Green Nanoparticles: The Future of Nanobiotechnology

Shilpi Srivastava • Atul Bhargava

Green Nanoparticles: The Future of Nanobiotechnology

Springer

Shilpi Srivastava
Amity Institute of Biotechnology
Amity University Uttar Pradesh,
Lucknow Campus
Lucknow, Uttar Pradesh, India

Atul Bhargava
Department of Botany
Mahatma Gandhi Central University
Motihari, Bihar, India

ISBN 978-981-16-7105-0 ISBN 978-981-16-7106-7 (eBook)
https://doi.org/10.1007/978-981-16-7106-7

This Springer imprint is published by the registered company Springer Nature Singapore Pte Ltd.
The registered company address is: 152 Beach Road, #21-01/04 Gateway East, Singapore 189721, Singapore

Preface

Nanobiotechnology has been proposed as the next general-purpose technology and engine for growth in the twenty first century. Applications employing nanotechnologies promise greater access to knowledge and information; new therapeutic interventions; improved environmental monitoring; greater safety and security; expanded communication capacities, and many other industrial and societal applications. The global nanotechnology market valued at nearly $1.05 billion in 2018, is expected to reach $2.23 billion by the year 2025 with an expected compound annual growth rate of 10.5% from 2019 to 2025. Millions of workers will be employed in nanotechnology-related industries, and many more will be part of supporting jobs like academics.

Nanoparticles are microscopic entities with at least one dimension less than 100 nm. These are gaining increasing attention due to their immense use in optics, electronics, and biomedical sciences. Synthesis of nanoparticles has been carried out through two approaches, viz. the top-down and the bottom-up approaches. In the top-down approach, a big component is broken down into smaller ones of desired size, while the bottom-up approach is primarily a build-up method that starts from atoms and is based on atomic transformations and molecular condensation. The physical methods are energy intensive, require costly vacuum systems or equipment to prepare nanoparticles besides having limitations like use of high temperature and not being environment-friendly. The chemical methods generally increase the particle reactivity and toxicity, and might harm human health and the environment due to the composition ambiguity and lack of predictability. Toxic chemicals accumulate on the surface of nanoparticles and nonpolar solvents that often limits their applications in biomedical applications. These production methods are expensive, labor-intensive, and are potentially hazardous to the environment as whole and living organisms in particular. There is an urgent need to develop environmentally friendly methods of synthesis of nanoparticles through techniques that are clean, biocompatible, nontoxic, and environment-friendly. Within the realm of green nanotechnology, biological synthesis of nanoparticles forms the most important component and has captured the fascination of the researchers the considering long-term sustainability goals. Use of living organisms or their biomass could be an alternative to the conventional chemical and physical methods of nanoparticle synthesis. Considering the vast potentiality of microbes and plants as sources, the biological synthesis can serve as a green technique for the synthesis of nanoparticles as an alternative to conventional methods. The biocompatibility of biologically synthesized nanoparticles offers very interesting applications in biomedicine and related fields.

Thus, the time is ripe for a comprehensive book which not only deals with the fundamental concepts on green synthesis of nanoparticles but also incorporates all the latest developments in this branch of science. The book describes the recent developments with respect to synthesis, characterization, and applications of nanoparticles in an easy-to-read, succinct format, and guides the readers to these technologies in a simple way. Spread over 16 chapters, the book gives an overview of the latest research and developments in the above-mentioned areas with added coverage of recent researchers in the field. Different groups of plants used in bionanofabrication have been dealt with in separate chapters. The topics are presented in a lucid, methodical way and care has been taken to keep the level of discussion simple and free from technical jargon.

The book will cater to advance undergraduate and post-graduate readers in nanobiotechnology, and practicing nanotechnologists. The book will also be of interest to researchers working in nanobiotechnology, nanosciences, pharmaceutical sciences, medical sciences, and general biology.

Lucknow, India
Motihari, India

Shilpi Srivastava
Atul Bhargava

Acknowledgments

Shilpi Srivastava

I am deeply indebted to my parent organization Amity University Uttar Pradesh, and Dr. Aseem Chauhan, Chairman, Amity University Uttar Pradesh (Lucknow Campus), for providing me a suitable forum to fulfill my ambitions. I owe my deepest sense of veneration and gratitude to Dr. Balwinder Shukla, VC, and Dr. Sunil Dhaneshwar, Pro VC, AUUP, Lucknow Campus, India, for giving me an opportunity to be a part of their prestigious and renowned institution without which this would not have been possible.

I also wish to thank Dr. J.K. Srivastava, Director, Amity Institute of Biotechnology, Lucknow, for his constant support and encouragement. I thank my colleagues and friends at Amity University namely, Dr. Prachi Srivastava, Dr. Rachna Chaturvedi, Dr. Jyoti Prakash, Dr. Gurjeet Kaur, Dr. Vineet Awasthi, Dr. Ruchi Yadav and Dr. Garima Awasthi for their support and psycho-stimulant company.

Atul Bhargava

I am highly grateful to Dr. Sanjeev Kumar Sharma, Vice Chancellor, Mahatma Gandhi Central University, Motihari, for his inspiring leadership and facilities provided in the University which helped me in accomplishing this task.

My sincere thanks are due to Dr. Anand Prakash, Campus Director (Chanakya Parisar), Mahatma Gandhi Central University for his never-ending support and affection showered on me. I am blessed and fortunate enough to have my support system in the form of my wonderful faculty colleagues Dr. Brijesh Pandey, Dr. Shahana Majumder, Dr. Ram Prasad, Dr. Akhilesh Kumar Singh, Dr. S.P. Singh, and Dr. Saurabh Singh Rathore for creating a lively atmosphere and relentless support during the writing of this text.

On a very personal note, I am grateful to my brother Mr. Akhilesh Bhargava, sister-in-law Dr. Meenakshi Bhargava, my niece Ms. Anushka Sharma, and my son Mr. Akshay Bhargava for their patience and perseverance, as well as their constant support during the writing of this book. They wholeheartedly supported me during my overburdened schedule and were with me during thick and thin.

Shilpi Srivastava and Atul Bhargava

We express our appreciation for the editorial guidance of Dr. Bhavik Sawhney (Editor, Biomedicine, Springer Nature) and Veena Perumal (Production Editor,

Springer Nature) whose efforts have helped us in shaping the text to bring out the best to the readers.

A final word of gratitude to all those who did not find mention above but have contributed somehow in the publication of this book.

Contents

1 Green Nanotechnology: An Overview 1
1.1 Introduction 1
1.2 Nanobiotechnology 4
1.3 Historical Perspectives 5
1.4 Generation Map of Nanotechnology 7
1.5 Green Nanotechnology 8
References 12

2 The Nanodomain 15
2.1 Introduction 15
2.2 Types of Nanomaterials 16
2.2.1 Natural Nanomaterials 17
2.2.2 Engineered Nanomaterials 18
2.3 Types of Nanomaterials 23
References 25

3 Tools and Techniques Used in Nanobiotechnology 29
3.1 Introduction 29
3.2 Microscopy 30
3.2.1 Scanning Electron Microscope (SEM) 33
3.2.2 Transmission Electron Microscope (TEM) 35
3.2.3 Scanning Tunneling Microscope (STM) 38
3.2.4 Atomic Force Microscope (AFM) 39
3.3 Spectroscopic Techniques 42
3.3.1 Ultraviolet-Visible Spectroscopy 42
3.3.2 Fourier Transform Infrared (FTIR) Spectroscopy 43
3.3.3 X-Ray Photoelectron Spectroscopy (XPS) 45
3.3.4 Energy Dispersive X-Ray Analysis (EDAX) 45
3.4 Nuclear Magnetic Resonance (NMR) 46
3.4.1 Advantages of NMR 46
3.4.2 Disadvantages of NMR 47
3.5 X-Ray Diffraction (XRD) 47
3.5.1 Advantages of X-Ray Diffraction 47
3.5.2 Disadvantages of X-Ray Diffraction 48

3.6 Dynamic Light Scattering (DLS) 48
3.6.1 Advantages of DLS Technique 48
3.6.2 Disadvantages of Dynamic Light Scattering 49
3.7 Zeta Potential 49
References 50

4 Green Synthesis of Nanoparticles 53
4.1 Introduction 53
4.1.1 Top-Down Approach 53
4.1.2 Bottom-Up Approach 54
4.2 Methods of Nanoparticle Synthesis 55
4.3 Physical Methods 55
4.3.1 High-Energy Ball Milling 55
4.3.2 Inert Gas Condensation (IGC) 57
4.3.3 Physical Vapor Deposition 58
4.3.4 Laser Pyrolysis 58
4.3.5 Flame Spray Pyrolysis 60
4.3.6 Electrospraying 60
4.3.7 Melt Mixing 61
4.4 Chemical Methods 62
4.4.1 Sol-Gel Synthesis 62
4.4.2 Microemulsion Method 63
4.4.3 Hydrothermal Synthesis 65
4.4.4 Polyol Synthesis 66
4.4.5 Chemical Vapor Deposition 67
4.4.6 Chemical Vapor Synthesis/Chemical Vapor Precipitation 67
4.4.7 Wet-Chemical Method 68
4.4.8 Advantages and Disadvantages of Chemical Methods 68
4.5 Biological Methods 69
References 70

5 Biological Synthesis of Nanoparticles: Bacteria 77
5.1 Introduction 77
5.2 Bacteria in Green Synthesis of Nanoparticles 79
5.3 Gram-Positive Bacteria 81
5.4 Gram-Negative Bacteria 84
5.5 Bacterial Biosurfactants in Nanoparticle Production 88
5.6 Mechanistic Aspects of Nanoparticle Production by Bacteria 90
5.7 Future Prospects 91
References 91

6 Biological Synthesis of Nanoparticles: Fungi 101
6.1 Introduction 101
6.2 Fungal Classification 102
6.3 Fungi in Nanoparticle Production 103
6.4 Mechanism of Nanoparticle Synthesis by Fungi 104
6.5 Filamentous Fungi in Nanoparticle Production 106
6.6 Yeast in Nanoparticle Production 111
6.7 Mushrooms in Nanoparticle Production 113
6.8 Lichens in Nanoparticle Production 115
6.9 Advantages of Using Fungi in Nanoparticle Production 126
6.10 Future Prospects 126
References 126

7 Biological Synthesis of Nanoparticles: Algae 139
7.1 Introduction 139
7.2 Algal Classification 140
7.3 Use of Algae in Nanoparticle Synthesis 142
7.4 Mechanism of Nanoparticle Synthesis by Algae 144
7.5 Blue-Green Algae or Cyanobacteria in Nanoparticle Production 144
7.6 Green Algae in Nanoparticle Production 150
7.7 Diatoms in Nanoparticle Production 150
7.8 Brown Algae in Nanoparticle Production 156
7.9 Red Algae in Nanoparticle Production 159
7.10 Future Prospects 161
References 162

8 Biological Synthesis of Nanoparticles: Bryophytes 173
8.1 Introduction 173
8.2 Evolutionary Trends in Bryophytes 174
8.3 Bryophytes in Nanoparticle Production 175
8.4 Future Prospects 178
References 178

9 Biological Synthesis of Nanoparticles: Pteridophytes 181
9.1 Introduction 181
9.2 Phylogenetic Relationships in Pteridophytes 182
9.3 Pteridophytes in Nanoparticle Production 182
9.4 Future Prospects 192
References 192

10 Biological Synthesis of Nanoparticles: Gymnosperms 195
10.1 Introduction 195
10.2 Phylogenetic Studies in Gymnosperms 196
10.3 Gymnosperms in Green Synthesis of Nanoparticles 196
10.3.1 Conifers in Nanoparticle Production 198

10.3.2 Nanoparticle Synthesis from Other Gymnosperms . . . 203
10.4 Future Prospects . . . 211
References . . . 212

11 Biological Synthesis of Nanoparticles: Angiosperms: Monocotyledons . . . 217
11.1 Introduction . . . 217
11.2 Origin of Monocotyledonous Plants . . . 217
11.3 Monocots for Green Synthesis of Nanoparticles . . . 220
References . . . 226

12 Biological Synthesis of Nanoparticles: Dicotyledons . . . 231
12.1 Introduction . . . 231
12.2 Dicotyledonous Plants in Green Synthesis of Nanoparticles . . . 232
12.2.1 Biological Synthesis of Nanoparticles Using Leaves . . . 232
12.2.2 Biological Synthesis of Nanoparticles Using Roots . . . 235
12.2.3 Biological Synthesis of Nanoparticles Using Fruits . . . 237
12.2.4 Biological Synthesis of Nanoparticles Using Seeds . . . 241
12.2.5 Biological Synthesis of Nanoparticles Using Flowers . . . 243
12.2.6 Biological Synthesis of Nanoparticles Using Bark . . . 246
12.2.7 Biological Synthesis of Nanoparticles Using Other Plant Parts . . . 246
12.3 Future Perspectives . . . 250
References . . . 250

13 Biological Synthesis of Nanoparticles: Animal Systems . . . 261
13.1 Introduction . . . 261
13.2 Insect Wings . . . 262
13.3 Chitosan . . . 263
13.4 Honeybee . . . 264
13.4.1 Honey . . . 264
13.4.2 Propolis . . . 265
13.5 Spider Silk . . . 268
13.6 Paper Wasp Nest . . . 270
13.7 Wasp Venom . . . 270
13.8 Silk Moth and Its Derivatives . . . 271
13.8.1 Silk Fibroin . . . 271
13.8.2 Silk Sericin . . . 272
13.9 Eggs . . . 274

13.9.1 Egg White . . . 275
13.9.2 Yolk . . . 277
13.10 Earthworms . . . 277
13.11 Milk . . . 279
13.12 Future Prospects . . . 279
References . . . 279

14 Applications of Biosynthesized Nanoparticles . . . 285
14.1 Introduction . . . 285
14.2 Insecticidal Activities . . . 286
14.2.1 Nanoparticles Against Storage Pests . . . 286
14.2.2 Nanoparticles Against Mosquito Vectors . . . 286
14.3 Environmental Applications . . . 293
14.3.1 Remediation of Pesticides . . . 293
14.3.2 Remediation of Dyes . . . 294
14.3.3 Heavy Metal Sensing and Remediation . . . 295
14.3.4 Removal of Antibiotics . . . 299
14.4 Nanoparticles in Agriculture . . . 299
14.5 Green Nanoparticles in Medicine . . . 302
14.5.1 Antimicrobial Activity . . . 302
14.5.2 Anticancer Activity . . . 304
14.5.3 Anti-inflammatory Activity . . . 308
14.5.4 Drug Delivery . . . 308
14.5.5 Bioimaging . . . 310
References . . . 310

15 Toxicity Aspects of Biologically Synthesized Nanoparticles . . . 325
15.1 Introduction . . . 325
15.2 Health Risks . . . 326
15.2.1 Pulmonary and Respiratory Effects . . . 326
15.2.2 Cutaneous Toxicity . . . 327
15.2.3 Cardiovascular Toxicity . . . 328
15.2.4 Genotoxicity and Carcinogenicity . . . 334
15.3 Environmental Hazards . . . 335
15.4 Toxicity Toward Plant Forms . . . 338
15.5 Ethical Issues . . . 338
References . . . 340

16 Nanobiotechnology: Present Status and Future Prospects . . . 345
16.1 Introduction . . . 345
16.2 Present Status and the Expanding Market . . . 346
16.3 Challenges in Nanobiotechnology . . . 347
16.3.1 Nanomedicines . . . 347
16.3.2 Food Industry . . . 351
16.3.3 Agriculture . . . 351
16.4 Thrust Areas . . . 351
References . . . 352

Green Nanotechnology: An Overview

1

Abstract

Nanotechnology lays emphasis on the production and usage of structures and systems by manipulation of atoms and molecules at the nanoscale. The primary concept of nanotechnology was put forward by Richard Feynman in 1959 while presenting his famous talk at the annual meeting of the American Physical Society. In this lecture Feynman dwelt on assembling atomic blocks at the molecular level. However, the term nanotechnology came into existence much later and was coined in 1974 by Norio Taniguchi, the Japanese scientist, who first used it while hypothesizing the idea of precision manufacturing at the scale of nanometers. Nanotechnology as a discipline was firmly established by the contributions of Richard Smalley, Eric Drexler, and Chad Mirkin. Green nanotechnology is aimed at sustainable production and use of nanomaterials for betterment of human society.

1.1 Introduction

The emerging technologies of the late twentieth and early twenty-first century, namely, information technology along with biotechnology, have altered the production methods as well as utilization of products in the industrial and service sectors and even for societal benefits to a significant extent. Applied science has taken giant leaps to a point where we have started manipulating substances at the atomic level rather than controlling them at the molecular level. New terms like nanoscience, nanotechnology, and nanobiotechnology have become quite popular in the last 30 years and have revolutionized scientific research and development. These terms have often been interchangeably used due to which a need arose for differentiating nanoscience and nanotechnology. Nanoscience is concerned with the manipulation of matter at the atomic, molecular, and supramolecular levels. In other words,

S. Srivastava, A. Bhargava, *Green Nanoparticles: The Future of Nanobiotechnology*,
https://doi.org/10.1007/978-981-16-7106-7_1

nanoscience is the foundation for nanotechnology. Nanotechnology is the branch of science that lays emphasis on the design, production, and usage of structures and systems by manipulation of atoms and molecules at the nanoscopic scale (nanoscale).

Extensive research in nanotechnology is projected to have an appreciable impact in health care, agriculture, and environmental restoration that will lead to a better understanding of nature and aims to have a positive impact on sustainability and human potential. It has been used in various sectors like food, information technology, communications, environmental restoration, energy, and biomedical science. After the advent of research at the nanoscale, various terms have come into prominence which needs to be defined in a clear way.

The term "nanotechnology" was first introduced by Professor Norio Taniguchi in 1974. The Office of Technology Assessment at the German Parliament (TAB) defines nanotechnology as the fabrication and exploitation of structures having at least one dimension in the nanometer scale. According to Goodsell (2004), nanotechnology refers to the "engineering and manufacturing at nanometer scales with atomic precision." The term is often interchangeably used with "molecular nanotechnology." The National Nanotechnology Institute (NNI) defined nanotechnology as the "understanding and control of matter at dimensions of roughly 1 to 100 nm, where unique phenomena enable novel applications." NNI considers a technology within the nanorealm only when it encompasses the following:

- Substances having size range of approximately 1–100 nm.
- Creation and use of structures and systems having unique properties due to their minute size.
- Control or manipulate articles at the atomic scale.
- The various definitions of nanotechnology have been provided in Box 1.1.

Box 1.1 Definitions of nanotechnology

Definition	Source
"The processing of separation, consolidation, and deformation of materials by one atom or one molecule"	Norio Taniguchi
"The understanding and control of matter at dimensions of roughly 1 to 100 nanometers, where unique phenomena enable novel applications"	US National Nanotechnology Initiative (NNI)
"The design, characterization, production and application of structures, devices and systems by controlling shape and size at the nanometer scale"	Royal Society and Royal Academy of Engineering
"The art and science of manipulating matter at the atomic or molecular scale and may lead	National Center For Environmental Research

(continued)

Box 1.1 (continued)

Definition	Source
to significant improvements in technologies for protecting the environment"	
"The creation of functional materials, devices, and systems through control of matter on the nanometer (1 to 100 nm) length scale and the exploitation of real properties and phenomena developed at that scale"	Los Alamos National Laboratory
"An interdisciplinary S&T that encompasses IT technology, the environmental sciences, life sciences, material sciences, etc. It is for controlling and handling atoms and molecules in the order of nanometer (1/1000000000), enabling discovery of new functions by taking advantage of its material characteristics unique to nano size, so that it can bring technological innovation in various fields"	Japan: Second Science and Technology Basic Plan (2001–2005)
"The production and application of structures, devices and systems by controlling the shape and size of materials at nanometre scale. The nanometre scale ranges from the atomic level at around 0.2 nm (2 Å) up to around 100 nm"	European Medicine Agency (EMA)
"The term nanotechnology covers entities with a geometrical size of at least one functional component below 100 nanometers in one or more dimensions susceptible of making physical, chemical or biological effects available which are intrinsic to that size"	European Patent Office
"Nanotechnology is the development and use of techniques to study physical phenomena and develop new devices and material structures in the physical size range of 1–100 nanometres (nm), where 1 nanometre is equal to one billionth of a meter"	Nanotechnology Initiative Division, Ministry of Electronics and Information Technology, Government of India
"Understanding and control of matter and processed at the nanoscale, typically, but not exclusively, below 100 nanometers in one or more dimensions where the onset of size-dependent phenomena usually enables novel applications"	ISO TCC 229 in 2007

1.2 Nanobiotechnology

Biotechnology involves the manipulation of living organisms in production of economically important products. On the other hand, nanobiotechnology involves manipulation, synthesis, and engineering of products at the nanoscale range through utilization of living forms. This multi-strategic discipline is the converging point of a diverse array of technologies having knowledge base in engineering, biology, agriculture, physics, and chemistry (Fig. 1.1). This multidisciplinary field

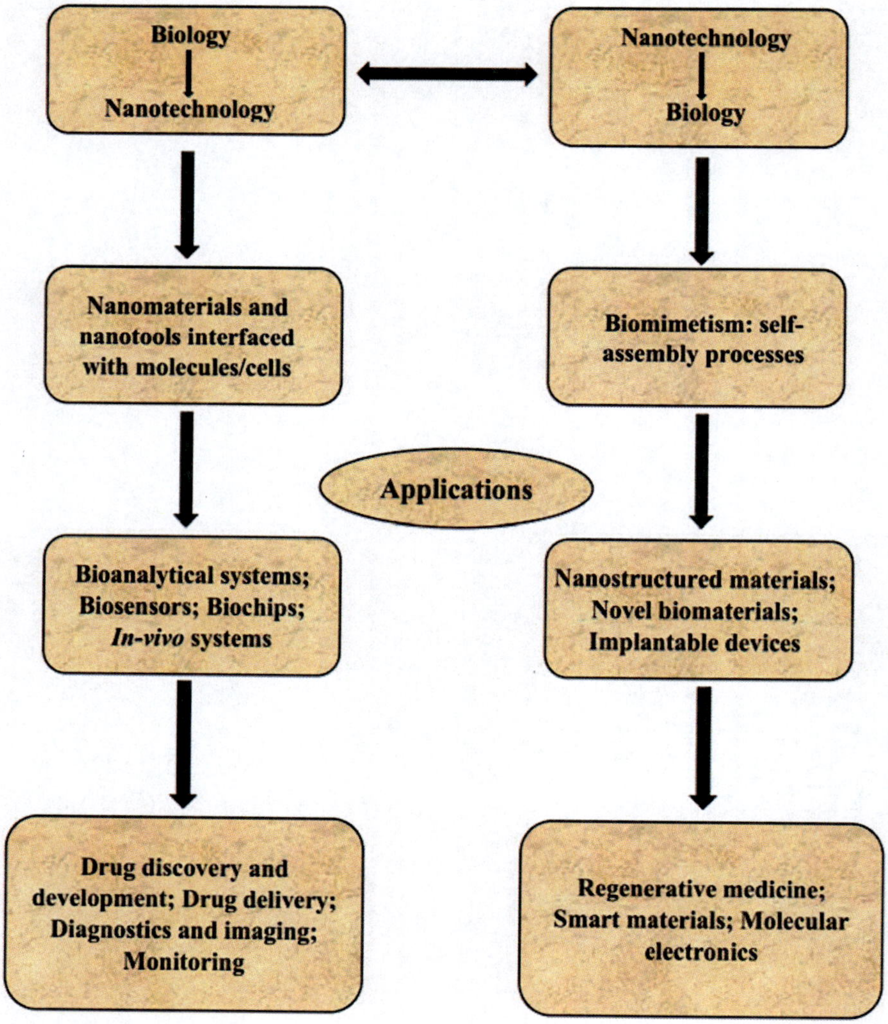

Fig. 1.1 Diagrammatic representation of nanobiotechnology and its applications

amalgamates nanotechnology and biotechnology to produce novel nanostructures having applications in industries, agriculture, and clinical science. The blending of nanotechnology and biology is advantageous in two ways:

- Nanotechnology introduces novel tools for better understanding of the biological processes.
- Biology puts into use the self-assembly and synthesis concepts for revolutionizing the manipulation and use of matter at the nanoscale.

The biological approaches makes use of tools and living entities for fabrication of nanostructures that mimic biological structures and are utilized in different developmental sectors. The primary difference between "nanobiology" and "nanobiotechnology" pertains to usage of the term "technology." Any structure built using human skill and contribution forms a part of technology and encompasses diverse disciplines. One major difference between nanotechnology and nanobiotechnology is the utilization of top-down approach in the former and bottom-up approach in the latter. The confluence of nanoscience with biotechnology conceptualizes the development of novel products with minimum environmental damage. However, the manufacturing, use, and disposal of these materials, along with regulations for their safe usage, are of utmost importance and need to be properly taken care of.

1.3 Historical Perspectives

Although unknowingly, nanoparticles have been found to be useful in human society for a long time. The earliest use of nanoparticles was in the glaze exhibited in Chinese porcelain. A Roman glass cage cup known as the Lycurgus Cup of fourth century AD used nanosized gold clusters to create different colors on illumination from different angles. This object present in the British Museum collection is the oldest synthetic nanomaterial known to date. Dichroic glass refers to a specific type of glass that displays different colors according to the lighting conditions. Thus, the Lycurgus Cup displays two types of colors, i.e., it appears green when lit from the front and reddish-purple when light shines through the glass. However, the cause of this effect was not known to those who were exploiting this. TEM studies in the 1990s showed that dichroism was due to the presence of nanoparticles having size ranging from 50 to 100 nm. These nanoparticles were silver-gold alloy having 330 ppm silver and 40 ppm gold, along with some amount of copper (10%). Other notable examples include the glittering "luster" ceramic glazes used during the Middle Ages in Middle East Asia and Europe that comprised of silver and copper nanoparticles, the Renaissance pottery fabricated during the sixteenth century, and the "Damascus" sabre of the seventeenth century containing nanoscale wires of cementite (iron-carbon compound). Carbon black is another common example of a nanoparticulate material that has been used for decades.

Nano concept was initially put forward at the start of the twentieth century by Richard Zsigmondy, the Hungarian chemist, who first used the term "nanometer." However, it was the theoretical physicist Richard Feynman (Box 1.2) who first elaborated the primary concept of nanotechnology in the year 1959 in his famous lecture entitled "There's Plenty of Room at the Bottom" at a congregation of the American Physical Society. Richard Feynman during his 1959 lecture mooted the idea of manipulating atoms for diverse uses. Feynman stated that "The principles of physics, as far as I can see, do not speak against the possibility of maneuvering things atom by atom." Thus, the cornerstone of nanoscale technology was laid by Feynman when he conceptualized the prospects of organizing atoms for creating novel products both at atomic and molecular level. In 1964, Glenn T. Seaborg, an American Chemist, identified, isolated, and obtained patents for two transuranium elements, viz., americium and curium. However, the term nanotechnology came into existence much later and was coined in 1974 by Norio Taniguchi, a Japanese scientist, who used it while hypothesizing the idea of precision manufacturing at the scale of nanometers. Nanotechnology as a discipline was firmly established by the works of Richard Smalley, Eric Drexler, and Chad Mirkin.

Box 1.2 Richard Feynman: contributions in nanosciences
Richard Feynman (1918–1988).

He was born in New York City in 1918 and did his B.Sc in 1939 from Massachusetts Institute of Technology. Feynman obtained his doctoral degree from Princeton University in 1942. He held the position of Professor (Theoretical Physics) at Cornell University (1945–1950) and later served as Professor at the California Institute of Technology for about nine years. Feynman made monumental contribution in quantum mechanics, corrected the inaccuracies of quantum electrodynamics, and pioneered the research in quantum computing. He worked at the US atomic bomb project during World War II and is also well known for Feynman diagrams and the Feynman Lectures on Physics. Feynman was ranked as seventh greatest physicist of all time and was bestowed the prestigious Albert Einstein Award and Lawrence Award. In 1965, Feynman was jointly awarded the Nobel Prize in Physics with Julian Schwinger (USA) and Shin'ichirō Tomonaga (Japan) for their monumental work in quantum electrodynamics that revolutionized the physics of elementary particles. In 1978, Feynman was diagnosed with liposarcoma, a rare form of cancer, and later for ruptured duodenal ulcer that caused kidney disfunction. After a long battle with multiple disorders, Feynman died in 1988.

Richard E. Smalley, an American scientist at Rice University, pioneered the field of nanotechnology and along with Robert Curl (Rice University) and Harold W. Kroto (University of Sussex) shared the Nobel Prize in 1996 for the development of carbon-composed clusters termed as buckminsterfullerene (C_{60}) or buckyballs. For synthesis of fullerenes, graphite was exposed to an

(continued)

Box 1.2 (continued)
intense lase pulse which on condensation and separation through chromatographic techniques formed structures having 60 and 70 carbon atoms. About 75% had 60 carbon atoms arranged in a three-dimensional hollow spherical cage comprising of 60 vertices and 32 faces. These structures with unique properties were named fullerenes in honor of the famous American architect R. Buckminster Fuller who is credited for the invention of the famous hemispherical geodesic dome. Upon his death, Smalley was honored and credited as the "Father of Nanotechnology" by the US Senate.

K. Eric Drexler is often described as "the founding father of nanotechnology." Drexler received his doctoral degree from Massachusetts Institute of Technology in the year 1991 which was the first degree of its kind in the world. His PhD thesis was published with some revisions as the book entitled *Nanosystems: Molecular Machinery Manufacturing and Computation* in the year 1992. He played an important role as founder of the Foresight Institute and served as the Chief Technical Advisor at Nanorex, a company associated with open-source design software for structural DNA nanotechnologies. Drexler introduced an idea of using machines that work at the molecular scale to structure matter from the bottom up in his famous book *Engines of Creation: The Coming Era of Nanotechnology*. Drexler's works made nanotechnology a burning topic among the researchers and brought to limelight the engineering approach to nanotechnology and nanosystems.

Chad A. Mirkin, a well-known nanoscience expert, is currently the Director of the International Institute for Nanotechnology located at Northwestern University, Illinois, USA. He has over 100 awards, is a part of about 800 manuscripts with 1200 patents, and has over 160,000 citations to his work. Mirkin was born in 1963 in Phoenix, Arizona, and received his doctorate from Pennsylvania State University, USA, in 1989. He has contributed in the development of nanotechnology-based biosensing systems for use in medical diagnostics and is credited with the invention of dip-pen nanolithography. His research work focused on designing novel ligands and molecule-based electronic devices, self-assembled monolayers, catalysis, and optics which provided a strong base for furthering the uses of nanotechnology in diverse areas.

1.4 Generation Map of Nanotechnology

The huge research potential in nanotechnology necessitated the formulation of a timeframe for nanotechnology development. Mihail Roco put forward a roadmap for development of nanotechnology in various phases or generations (Fig. 1.2). The first generation mainly comprised of passive nanostructures and lasted till the year 2000.

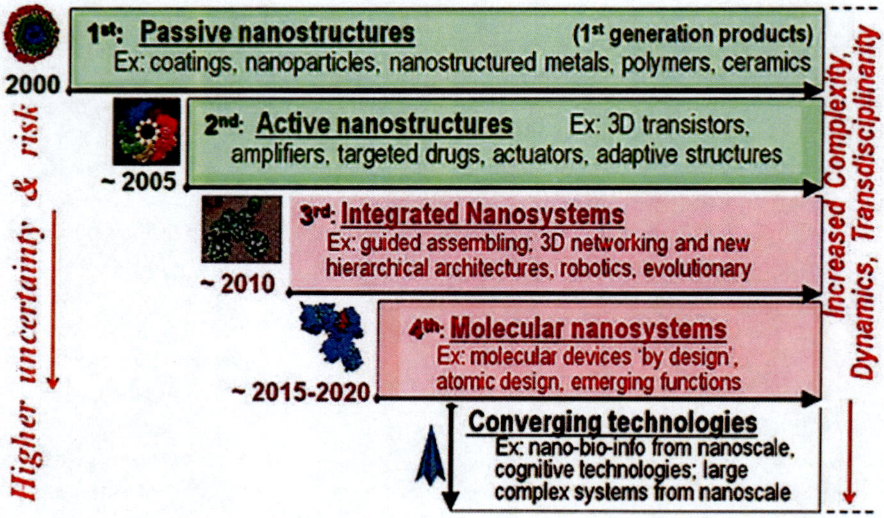

Fig. 1.2 Projected stages for the beginning of industrial prototyping and nanotechnology commercialization over time (Reprinted by permission from Springer Nature)

Contact and dispersed nanostructures like aerosols, colloids, nanoparticle-reinforced composites, nanostructured metals, polymers, and ceramics were the chief elements of the first generation. The second generation envisaged the development of active nanostructures for multitask uses like actuators, drug delivery devices, 3-D transistors, and bio-based sensors. The third generation was speculated to be emerging around 2010 and was likely to feature nanosystems having loads of interacting components. Examples of third-generation nanostructures include guided assemblies, three-dimensional (3-D) networking, and hierarchical architectures. The coming years are likely to witness the development of the first integrated nanosystems that will function like a mammalian cell with hierarchical systems within systems. Table 1.1 depicts the important discoveries in nanobiotechnology.

1.5 Green Nanotechnology

The use of "green" and environmentally friendly processes in nanomaterial production has been an upcoming trend which is gaining popularity due to several limitations and the environmental impact of the physical and chemical methods. The concept of green and sustainable chemistry arose in the early 1990s. The term "green chemistry" was coined in 1991 by P.T. Anastas, popularly known as the "Father of Green Chemistry," to focus on the design and fabrication of products that are ecofriendly and non-hazardous (Fig. 1.3).

Green chemistry aims at the following:

Table 1.1 Important landmarks in nanotechnology

Year	Achievement
1900	Max Planck proposes energy quantization
1927	Werner Heisenberg developed the uncertainty principle
1931–1933	Electron microscope built by physicist Ernst Ruska and Max Knoll, the electrical engineer
1938	First commercial electron microscope developed by Siemens
1939	Transmission electron microscope developed by Siemens
1959	Feynman describes building molecular machines with atomic precision
1965	Moore's law
1974	Norio Taniguchi coined the term "nanotechnology"
1977	Drexler originates molecular nanotechnology concepts at MIT
1981	Eric Drexler writes *Engines of Creation* a treatise on molecular engineering; scanning tunneling microscope developed by Gerd Binnig and Heinrich Rohrer at IBM-Research at Zurich
1985	Buckyballs discovered by Kroto, Curl, and Smalley
1986	AFM invented by Gerd Binnig, Calvin Forrest Quate, and Christoph Gerber
1987	First protein engineered
1989	Don Eigler and Erhard Schweizer developed the IBM logo *by placing 35 individual xenon atoms* on chilled nickel crystal using a scanning tunneling microscope
1990	First journal on nanotechnology
1991	Carbon nanotube invented by the Japanese physicist Sumio Iijima Japan's MITI announces the establishment of bottom-up "atom factory" and commits $200 million
1992	C.T. Kresge and colleagues at Mobil Oil Corporation discovered *Mobil* Composition of Matter-*41 (MCM-41) and Mobil* Composition of Matter-*48 (MCM-48), both being* nanostructured catalytic materials
1993	First Feynman Prize in Nanotechnology awarded to Charles Musgrave Controlled synthesis of nanocrystals successfully accomplished by Moungi Bawendi of MIT
1995	Nadrian C. Seeman conferred Feynman prize for developing the concept of DNA nanotechnology
1996	$250,000 Feynman Grand Prize announced by Palo Alto, CA—Foresight Institute. NASA initiates research on computational nanotechnology First conference on nanobiotechnology entitled "Biological Approaches and Novel Applications for Molecular Nanotechnology" held in San Diego, CA
1997	Zyvex, the first company for production of nano-engineered materials was setup. first nanorobotic system designed
1998	First *two-dimensional DNA crystals* developed by the American nanotechnologist and crystallographer Nadrian C. Seeman.
1999	First nanomedicine book published Chad Mirkin invented *dip-pen nanolithography*
2000	US National Nanotechnology Initiative (NNI) announced by US President Bill Clinton First DNA motor created by Lucent and Bell Labs
2001	A center for military applications conceptualized by the USA

(continued)

Table 1.1 (continued)

Year	Achievement
2003	Gold nanoshells developed by Naomi J. Halas and co-workers at Rice University Drexler-Smalley debate on molecular nanotechnology is published in Chemical and Engineering News, a magazine published by the American Chemical Society
2004	First conference held on advanced nanotechnology Establishment of first center for nanomechanical systems The term "nanotoxicology" was first used
2005	Erik Winfree and Paul Rothemund put forward theory on DNA computing and develop "algorithmic molecular self-assembly" Metallic nanowires synthesized using a ray of electrons
2006	First report on the development of thin-film nanotubes by evaporation Scientists at Rice University led by James Tour constructed the world's first "nanoscale car" Fabrication of a 3 nm metal–oxide–semiconductor field-effect transistor (MOSFET), the world's smallest nanoelectronic device accomplished
2007	Russia resolves 1 billion US dollar investment in nanotechnology MIT research group led by Angela Belcher develop a *lithium-ion battery by manipulating genes of a common virus*
2008	NNI Strategy for Nanotechnology-Related Environmental, Health, and Safety (EHS) Research published.
2010	Initiation of DNA-based "robotic" assembly Michael S. Strano at MIT published the first report in Nature Materials on the discovery of thermopower waves using carbon nanotubes
2011	First reports on the development of programmable nanowire circuits DNA molecular robots developed
2012	NNI launches Nanosensors and the Nanotechnology Knowledge Infrastructure (NKI)
2013	World's first carbon nanotube computer developed at Stanford University
2014	Release of 2014 Strategic Plan by NNI.
2016	Nobel Prize in Chemistry jointly awarded to Jean-Pierre Sauvage, Sir J. Fraser Stoddart, and Bernard L. Feringa for developing molecules with controllable movements (molecular machines)
2018	World's smallest tic-tac-toe game board developed at Caltech using dynamic DNA origami tiles
2019	Chad Mirkin conferred the Kabiller Prize in Nanoscience and Nanomedicine

1. Making better use of resources for the development of an efficient chemical process.
2. Greening extraction procedures.
3. Greening sample digestion and derivatization.
4. Reduction in the hazards and wastes produced in the preparation and/or handling of chemicals.
5. Preparation of materials through designing or redesigning manufacturing processes that are eco-friendly.
6. Replacing toxic reagents and products with those having similar properties and applications but have less unwanted effects on the environment.

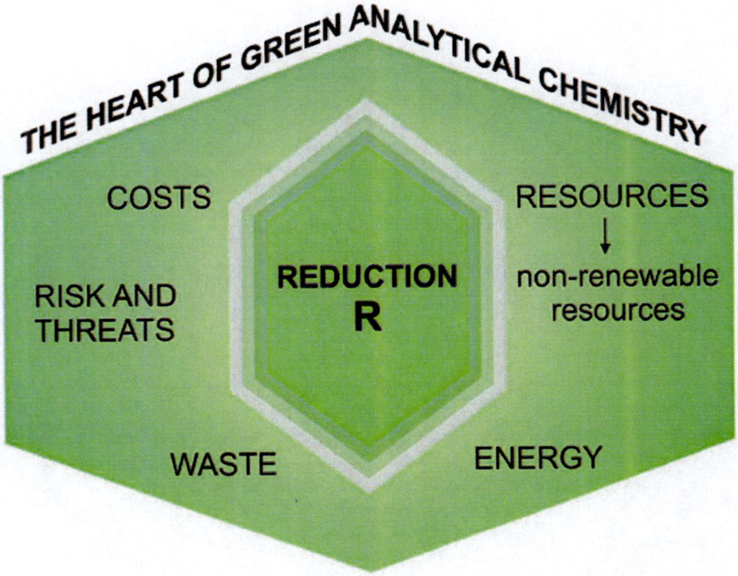

Fig. 1.3 The heart of green analytical chemistry (Reprinted with permission from Płotka-Wasylka et al. 2019; Springer Nature)

7. Reduction in the energy requirement either by the use of faster processes or by using renewable energy systems having less energy cost but similar efficiency.
8. Reduction in the toxicity for a compound or substance.
9. Reducing the costs and time.
10. Encouraging the development and usage of materials in consonance with the thought of sustainable development.

Green nanotechnology appears to be an offshoot of green chemistry and relies on application of environmentally benign approaches for the development of techniques and materials for diverse applications that can go a long way in enhancing environmental sustainability (Fig. 1.4). It utilizes the principles of green chemistry and eco-friendly engineering with the aim to effectively minimize the environmental and health risks associated with the manufacture and use of nanomaterials. Thus, green nanotechnology is aimed at mitigating the health hazards of nanomaterials, reduces the risks involved in nanofabrication, and promotes sustainable use of nanomaterials. The use of green nanotechnology effectively scales down the threats associated with the synthesis of unwanted chemical intermediates during different stages of use of nanotechnology, thus making the technology more environmentally friendly.

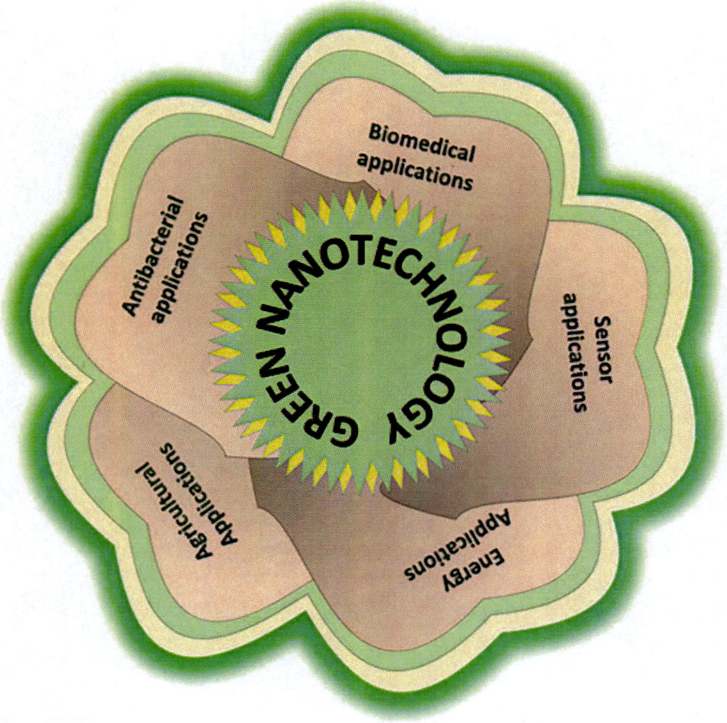

Fig. 1.4 Diverse applications of green nanotechnology (Reprinted with permission from Ganachari et al. 2019; Springer Nature)

References

Anastas PT, Warner JC (1999) Green chemistry: theory and Practice. Oxford University Press, Oxford

Anu Mary Ealia S, Saravanakumar MP (2017) A review on the classification, characterisation, synthesis of nanoparticles and their application. IOP Conf Ser: Mater Sci Eng 263:032019

Barber DJ, Freestone IC (1990) An investigation of the origin of the colour of the Lycurgus Cup by analytical transmission electron microscopy. Archaeometry 32:33–45

Bassi M, Santinello I, Bevilacqua A, Bassi P (2013) Nanotecnologie: una grande rivoluzione che parte dal piccolo [Nanotechnology: a big revolution from the small world]. Urologia 80:46–55

Bayda S, Adeel M, Tuccinardi T, Cordani M, Rizzolio F (2019) The history of nanoscience and nanotechnology: from chemical-physical applications to nanomedicine. Molecules 25:112

Drexler EK (1986) Engines of creation: the coming era of nanotechnology. Anchor Press, Garden City, NY

Drexler EK, Peterson C, Pergamit G (1991) Unbounding the Future: The Nanotechnology Revolution. William Morrow and Company, New York

Freestone I, Meeks N, Sax M, Higgitt C (2007) The Lycurgus cup—a Roman nanotechnology. Gold Bull 40:270–277

Ganachari SV, Yaradoddi JS, Somappa SB, Mogre P, Tapaskar RP, Salimath B, Venkataraman A, Viswanath VJ (2019) Green nanotechnology for biomedical, food, and agricultural applications. In: Martínez L, Kharissova O, Kharisov B (eds) Handbook of ecomaterials. Springer, Cham, pp 2681–2698

Goodsell D (2004) Bionanotechnology: lessons from nature. Wiley-Liss, New York

Hulla JE, Sahu SC, Hayes AW (2015) Nanotechnology: history and future. Hum Exp Toxicol 34(12):1318–1321

Isac-García J, Dobado JA, Calvo-Flores FG, Martínez-García H (2016) Green chemistry. In: Isac-García J, Dobado JA, Calvo-Flores FG, Martínez-García H (eds) Experimental organic chemistry. Academic Press, Amsterdam, pp 409–415

Karn B, Bergeson L (2009) Green nanotechnology: straddling promise and uncertainty. Nat Res Environ 24:2

Karn B, Wong SS (2013) Ten years of green nanotechnology. In: Sustainable nanotechnology and the environment: advances and achievements. American Chemical Society, Washington DC

Kharissova OV, Dias HV, Kharisov BI, Pérez BO, Pérez VM (2013) The greener synthesis of nanoparticles. Trends Biotechnol 31:240–248

Kreuter J (2007) Nanoparticles da historical perspective. Int J Pharm 331:1–10

Li CJ, Trost BM (2008) Green chemistry for chemical synthesis. Proc Natl Acad Sci USA 105: 13197–13202

Narayanan KB, Sakthivel N (2010) Biological synthesis of metal nanoparticles by microbes. Adv Colloid Interface Sci 156:1–13

National Nanotechnology Initiative (NNI) (n.d.). www.nano.gov

Niemeyer C (2004) Nanobiotechnology: concepts, applications and perspectives. John Wiley & Sons, New York

OECD (2009) Nanotechnology: an overview based on indicators and statistics. STI Working Paper 2009/7, OECD

Pasut G (2019) Grand Challenges in Nano-Based Drug Delivery. Front Med Technol 1:1

Piri-Moghadam H, Gionfriddo E, Grandy JJ, Alam MN, Pawliszyn J (2018) Development and validation of eco-friendly strategies based on thin film microextraction for water analysis. J Chromatogr A 1579:20–30

Płotka-Wasylka J, Gałuszka A, Namieśnik J (2019) Green analytical chemistry: summary of existing knowledge and future trends. In: Płotka-Wasylka J, Namieśnik J (eds) Green analytical chemistry. Green chemistry and sustainable technology. Springer, Singapore, pp 431–449

Poole CP, Owens FJ (2003) Introduction to Nanotechnology. John Wiley & Sons, New York

Pradell T, Climent-Font A, Molera J, Zucchiatti A, Ynsa MD, Roura P, Crespo D (2007) Metallic and nonmetallic shine in luster: An elastic ion backscattering study. J Appl Phys 101:103518

Reibold M, Paufler P, Levin AA, Kochmann W, Pätzke N, Meyer DC (2006) Materials: carbon nanotubes in an ancient Damascus sabre. Nature 444:286

Roco MC (2011) The long view of nanotechnology development: the national nanotechnology initiative at 10 years. J Nanopart Res 13:427–445

Sajid M (2018) Dispersive liquid-liquid microextraction coupled with derivatization: a review of different modes, applications, and green aspects. Trends Anal Chem 106:169–182

Santamaria A (2012) Historical overview of nanotechnology and nanotoxicology. Methods Mol Biol 926:1–12

Sharma VK, Yngard RA, Lin Y (2009) Silver nanoparticles: green synthesis and their antimicrobial activities. Adv Colloid Interfac 145:83–96

Singh H, Du J, Singh P, Yi TH (2018) Ecofriendly synthesis of silver and gold nanoparticles by *Euphrasia officinalis* leaf extract and its biomedical applications. Artif Cells Nanomed Biotechnol 46:1163–1170

Wagner FE, Haslbeck S, Stievano L, Calogero S, Pankhurst QA, Martinek KP (2000) Before striking gold in gold-ruby glass. Nature 407:691–692

Yu H, Park JY, Kwon CW, Hong SC, Park KM, Chang PS (2018) An overview of nanotechnology in food science: preparative methods, practical applications, and safety. J Chem 5427978:1–10

The Nanodomain

2

Abstract

Nanomaterials are any organic, inorganic, or mixed (organometallic) materials that bear specific properties owing to their miniscule size, which is in the range of 1–100 nanometers (nm). Many types of nanomaterials are known, namely, natural, incidental, engineered, and anthropogenic. Another classification based on the composition of nanostructures divides them into inorganic and organic nanomaterials. The engineered nanomaterials show greater biological activity as compared to larger entities having the same composition and also show greater uptake in living beings. The chief features that influence the toxicity of engineered nanostructures are their size, shape, aspect ratio, surface functionalization, solubility, binding with receptors and proteins, and the propensity to agglomerate. The engineered nanomaterials are sometimes considered as potential threat to soil, animal, and human health which has ignited a debate on the consequences of their indiscriminate use.

2.1 Introduction

The prefix of nanotechnology has been derived from the Greek word "nanos," meaning dwarf. A nanometer (nm) is about 10^{-9} of a meter which is quite small in comparison to the human hair and red blood cell. The diameter of human DNA (2.5 nm) is within the size range of nanometer scale (1–100 nm) commonly termed as the nanoscale. The *Betacoronavirus* SARS-CoV-2 has diameter centered around 100 nm, while amyloids are 7–13 nm in diameter. Soluble proteins like the Rubisco monomer have a diameter of 3–6 nm, while submicelles of casein protein have a diameter of 10 nm. The wavelength of visible light ranges between 400 and 700 nm. Table 2.1 sheds light on the size of some objects in the nanometer range.

S. Srivastava, A. Bhargava, *Green Nanoparticles: The Future of Nanobiotechnology*, https://doi.org/10.1007/978-981-16-7106-7_2

Table 2.1 Comparisons of scale from macro to molecular

Size (nm)	Terminology	Example
$> 10^4$	Macro	Normal matter
10^3–10^4	Micro	Living cells, MEMS Microfluidic channels
1–100	Nano	Proteins, DNA, inorganic nanoparticles
0.5–1.0	Molecular	Pores in zeolites
0.1–0.5	Atomic/molecular	Chemical bonds

A nanometer is a billionth of a meter, or about 80,000 times thinner than a human hair. Living beings are built up of cells that are quite small in size. The different components of the cells are even smaller and are in the sub-micron size domain. A comparison of some nanosized materials in nature has been provided in Table 2.2. In a normal room, there are about 15,000 nanoparticles per cubic centimeter in the air; in a forest around 50,000 and in a large city, their number may reach up to 100,000 nanoparticles.

2.2 Types of Nanomaterials

The European Commission (EC) defines nanomaterial as "a natural, incidental or manufactured material containing particles, in an unbound state or as an aggregate or as an agglomerate and where, for 50% or more of the particles in the number size distribution, one or more external dimensions is in the size range 1–100 nm." Nanomaterials are any organic, inorganic, or mixed particles having at least one external dimension measuring 1–100 nm (Fig. 2.1). Due to their miniscule size, nanomaterials possess specific chemical, physical, and/or electrical properties.

Different types of nanomaterials have been aptly defined by Hochella et al. (2019). These are the following:

1. Natural nanomaterials—Nanomaterials generated in nature by different physical, chemical, and biological processes, without any human interference.
2. Incidental nanomaterials—Nanomaterials that are generated unintentionally in nature due to human-related activities such as mining, tillage, and demolition/ construction.
3. Engineered nanomaterials—Nanomaterials that are intentionally produced and released by human beings into the environment. Some examples include quantum dots, dendrimers, carbon nanotubes, liposomes, and fullerene just to name a few.
4. Anthropogenic nanomaterials—These are comprised of both incidental and engineered nanomaterials.

However, majority of the studies have divided nanomaterials into two forms, viz., natural and engineered nanomaterials.

Table 2.2 Some natural and man-made nanosized entities in our environment

Size of molecule	Example
Natural objects	
1 nm	Width of an atom
7 nm	Microfilaments
9 nm	Diameter of F1 particle
2–12 nm	Width of DNA double helix
5–10 nm	Hemoglobin
13 nm	Diameter of ribulose 1,5-bisphosphate carboxylase-oxygenase (rubisco)
10–20 nm	Ribosome
25 nm	Diameter of microtubules
10–100 nm	Viruses
120 nm	Nuclear pore complex in vertebrates
100–200 nm	Mycoplasma
300 nm	Wavelength of UV light
800 nm	Width of dust particle
100–1000 nm	Diameter of peroxisome
100–1200 nm	Lysosome
1000 nm	Length of bacteria
1000–3000 nm	Mitochondria
3000 nm	Diameter of *Chlamydomonas* cell
3000–4000 nm	Width of *spirulina* trichome
3000–10,000 nm	Nucleus
5000–10,000 nm	Chloroplast
7000–8000 nm	RBCs
75,000–100,000	Width of human hair
Manmade objects	
1 nm	Carbon buckyball
1.3 nm	Carbon nanotube
1.5–10 nm	Dendrimer
35 nm	Zone plate x-ray "lens"
100 nm	Width of wire
70–350 nm	Liposome
10–100 μm	Microelectromechanical systems (MEMS) devices
1,000,000 nm	Head of a pin

2.2.1 Natural Nanomaterials

Natural nanoparticles are produced in different environmental zones by various processes mentioned previously (Fig. 2.2) and evolve chemically and physically through diverse complex mechanisms like weathering of rocks, forest fires, photo-oxidation, bio-assimilation, redox and precipitation reactions, biomineralization, soil erosion, sand storms, aggregation, and gas-solid nucleation. These nanosized entities

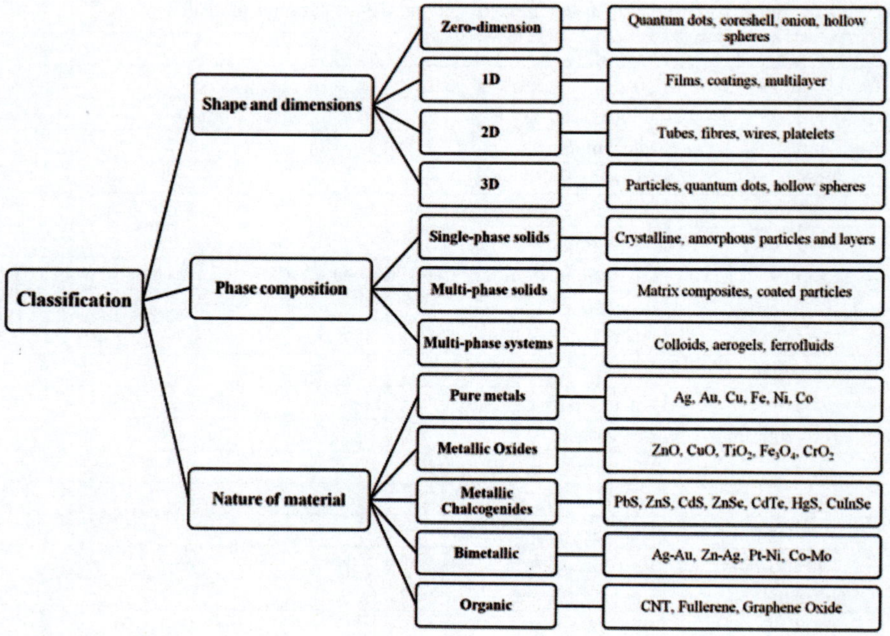

Fig. 2.1 Classification of nanomaterials (Reprinted with permission from Khanna et al. 2019)

move through different environmental compartments via the biogeochemical cycles and interact with environmental components. The anthropogenic nanoparticles are generated through human activity, some of which have been mentioned below:

1. Dust generation through mining, tillage, demolition, and construction work.
2. Atmospheric release and nucleation.
3. Release of waste water into the biosphere.
4. Insufficient treatment and spread of sludge from sewage treatment plants.
   Nanoscopic ash and soot are representatives of natural nanoparticles that are produced as a result of volcanic activity or other types of combustion. The volcanic ash clouds contain different polydisperse micro- and nanoparticles ranging from 100 to 200 nm in size that are composed of silicate and iron compounds and are found suspended in air.

2.2.2 Engineered Nanomaterials

Engineered nanomaterials, with size of 100 nm or less, are intentionally produced or designed from bulk materials by human beings and have specific properties associated with their unique dimensions. These are released into the biosphere during the production, consumption, and disposal of nano-enabled products. An enormous number of engineered nanomaterials having different shape, size, and

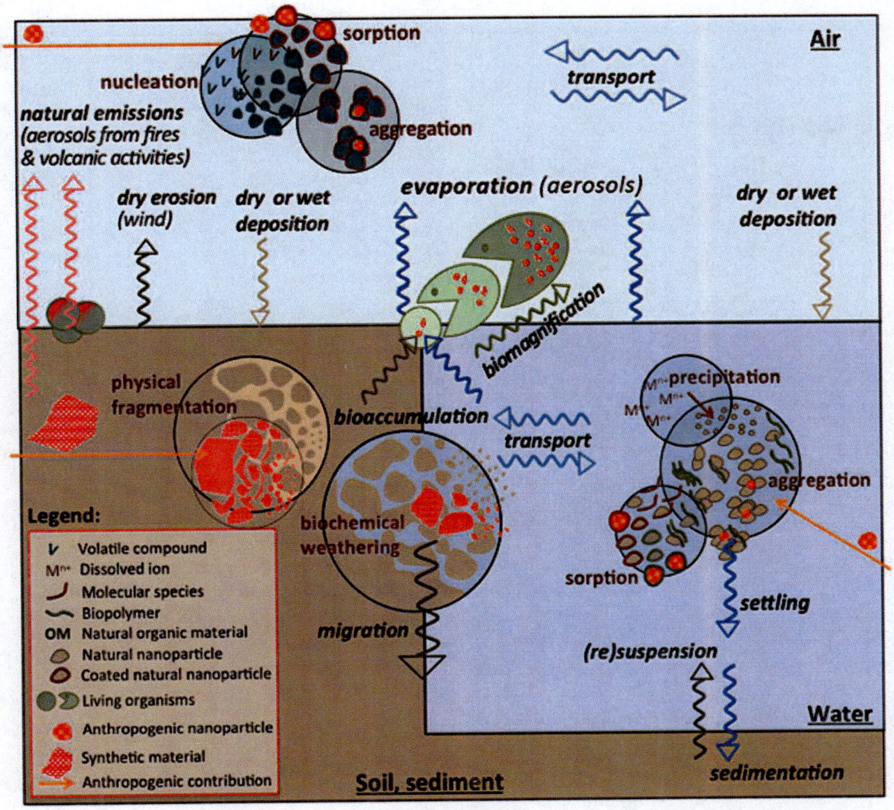

Fig. 2.2 Nano-sized particles in the environmental compartments (Reprinted with permission from Lespes et al. 2020)

composition have been developed (Fig. 2.3). For example, more than 50,000 types of carbon nanotubes (CNTs) have been produced, and about 5000 patents have been awarded for the CNTs. The engineered nanomaterials exhibit distinct characteristics at cellular, atomic, and molecular levels which help in their utilization for different purposes.

Even the *Bhasma* (ash obtained through incineration), a metallo-mineral preparation used in the Indian Ayurvedic system of medicine, is considered a mixture of biologically produced nanoparticles. The starter material undergoes purification through a complex process, followed by the reaction phase wherein another mineral or herbal component is incorporated in the mixture. Then, the mixture is incinerated in a furnace which minimizes the toxicity of the metals leading to the production of nontoxic biologically active nanoparticles. Examples are *Swarna Bhasma* (gold), *Heerak Bhasma* (diamond), *Trivang Bhasma* (tin, lead, and zinc), *Abhrak Bhasma* (mica), *Akik Bhasma* (silicon dioxide), *Naga Bhasma* (lead), *Kansya Bhasma* (bronze), *Loha Bhasma* (iron), *Shankha Bhasma* (conch shell), and *Tamra Bhasma* (copper). Electron microscopic studies have shown that the *Swarna Bhasma*

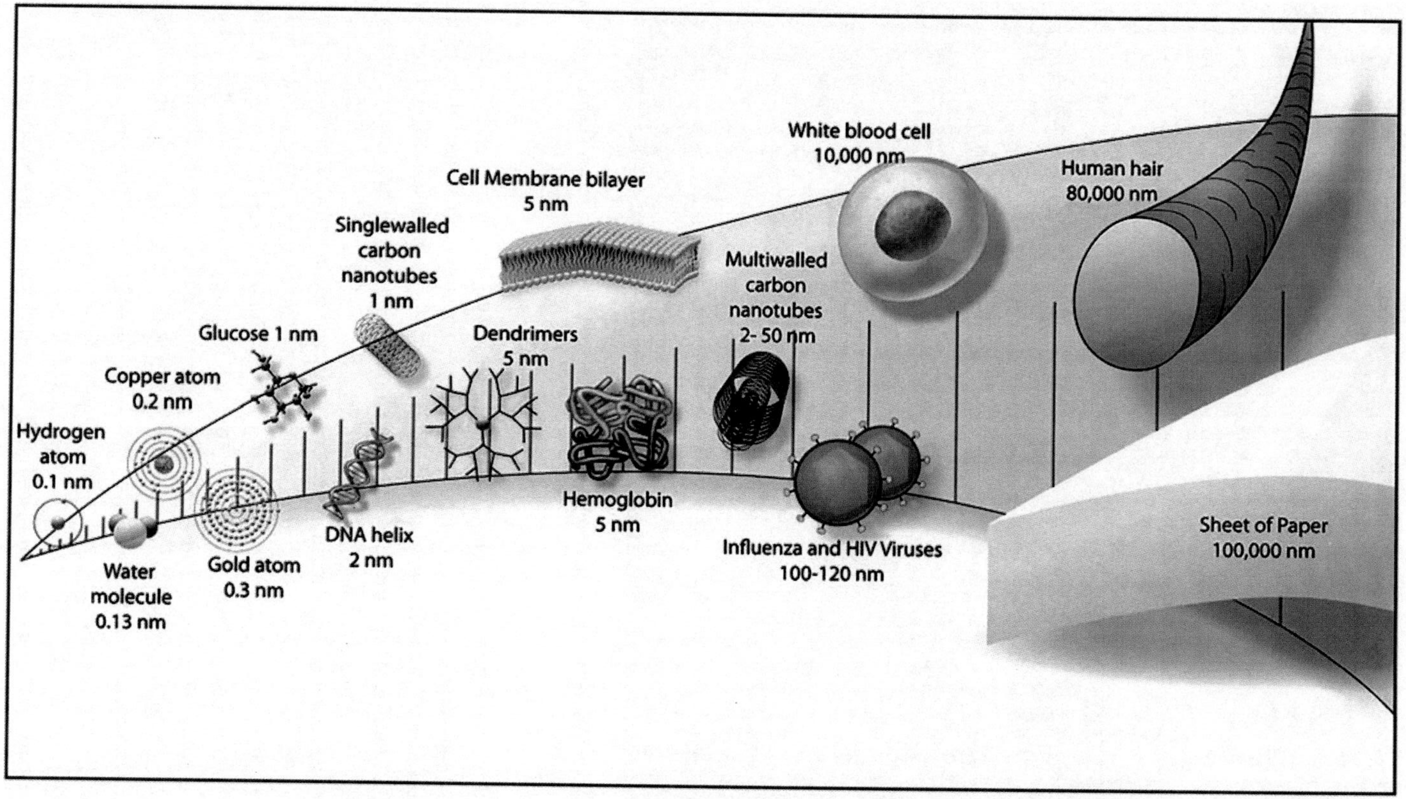

Fig. 2.3 Morphology and size of some engineered nanomaterials and some common materials (Reprinted with permission from Yokel and MacPhail, 2011)

Table 2.3 Role of engineered nanoparticles in soil restoration (Reprinted with permission from Bandala and Berli 2019)

ENM type	Soil application	Notes	Reference
Natural iron oxide and silver NPs	Metal removal (Pb, Cu, Sb)	NMs were tested jointly with biochar for soil restoration	Rajapaksha et al. (2015)
SAMMS[a]	Extraction of PAHs	The sequestration of PAHs in soil using SAMMS was demonstrated	Brandl et al. (2015)
Emulsified zero-valent iron (ZVI) NPs	Trichloroethylene (TCE) removal	TCE ranging 0.439–1.18 mg L^{-1} was removed 65–85% in 90 days	Viskutyte and Varma (2014)
ZVI NPs	Hg, Ni, Cd, Pb, and Cr removal	Nano ZVI was found to reduce and adsorb metals from soil	Rabbani et al. (2015)
Colloidal ZVI	Cd removal from aqueous soil solution	Use of carboxymethyl cellulose stabilized ZVI to improve colloid stability	Nasiri et al. (2013)
Iron nano-oxides	As removal	Nano-goethite, nanomahemite, and nanomagnetite was used to trap as in soil samples	Zhang et al. (2010); Shipley et al. (2011)
Nano-manganese oxides	As removal/ oxidation	Different Mn oxides were used for as removal/oxidation in soil	Watanabe et al. (2013); Villalobos et al. (2014)
Metal-NPs	Organic and inorganic pollutants removal	Hetero- and homogeneous systems are used for abiotic soil remediation	Floris et al. (2017)
Other NMs	Cr adsorption and reduction	Cr and As are immobilized/ reduced in soil	Martínez-Fernández et al. (2017)

[a]Self-assembled monolayers on mesoporous supports.

comprises of gold nanoparticles having size of 56–57 nm, while the *Bhasma* containing mercury compounds contain mercury sulfide in the range of 25–50 nm.

The engineered nanomaterials enter the biosphere through various means, with some nanomaterials being more prevalent in the biosphere than others. These entities are biologically more active than bulk particles having the same composition and also show greater uptake in living beings. The chief features influencing the toxicity of engineered nanomaterials are their size, shape, aspect ratio (length/diameter), surface coating, solubility, surface reactivity, interaction with proteins, receptor binding, and propensity to agglomerate.

Engineered nanomaterials are increasingly being explored for improving soil characteristics and their effectiveness in promoting sustainable agriculture (Table 2.3). The small size coupled with large surface area of the engineered nanomaterials makes them highly reactive and confers unique properties like enhanced cation exchange, sustained nutrient release that find potential applications in soil remediation and food industry, improved efficacy of traditional plant disease

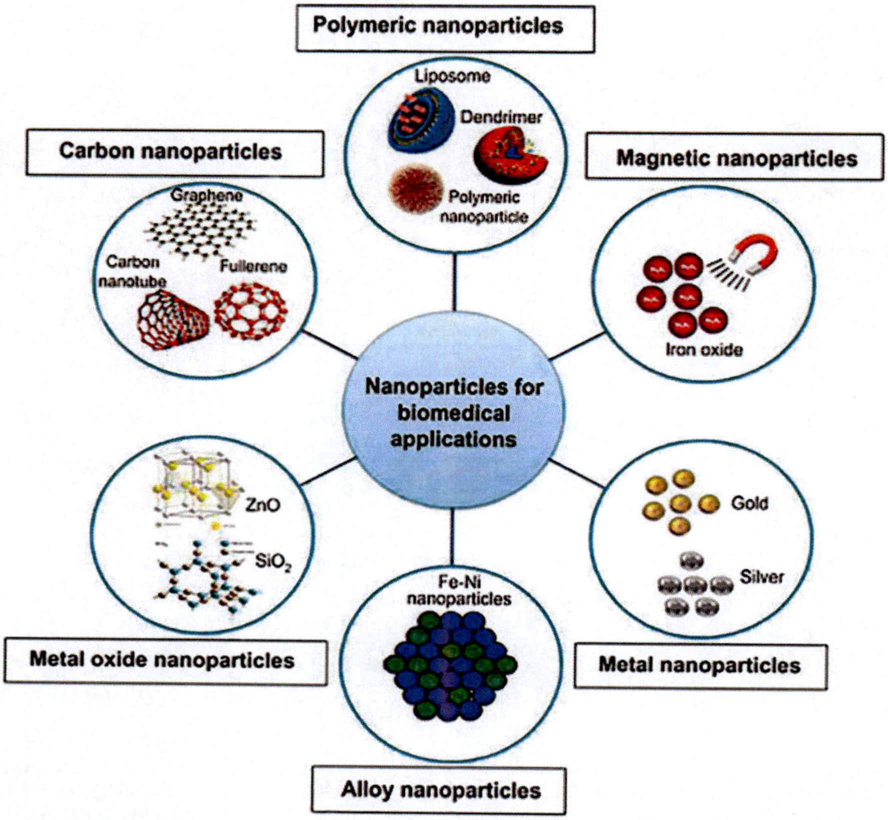

Fig. 2.4 Diverse nanoparticles used in biomedical applications (Reprinted with permission from Syama and Mohanan 2018)

management strategies, as well as beneficial effects on crop yield. Some notable examples include nanoclays, nanosized sulfonated polyaniline (nSPANI), chitosan, and zein nanoparticles that act as suitable carriers to encapsulate natural insecticides and aid in their controlled release which aids in increasing crop productivity and lessens environmental degradation. Engineered nanomaterials also help in the reclamation of water and energy generation.

A promising area that is gaining popularity is the use of engineered nanomaterials in biomedical research due to their biocompatibility, chemical stability, high saturation magnetization, and ease of functionalization which enhances their stability. However, nanomaterials require adequate surface treatment to tune them for biomedical applications. Figure 2.4 depicts the different types of nanoparticles used in biomedical research.

However, indiscriminate use of engineered nanomaterials in increasing agricultural productivity poses immense risk to edaphic and human systems. The harmful effects of nanomaterials range from damage to beneficial soil microorganisms, root growth inhibition, impaired cell division, and reduced seed germination. With respect to their effects after entry in animal systems including human beings, nanoparticles generate reactive oxygen species (ROS) inside the cells leading to oxidization of several biomolecules and resultant alteration in cell signaling. Therefore, surface functionalization of nanomaterials is of immense importance before using them for biomedical applications.

2.3 Types of Nanomaterials

Nanotechnology enables the fabrication, modification, and subsequent use of various types of nanomaterials. The nanomaterials can be classified using different criteria like the nanometer range dimensions or their composition into the following three types:

1. Zero-dimension structures: These are nanomaterials of ultra-small size which are confined in three dimensions, all of which are in nanometer range. These nanomaterials are equipped with novel properties like high photoluminescence (PL) quantum efficiency and chemiluminescence as compared to their bulk counterparts. Examples of zero-dimensional structures are graphene quantum dots, fullerenes, inorganic quantum dots, nanoporous silicon, nanograins, nanoshells, nanocapsules, nanorings, magnetic nanoparticles, activated carbon, and quasicrystals.
2. One-dimensional structures: These structures are confined to two dimensions known as one-dimensional structures, i.e., one dimension is outside the nanoscale. Some examples of one-dimensional structure are nanorods, nanofilaments, quantum wires, and nanowires. Carbon nanotubes (CNTs) were the first one-dimensional structures to be synthesized by Sumio Iijima in 1991 using a simple arc-evaporation method. These are cylindrical graphene sheets of sp^2-bonded carbon atoms. CNTs have properties like large surface area and exceptional mechanical strength (hundred times that of steel) along with high elasticity due to which the CNTs can bend, twist, and buckle without significant damage to its structure. Other important features include high thermal (750 °C at normal pressure) properties, unique electrical conductivity, and high surface-area-to-volume ratio making them ideal for industrial applications like vehicle manufacturing, electrochemical capacitors, solar cells, nano-electronics, optics, material applications, and lithium-ion batteries. Another example of such structure is nanowires, viz., cylindrical structures having a diameter $\leq$100 nm but more than 1000 times longer than their diameter. Nanowires show immense potential for use in electronics, opto-electronics, and nano-electromechanical devices like transistors, memory devices, electronic circuits, biomolecular nanosensors, field emitters, and quantum devices.
3. Two-dimensional structures: These were first reported by Novoselov et al. (2004) who successfully synthesized graphene by exfoliating from graphite. Two-dimensional nanomaterials are few atoms thick having sheet-like

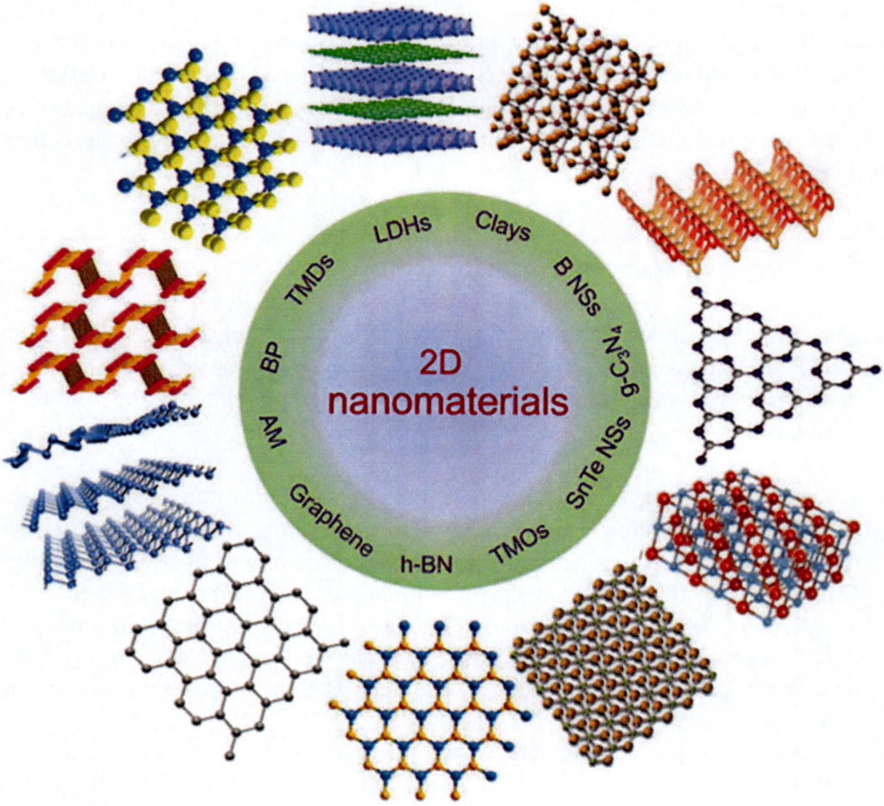

Fig. 2.5 Some examples of 2-D nanomaterials (Reprinted with permission from Hu et al. 2019)

morphology, large surface, and anisotropic physical/chemical properties. These types of nanomaterials can be amorphous or crystalline, having a high degree of chemical functionality, and comprise of different components integrated in a matrix of metals, ceramics, or polymeric substances. They are the thinnest material known with strong atomic and chemical bonds along with high theoretically specific surface area. Some examples of two-dimensional nanomaterials include graphene, boron nitride nanosheets, transition metal dichalcogenides (TMDs), transition metal oxides (TMOs), MXenes, super lattices, and quantum wells (Fig. 2.5). The unique physical, electronic, optical, chemical and surface properties make 2-D nanostructures ideal for diverse use in optoelectronics, catalysis/electrocatalysis, energy storage facilities, biosensing, supercapacitors, and biomedical applications especially drug delivery, photothermal therapy, photodynamic therapy, and tissue engineering.

Another arrangement is based on whether the nanomaterials are organic, nonorganic, or combination of both.

1. Organic nanomaterials: Organic nanomaterials are specialized structures where natural or synthetic organic molecules act as templates. The organic nanomaterials are more biodegradable and used for the production of bionanocomposites by association with a biodegradable polymer. A good example of organic nanomaterials is polymeric nanoparticle of 1–1000 nm consisting of chitosan, alginic acid, and albumin forming a matrix surrounded by the surfactants. These are structured via spontaneous sophisticated self-assembly process during which the medicinally important molecule is entrapped within the polymer nanoparticle core. Polymeric nanoparticles offer numerous advantages like controlled release to the desired site, conferring stability to the labile molecules, and modification of surfaces with ligands for targeted drug delivery. Nanoemulsions, another organic nanomaterial, are a colloidal particulate system which are solid spheres having a negatively charged amorphous and lipophilic surface. Nanoemulsions are formed either by low or high energy emulsification with the stability and functional properties influenced by many factors. Nanoemulsions find important applications in the food industry, namely, as coloring agent, ingredient of coating in food packaging, encapsulation of flavor and nutraceuticals in medical science for drug delivery, and biological or diagnostic agents. Other organic nanomaterials are liposomes, solid-lipid nanoparticles, and carbon-based nanomaterials.
2. Inorganic nanomaterials: Inorganic nanomaterials are synthesized by the precipitation of inorganic substances comprising of elemental metals, metal oxides (TiO_2, Al_2O_3, ZrO_2, MnO, Fe_3O_4, and CeO_2), and metal salts. Some attractive inorganic nanoparticles include metals, silica, inorganic dendrimers, organic-inorganic hybrids, and bioinorganic hybrids. Inorganic nanoparticles are non-toxic and hydrophilic and can be either crystalline or amorphous solids at ambient temperature with spherical or nonspherical shape. Their unique characteristics enable their utilization in several physical, biological, and biomedical (disease diagnosis, imaging, and drug delivery) applications. Inorganic nanoparticles are utilized in clinical science as drugs, imaging agents, and antiseptics and in food industry for biofortification, browning inhibition, detection, and prevention of food contamination, food processing, and biodegradable food packaging. Quantum dots (QDs), also known as "artificial atoms" are man-made crystalline semiconductor nanomaterials that exhibit quantum mechanical behavior due to the specific size of their energy band gaps. These are formed either by single-element materials (Se or Ge) or from compound semiconductors (CdSe, CdTe, or Pb) in which all three dimensions fall in the 1–10 nm range.

References

Albalawi F, Hussein MZ, Fakurazi S, Masarudin MJ (2021) Engineered nanomaterials: the challenges and opportunities for nanomedicines. Int J Nanomedicine 16:161–184

Aswathanarayan JB, Vittal RR (2019) Nanoemulsions and their potential applications in food industry. Front Sustain Food Syst 3:95

Babu VJ, Vempati S, Uyar T, Ramakrishna S (2015) Review of one-dimensional and two-dimensional nanostructured materials for hydrogen generation. Phys Chem Chem Phys 17(5):2960–2986

Bajpai VK, Kamle M, Shukla S, Kumar Mahato D, Chandra P, Hwang SK, Kumar P, Huh YS, Han YK (2018) Prospects of using nanotechnology for food preservation, safety, and security. J Food Drug Anal 26(4):1201–1214

Bandala ER, Berli M (2019) Engineered nanomaterials (ENMs) and their role at the nexus of food, energy, and water. Mater Sci Energy Technol 2:29–40

Borthakur P, Boruah PK, Sharma B, Das MR (2016) Nanoemulsion: preparation and its application in food industry. Emulsions 3:153–191

Brandl F, Bertrand N, Lima EM, Langer R (2015) Nanoparticles with photo induced precipitation for the extraction of pollutants from water and soil. Nat Commun 6:7765

Chen Y, Li CP, Chen H, Chen Y (2006) One-dimensional nanomaterials synthesized using high-energy ball milling and annealing process. Sci Technol Adv Mater 7:839–846

Date AA, Desai N, Dixit R, Nagarsenker M (2010) Self-nanoemulsifying drug delivery systems: formulation insights, applications and advances. Nanomedicine 5:1595–1616

De la Rosa G, García-Castañeda C, Vázquez-Núñez E, López-Camacho PY, Basurto-Islas G, Castro-Beltrán R, Alba-Rosales JE (2019) Synthesis and production of engineered nanomaterials for laboratory and industrial use. In: Marmiroli N, White J, Song J (eds) Exposure to Engineered Nanomaterials in the Environment. Elsevier, Amsterdam, pp 3–30

Donsi F (2018) Applications of nanoemulsions in foods. In: Jafari SM, McClements DJ (eds) Nanoemulsions: formulation, applications, and characterization. Academic Press, Cambridge, pp 349–377

Erickson HP (2009) Size and shape of protein molecules at the nanometer level determined by sedimentation, gel filtration, and electron microscopy. *Biol Proced Online* 11:32–51

Fathi M, Donsi F, McClements DJ (2018) Protein-based delivery systems for the nanoencapsulation of food ingredients. Compr Rev Food Sci Food Safety 17(4):920–936

Floris B, Galloni P, Sabuzi F, Conte V (2017) Metal systems as tools for soil remediation. Inorg Chim Acta 455(2):429–445

Fryd MM, Mason TG (2012) Advanced nanoemulsions. Annu Rev Phys Chem 63:493–518

Goindi S, Kaur A, Kaur R, Kalra A, Chauhan P (2016) Nanoemulsions: an emerging technology in the food industry. Emulsions 3:651–688

Griffin S, Masood MI, Nasim MJ, Sarfraz M, Ebokaiwe AP, Schäfer KH, Keck CM, Jacob C (2017) Natural nanoparticles: A particular matter inspired by nature. Antioxidants (Basel) 7(1):3

Hochella MF Jr, Mogk DW, Ranville J, Allen IC, Luther GW, Marr LC, McGrail BP, Murayama M, Qafoku NP, Rosso KM, Sahai N, Schroeder PA, Vikesland P, Westerhoff P, Yang Y (2019) Natural, incidental, and engineered nanomaterials and their impacts on the Earth system. Science 363(6434):eaau8299

Hu T, Mei X, Wang Y, Weng X, Liang R, Wei M (2019) Two-dimensional nanomaterials: fascinating materials in biomedical field. Sci Bull 64:1707–1727

Jaiswal M, Dudhe R, Sharma PK (2015) Nanoemulsion: an advanced mode of drug delivery system. 3. Australas Biotechnol 5(2):123–127

Janković NZ, Plata DL (2019) Engineered nanomaterials in the context of global element cycles. Environ Sci Nano 6:2697–2711

Jin W, Xu W, Liang H, Li Y, Liu S, Li B (2016) Nanoemulsions for food: properties, production, characterization, and applications. Emulsions 3:1–36

Jun Y-S, Kim D, Neil CW (2016) Heterogeneous nucleation and growth of nanoparticles at environmental interfaces. Acc Chem Res 49:1681–1690

Khanna P, Kaur A, Goyal D (2019) Algae-based metallic nanoparticles: Synthesis, characterization and applications. J Microbiol Methods 163:105656

Lalwani G, Henslee AM, Farshid B, Lin L, Kurtis Kasper F, Qin Y-X, Mikos AG, Sitharaman B (2013) Two-dimensional nanostructure-reinforced biodegradable polymeric nano-composites for bone tissue engineering. Biomacromolecules 14:900–909

Lee S-H, Gordon H, Yu H, Lehtipalo K, Haley R, Li Y, Zhang R (2019) New particle formation in the atmosphere: from molecular clusters to global climate. J Geophys Res Atmos 124:7098–7146

Lespes G, Faucher S, Slaveykova VI (2020) Natural nanoparticles, anthropogenic nanoparticles, where is the frontier? Front Environ Sci 8:71

Martínez-Fernández D, Vitkova M, Michalkova Z, Komarez M (2017) Engineered nanomaterials for phytoremediation of metal/metalloids contaminated soils: implications for plant physiology. In: Ansari AA, Gill SS, Gill R, Lanza GR, Newman (eds) Phytoremediation: management of environmental contaminants. Springer, Switzerland

Mas-Ballesté R, Gómez-Navarro C, Gómez-Herrero J, Zamora F (2011) 2D materials: to graphene and beyond. Nanoscale 3:20–30

McClement DJ, Xiao H (2012) Potential biological fate of ingested nanoemulsions: influence of particle characteristics. Food Funct 3(3):*202–220*

McClements DJ, Rao J (2011) Food-grade nanoemulsions: formulation, fabrication, properties, performance, biological fate, and potential toxicity. Crit Rev Food Sci Nutr 51:285–330

Mortimer M, Holden PA (2019) Fate of engineered nanomaterials in natural environments and impacts on ecosystems. In: Marmiroli N, White J, Song J (eds) Exposure to engineered nanomaterials in the environment. Elsevier, Amsterdam, pp 61–103

Nasiri J, Gholami A, Panahpour E (2013) Removal of cadmium from soil resources using stabilized zero-valent iron nanoparticles. J Civ Eng Urban 3(6):338–341

Nile SH, Baskar V, Selvaraj D, Nile A, Xiao J, Kai G (2020) Nanotechnologies in food science: applications, recent trends, and future perspectives. Nano-Micro Lett 12:45

Novoselov KS, Geim AK, Morozov SV, Jiang D, Zhang Y, Dubonos SV, Grigorieva IV, Firsov AA (2004) Electric field effect in atomically thin carbon films. Science 306(5696):666–669

Pal D, Sahu CK, Haldar A (2014) Bhasma: the ancient Indian nanomedicine. J Adv Pharm Technol Res 5(1):4–12

Pal SK (2015) The Ayurvedic bhasma: the ancient science of nanomedicine. Recent Patents Nanomed 5:12

Rabbani MM, Ahmed I, Park SJ (2015) Application of nanotechnology to remediate contaminated soils. In: Hasegawa H, Rahman HM, Rahman MA (eds) Environmental remediation technologies for metal-contaminated soils. Springer, New York, pp 219–229

Rajapaksha AU, Ahmad M, Vithanage M, Kim K, Chang JY, Lee SS, Ok YS (2015) The role of biochar, natural iron oxides, and nanomaterials as soil amendments for immobilizing metals in shooting range soil. Environ Geochem Health 37:931–942

Sarkar PK, Chaudhary AK (2010) Ayurvedic Bhasma: The most ancient application of nanomedicine. J Sci Ind Res 69:901–905

Sharma VK, Filip J, Zboril R, Varma RS (2015) Natural inorganic nanoparticles- formation, fate, and toxicity in the environment. Chem Soc Rev 44:8410–8423

Shipley HJ, Engates KE, Guettner AM (2011) Study of iron oxide nanoparticles in soil for remediation of arsenic. J Nanopart Res 13(6):2387

Syama S, Mohanan P (2018) The promising biomedical applications of engineered nanomaterials. In: Hussain CM (ed) Handbook of nanomaterials for industrial applications. Elsevier, Amsterdam, pp 530–542

Tiwari JN, Tiwari RN, Kim KS (2012) Zero-dimensional, one-dimensional, two-dimensional and three-dimensional nanostructured materials for advanced electrochemical energy devices. Prog Mater Sci 57:724–803

Villalobos M, Escobar-Quiroz IN, Salazar-Camacho C (2014) The influence of particle size and structure on the sorption and oxidation behavior of birnessite: I. Adsorption of As (V) and oxidation of As(III). Geochim Cosmochim Acta 125:564–581

Viskutyte J, Varma RS (2014) Greener and sustainable remediation using iron nanomaterials. In: Sharma VK, Chang S, Doong R, Wu C (eds) Green catalysis for energy transformation and emission control. American Chemical Society, New York, pp 1–21

Wagner S, Gondikas A, Neubauer E, Hofmann T, von der Kammer F (2014) Spot the difference: engineered and natural nanoparticles in the environment-release, behaviour, and fate. Angewandte Chemie 53:12398–12419

Wan J, Lacey SD, Dai J, Bao W, Fuhrer MS, Liangbing H (2016) Tuning two-dimensional nanomaterials by intercalation: materials, properties and applications. Chem Soc Rev 45:6742–6765

Watanabe J, Tani Y, Chang J, Miyata N, Naitou H, Seyama H (2013) As (III) oxidation kinetics of biogenic manganese oxides formed by *Acremonium strictum* strain KR21 2. Chem Geol 347: 227–232

Yokel RA, MacPhail RC (2011) Engineered nanomaterials: exposures, hazards, and risk prevention. J Occup Med Toxicol 6:7

Zhang MY, Wang Y, Zhao DY, Pan G (2010) Immobilization of arsenic in soils by stabilized nanoscale zero-valent iron, iron sulfide (FeS), and magnetite (Fe_3O_4) particles. Chin Sci Bull 55(4–5):365–372

Zielińska A, Carreiró F, Oliveira AM, Neves A, Pires B, Venkatesh DN, Durazzo A, Lucarini M, Eder P, Silva AM, Santini A, Souto EB (2020) Polymeric nanoparticles: production, characterization, toxicology and ecotoxicology. Molecules 25(16):3731

Tools and Techniques Used in Nanobiotechnology

3

Abstract

The diverse application of nanoparticles in science, technology, and medicine requires an in-depth understanding of properties of nanomaterials. Detailed knowledge of the nanomaterials from the atomic/molecular to crystal structure and finally up to the microstructure should be available at the disposal of personnel for their use in different sectors. Alteration in structural and functional parameters of nanoparticles affects their physicochemical and physiological properties which are of prime importance while utilizing nanomaterial in diverse areas. Therefore, proper characterization of the nanomaterials is of prime importance and needs to be accurately elucidated. Several characterization techniques ranging from microscopy to spectroscopy are available that aid in the detailed analysis of different types of nanoparticles.

3.1 Introduction

Nanomaterials are polycrystalline solids with particle size ranging from 1 to 100 nm and are characterized by small size coupled with large surface area that imparts peculiar properties to the nanoparticles. Utilization of nanomaterials in electronics, medicine, and agriculture requires a proper and in-depth understanding of properties of nanomaterials. Detailed knowledge of the structure right from the atomic/molecular to crystal structure and finally up to the microstructure (mesoscopic scale and defect structure) should be available at the disposal of personnel for their use in different sectors. The following needs to be fully understood with respect to the nanostructures:

- Size range
- Porosity

S. Srivastava, A. Bhargava, *Green Nanoparticles: The Future of Nanobiotechnology*, https://doi.org/10.1007/978-981-16-7106-7_3

- Surface area
- Morphology
- Zeta potential
- Adsorption potential
- Agglomeration
- Distribution of conjugated moieties
- Impurities
- Molecular structure
- Chemical composition
- Melting and boiling point
- Vapor pressure
- Flash point
- pH
- Solubility

Alteration in the abovementioned parameters affects their physicochemical and physiological properties which are of prime importance while utilizing nanomaterial in diverse areas. Several accurate characterization techniques are in use for exploring various aspects of nanostructures (Table 3.1).

3.2 Microscopy

Man has always been passionate to explore the unexplored. Microscopy is the visualization of minute objects which cannot be seen with the naked eye with the use of instruments known as microscopes. The word microscope has been derived from the words "μικρός, *mikrós*," meaning "small," and "σκοπεῖν, skopeî," meaning "to see." A microscope must fulfill the following:

1. Production of an enlarged image of the object.
2. Segregation of details of the object.
3. Making the object sufficiently visible for human beings.

Microscopy has been in a period of rapid evolution, with its core technologies and techniques becoming diverse and more sensitive. The development of new imaging structures have made advances possible across a wide range of research disciplines, demonstrating that microscopy is as important now as it has ever been.

The credit for the development of a microscope, comparable to the compound microscope, goes to the Dutchmen Hans Lippershey and Zacharias Janssen. This instrument had an adjustable magnification between 3 and 9x. However, the word "microscope" was coined by Giovanni Faber in the year 1625 referring to an instrument invented by Galileo in 1609 by modifying one of his telescopes into microscope. The British microscopist Robert Hooke developed an instrument having an objective lens that was moved away or close to the specimen for better focusing. However, this instrument was marred by chromatic aberration, with all images

Table 3.1 Parameters needed to be determined and the corresponding characterization techniques (Reprinted from Mourdikoudis et al. 2018)

Entity characterized	Characterization techniques suitable
Size (structural properties)	TEM, XRD, DLS, NTA, SAXS, HRTEM, SEM, AFM, EXAFS, FMR, DCS, ICP-MS, UV-vis, MALDI, NMR, TRPS, EPLS, magnetic susceptibility
Shape	TEM, HRTEM, AFM, EPLS, FMR, 3D-tomography
Elemental-chemical composition	XRD, XPS, ICP-MS, ICP-OES, SEM-EDX, NMR, MFM, LEIS
Crystal structure	XRD, EXAFS, HRTEM, electron diffraction, STEM
Size distribution	DCS, DLS, SAXS, NTA, ICP-MS, FMR, superparamagnetic relaxometry, DTA, TRPS, SEM
Chemical state—Oxidation state	XAS, EELS, XPS, Mössbauer
Growth kinetics	SAXS, NMR, TEM, cryo-TEM, liquid-TEM
Ligand binding/composition/density/arrangement/mass, surface composition	XPS, FTIR, NMR, SIMS, FMR, TGA, SANS
Surface area, specific surface area	BET, liquid NMR
Surface charge	Zeta potential, EPM
Concentration	ICP-MS, UV-vis, RMM-MEMS, PTA, DCS, TRPS
Agglomeration state	Zeta potential, DLS, DCS, UV-vis, SEM, Cryo-TEM, TEM
Density	DCS, RMM-MEMS
Single particle properties	Sp-ICP-MS, MFM, HRTEM, liquid TEM
3-D visualization	3D-tomography, AFM, SEM
Dispersion of NP in matrices/supports	SEM, AFM, TEM
Structural defects	HRTEM, EBSD
Detection of NPs	TEM, SEM, STEM, EBSD, magnetic susceptibility
Optical properties	UV-vis-NIR, PL, EELS-STEM
Magnetic properties	SQUID, VSM, Mössbauer, MFM, FMR, XMCD, magnetic susceptibility

containing "halos" of blue or red in color. By the late 1600s, there was significant improvement in the magnifying power and image quality that was attributed to improvement in lenses. However, it was in 1667 that an English scientist, architect, and polymath scholar Robert Hooke developed scientific application of microscopes in his monumental work "Micrographia." The text contained illustrations with some having 50x magnifications. About a century later, the achromatic lens was invented by Charles Hall in the 1730s. Another 100 years later Joseph Lister solved the intriguing problem of spherical aberration by placing lenses at precise distances from each other. These two inventions were of utmost significance and led to a considerable improvement in the quality of image. Table 3.2 depicts the chronology of development of microscope and microscopy.

Table 3.2 Chronological development of microscopic techniques

Year (AD)	Achievement
1000	The first vision aid, called a reading stone, is invented
1021	Muslim scholar Ibn al-Haytham wrote his *book of optics*
1284	Salvino D'Armate invented the first wearable eye glasses
1590	Hans Lippershey and Zacharias Janssen invent the forerunner of the compound microscope
1609	Galileo Galilei develops a compound microscope with a convex and a concave lens
1625	Giovanni Faber coins the name "microscope"
1665	English physicist Robert Hooke publishes *Micrographia*
1676	Dutch scientist Antonie van Leeuwenhoek builds a simple microscope with one lens
1730	Achromatic lens was invented by Charles hall
1830	Joseph Jackson Lister reduces spherical aberration
1929	Philipp Ellinger and august Hirt invented the "epi-fluorescence" microscope
1931	Ernst Ruska and max knoll design and build the first transmission electron microscope (TEM)
1932	Development of phase contrast microscope
1942	Ernst Ruska builds the first scanning electron microscope (SEM)
1957	Marvin Minsky patents the principle of confocal imaging
1962	Osamu Shimomura, frank Johnson, and Yo Saiga discover green fluorescent protein (GFP) in the jellyfish *Aequorea victoria*
1967	Invention of dichroic mirrors
1972	Godfrey Hounsfield and Allan Cormack develop the computerized axial tomography (CAT) scan
1973	John Venables and CJ Harland observe electron backscatter patterns (EBSP) in the scanning electron microscope
1978	Thomas and Christoph Cremer develop the first practical confocal laser scanning microscope
1981	Gerd Binnig and Heinrich Rohrer invent the scanning tunneling microscope (STM) at IBM research-Zurich
1986	Nobel prize in physics is awarded jointly to Ernst Ruska, Gerd Binnig, and Heinrich Rohrer First experimental implementation of atomic force microscope (AFM) made by Binnig, Quate, and Gerber
1988	Electrochemical scanning tunneling microscope developed by Itaya and Tomita
1990	Denk used a laser in a confocal scanning microscope to image human kidney cells with two-photon excitation (2PE)
1993	Scanning near-field optical microscopy (SNOM) developed by Eric Betzig and Robert Chichester
1994	Stimulated emission depletion (STED) microscopy developed by Stefan W. hell and Jan Wichmann
1993–96	Stefan hell pioneers super-resolution microscopy
1997	Dynamics of neurons within a live rat brain were first observed by K. Svoboda
2004	Selective plane illumination microscopy pioneered and developed by Ernst Stelzer
2008	Osamu Shimomura, Martin Chalfie and Roger Tsien awarded Nobel prize in chemistry for discovery and development of green fluorescent protein (GFP)

(continued)

Table 3.2 (continued)

Year (AD)	Achievement
2014	Nobel prize in chemistry awarded to Eric Betzig, Stefan hell, and William Moerner for the development of super-resolved fluorescence microscopy
2017	Nobel prize in chemistry awarded jointly to Jacques Dubochet, Joachim frank, and Richard Henderson for developing "cryo-electron microscopy"

Microscopy is used for detailed study of the size, structure, and composition of different nanomaterials. Among the various microscopic techniques, electron microscopy stands has been the technique of choice for thorough analysis of nanomaterials due to high magnification which enables easy viewing of materials at nanometer range.

3.2.1 Scanning Electron Microscope (SEM)

SEM is an advanced analytical tool in which the object is scanned with a highly focused beam of electrons. The beam interacts with different atoms presented at the surface of the sample which provides sufficient details about the surface morphology and produces high-quality images with good spatial resolution. The electron bean used for scanning comprises of highly energetic electrons typically from 0.2 to 40 keV that leads to the image of the sample. The image yields information not just on the purity of the sample but also the degree of aggregation. The magnification obtained by SEM can range from 10 to about 3,000,000 times.

3.2.1.1 SEM Setup

The typical experimental setup of SEM has been provided in Fig. 3.1. The main components of SEM include the following:

- Electron gun for steady stream of electrons
- Lenses to produce clear and detailed images
- Low vacuum electron detectors
- Vacuum chamber where specimens are placed
- Computer and display to view the images

Typically, the source of electrons is an electron gun (either thermionic or field emission guns) which produces a continuous beam of electrons. The filament of the commonly used thermionic gun is made up of tungsten or lanthanum hexaboride, both being cheap and having high melting point and low vapor pressure. One or two condenser lenses positioned above the electron source and apertures focus these electrons to a very fine point of the dimension of nanometer. A pair of deflector plates align, tilt, shifts, and scan the electron beam for effective rastering over the specimen placed in the path of the electron beam. The maximum resolution obtained

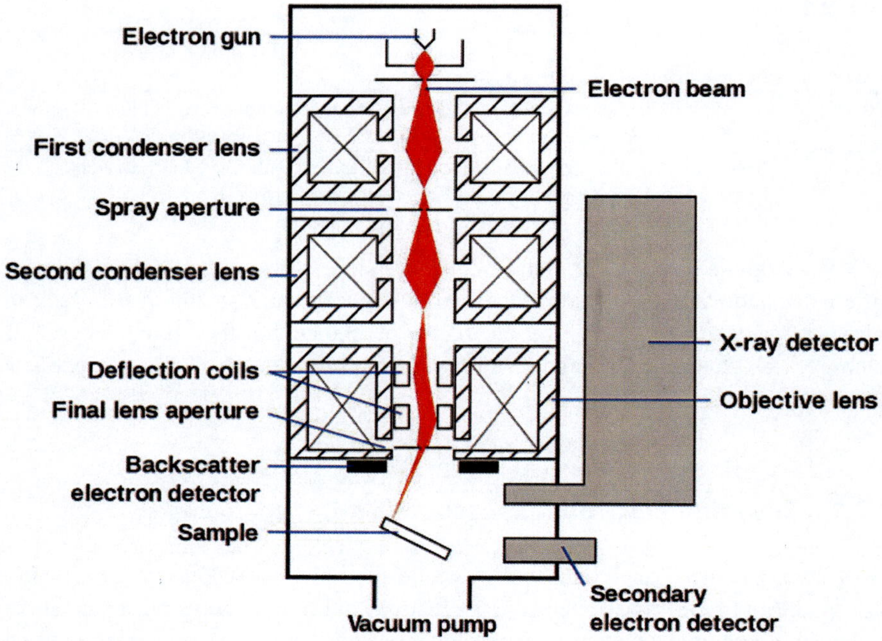

Fig. 3.1 Diagrammatic representation of a SEM setup (Source: Wikipedia)

in a SEM depends on factors like electron spot size and interaction volume of the beam with the object. A modified SEM technique, known as "environmental scanning electron microscopes," facilitates observation of wet and insulating specimens. This instrument, designed by Gerasimos Danilatos, is different from conventional SEM in that imaging can be carried out without destruction using an electron beam in a high chamber pressure atmosphere of water vapor.

3.2.1.2 Advantages of SEM

SEM provides the following advantages:

- SEM offers high-quality images with resolution (high lateral resolution of 1–10 nm) having dimensions as low as a few nanometers.
- Invaluable information about the structure and composition of the specimen.
- Complex sample preparation is not required.
- A large depth of field that enables greater part of the specimen to be focused at one time.
- Sample can be of any thickness, i.e., large bulky samples can be accommodated.
- Sample can be visualized as a three-dimensional image.
- SEM enables magnification of closely spaced samples and provides for higher resolution.
- User-friendly and easy to operate with proper training.
- Data can be generated in digital form.

3.2.1.3 Disadvantages of SEM

1. Cost and setup: The setup of SEM is expensive and huge and requires an area free of electrical and magnetic interference.
2. Maintenance: SEM requires hefty maintenance in terms of steady voltage, electromagnetic coils, and cold water supply.
3. This technique is confined to solid specimens that can withstand vacuum and can be accommodated inside the vacuum chamber. In majority of the cases, specimens should be solid and vacuum compatible. However, imaging of non-conductive and volatile vacuum-sensitive samples can be effectively carried out using higher pressures.
4. The coating of insulators with metals or carbon leads to production of artifacts which sometimes lead to inaccurate results (Fig. 3.2). The artifacts produced during SEM are of three types, viz., random, time-dependent distortion; non-random, time-independent spatial distortion; and non-random, time-dependent distortion.
5. SEM is a radiation-generating device where electrons are backscattered from the specimen. However, this risk has been minimized by designing of chamber that eliminates the escape of radiation.

3.2.2 Transmission Electron Microscope (TEM)

TEM has been the technique of choice for efficient characterization of nanomaterials. It provides reliable chemical information and high-resolution images down to the level of nanometers. TEM provides information on the size, morphology, and composition of the specimens. The first TEM was developed by Knoll and Ruska (Nobel prize in 1986 for this invention) in 1931 followed by its commercial version in 1939.

3.2.2.1 TEM Setup

TEM involves focusing of a high-energy coherent beam of electrons on an ultra-thin specimen with some of it being transmitted depending upon the thickness and transparency of the sample. A typical TEM system comprises of the following components:

1. Emission source or cathode
2. Specimen stage
3. Objective lens
4. Projector lens
5. Magnification system
6. Imaging device: normally a CCD (charge-coupled device) camera

The initial step in TEM involves the preparation of the specimen through various stages like fixation, dehydration, embedding, thin sectioning, and staining. The sample used in TEM should be quite thin and generally less than 150 nm so that

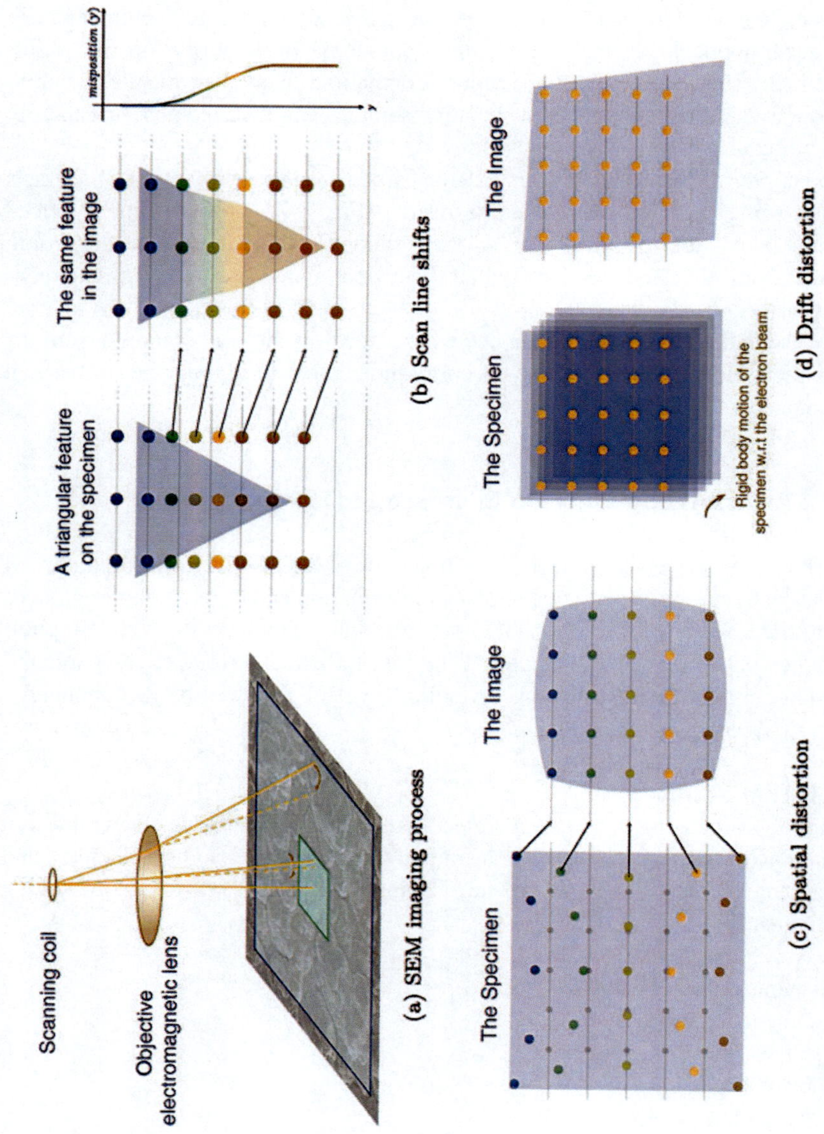

Fig. 3.2 Origin of SEM imaging artifacts (Reprinted from Maraghechi et al., 2019)

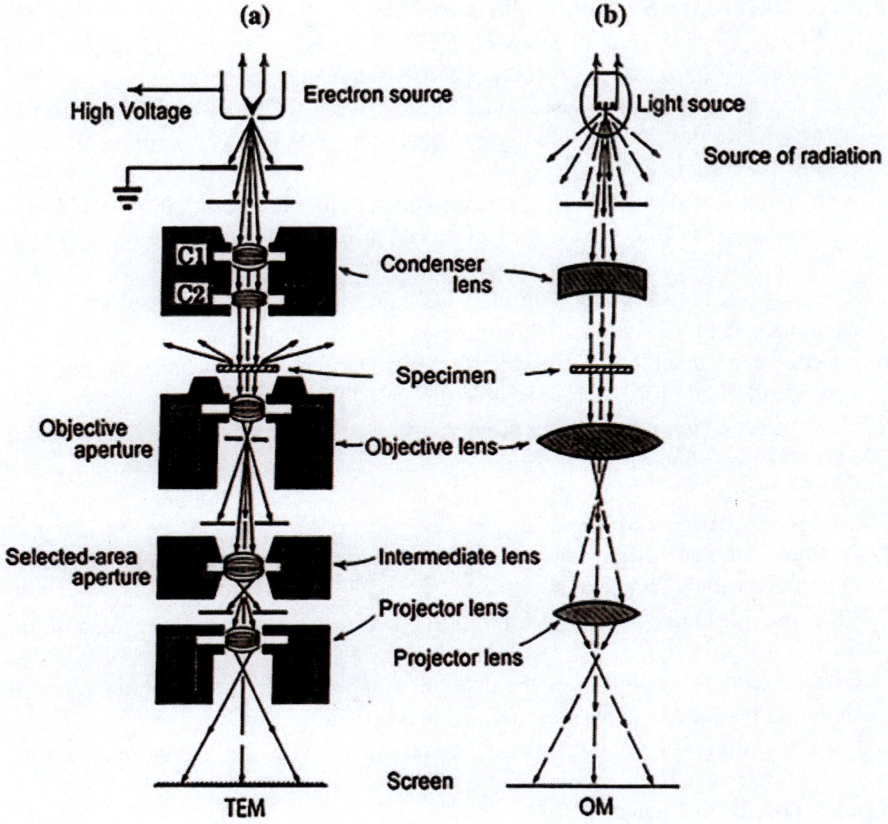

Fig. 3.3 Comparison of TEM (**a**) and an optical microscope (**b**) (Reprinted with permission from Tanaka 2017)

electrons can pass through it. Thereafter, a high voltage (40–200 keV) monochromatic beam of electrons is directed through the specimen of interest. During this transmission, electrons interact with the object to form a real, highly magnified image that is focused onto an imaging device. The different parts of the specimen are displayed in varied darkness according to density. The typical magnification obtained through TEM is around 1,000,000 x, while the resolution power is about 0.2 nm. TEM works on the same principle as light microscope but has about 1000 times more resolution. TEM is also similar to SEM with the difference that in TEM the beam passes through the sample. A number of TEM modifications are available like scanning TEM (STEM), high-resolution transmission electron microscopy (HRTEM), low-voltage electron microscope (LVEM), cryo-TEM, environmental TEM, and aberration corrected TEM.

3.2.2.2 Difference Between SEM and TEM

1. TEM uses a broad static beam, while SEM focusses the beam to a fine point and scans line by line over the specimen in a rectangular raster pattern.
2. The accelerating voltages used in SEM are quite low (30 kV) in comparison to TEM which operates at much higher values (60–300 kV) as it requires penetration of the sample.
3. TEM requires tedious sample preparation, while it is much simpler in SEM.
4. An object of any thickness can be observed in SEM, while it must be <150 nm thick in TEM.
5. The magnifications obtained through TEM are much higher (~50 million times) as compared to SEM (1–2 million times).
6. SEM offers a much larger view of the specimen which is in sharp contrast to a very small view of the sample visible through TEM.
7. TEM offers a two-dimensional projection image, while image obtained through SEM is three-dimensional.

3.2.2.3 Advantages of TEM

1. A phase-contrast image comprising of bright as well as dark field images is obtained through TEM.
2. TEM provides huge magnification of the specimens that can be over one million times.
3. Offers excellent resolution of the crystallographic structure of specimens, even showing the arrangement of atoms within a sample.
4. Provides information about surface features, shape, size, and structure.

3.2.2.4 Disadvantages of TEM

1. Many samples require extensive preparation for producing thin sample through which electrons can pass. Due to this, TEM imaging requires skilled personnel and ample time leading to low throughput of samples.
2. Requirement of an ultrahigh vacuum environment.
3. The structure of the sample may be altered during the preparation process. Some specimens, especially the biological materials, are damaged by the electron beam.
4. Requirement of trained manpower for operating it.
5. Live specimens cannot be observed since it requires placing the specimens in a vacuum chamber.
6. Setup is large and expensive.
7. Only those specimens can be studied that can fit into the vacuum chamber and withstand high pressure of the order of 10^{-4} Pa.

3.2.3 Scanning Tunneling Microscope (STM)

STM involves obtaining ultra-high resolution images of the specimen by measurement of the tunneling current between a conductive tip and the specimen on application of an electric potential. It was developed by Gerd Binnig and Heinrich

Rohrer in 1981 for which they were conferred with the Nobel Prize about 5 years later. The invention of STM heralded the era of nanotechnology with wide applications in electrochemistry, semiconductivity and molecular biology. STM works on the principle of quantum tunneling. This involves bringing a conducting tip (usually metallic or semiconducting) close to the surface of the specimen and application of a bias voltage between the two. This allows electrons to quantum mechanically tunnel between the object and the conducting tip and formation of an image due to variation in tunneling current. The tunneling current is a measure of location, applied voltage, and local density on the surface of specimen. A STM comprises of the following components:

1. Rigid metallic scanning tip: usually made of metal or semiconductors like tungsten, platinum-iridium, or gold
2. An xyz-piezoelectric scanner
3. Coarse positioning unit
4. Vibration isolation system
5. Computer

STMs are classified as either ambient STMs that can function in air or other gases at about ambient temperature, or ultra-high vacuum (UHV) STMs that operate under very high vacuum.

3.2.3.1 Limitations of STM

1. It is a surface-sensitive technique that requires an absolutely clean surface for high success rate.
2. This technique is only useful in analyzing conductor and semiconductor samples. Biomaterials are not efficiently analyzed since most of them are insulating.
3. Stable images of thick organic specimens are less frequently obtained through conventional STM since it requires considerable increase in electrical conductivity.

3.2.4 Atomic Force Microscope (AFM)

AFM, also known as scanning force microscope, is a form of high-resolution microscope having the ability to image nanosized objects. The AFM has been widely used for imaging and material characterization at the nanoscale. The AFM was developed in 1986 by Binnig, Quate, and Gerber. The original instrument had an ultra-small probe tip made of diamond at the end of a cantilever with interatomic van der Waals forces providing the necessary interaction. An STM placed above the cantilever acted as the second tip for detecting the vertical movements of the cantilever. The noncontact mode was introduced in 1987, while the first commercially available AFM, the Digital Instruments Nanoscope, was introduced in 1989. AFM has been established as an efficient measuring tool in nanobiotechnology and is expected to contribute a lot in the coming years. The technique can measure structures with vertical resolutions <0.1 nm and X-Y resolutions of around 1 nm.

3.2.4.1 Principle and Components

The instrument comprises of a cantilever having a sharp mounted tip (probe) made of silicon (Si), borosilicate glass, or silicon nitride (Si_3N_4) of 10–20 nm dimensions. The silicon nitride tips used for most applications are pyramidal in shape and durable and provide the sample with a dry surface. On the contrary, the silicon tips are sharp and conical in shape and present a wet surface which is more suited for biomolecular applications. The silicon tips efficiently measure flat surfaces but have the limitation that they do not penetrate the fissures present in small structures. Another disadvantage is the rapid wearing out of silicon tips which reduces image resolution. When the probe is brought close to the specimen surface, a deflection in cantilever is observed due to forces between the probe and specimen. The force imposed by the specimen on the probe is utilized in the formation of a 3-D image of the sample surface. Scanning can be done in the following modes:

1. Contact mode or variable-deflection mode: In this mode the deflection of cantilever is kept constant by repeated adjustment of the tip which is kept in hard contact with the surface. In most cases, this technique is used where the overall force is repulsive. This mode is quite attractive in specialized applications like small, high-speed atomic resolution scans. However, the contact has many drawbacks that limit its application. The downward force on the tip often modifies or induces damage in the soft surfaces like polymers as well as the tip. Therefore, care must be taken while using biological samples by placing them on a suitable substrate for imaging. Sometimes image artefacts are also produced which adds to its demerits.
2. Noncontact mode: In this mode the tip of cantilever never makes contact with the surface of the sample since it slowly scans across the surface at a few angstroms away without touching it. The van der Waals forces existing between the atoms on the sample surface and the cantilever decreases the resonance frequency of the cantilever and maintains a constant oscillation amplitude or frequency which is recorded by a scanner for constructing an image of the sample surface. The most common scanner utilizes a laser beam focused on the top of the cantilever and reflected onto photodetectors. The noncontact mode is much favorable for measurement of soft samples especially the biological ones and causes less or no damage to the sample.
3. Tapping mode: This is intermediate between the contact and the noncontact mode since it involves vibrating the cantilever up and down at or near its resonance frequency, i.e., above the surface of the specimen, the sample with its tip coming in contact with the surface only intermittently. The tapping mode helps in minimizing the shear forces associated with the tip movement and can be used on both wet and dry sample surfaces with equal efficiency. It also provides improved lateral resolution since the specimen is not dragged on the surface which lessens damage to soft samples. Tapping mode is also the most preferred

Table 3.3 Comparison of scanning electron microscopy (SEM) with atomic force microscopy (AFM)

	Scanning electron microscopy (SEM)	Atomic force microscopy (AFM)
1.	Provides a two-dimensional projection or a two-dimensional image of a sample	Provides a true three-dimensional surface profile
2.	Samples viewed require special treatment such as metal/carbon coatings that would irreversibly change or damage the sample	Samples do not require any special treatments
3.	Requires an expensive vacuum environment for proper operation	Can work perfectly well in ambient air or a liquid environment
4.	Low resolution. SEM can image an area on the order of square millimeters with a depth of field on the order of millimeters	High resolution. AFM can only image a maximum height of 10–20 μm and a maximum scanning area of about 150 × 150 mm
5.	Scanning of images is rapid	Relatively slow rate of scanning that often leads to thermal drift in the image
6.	No such phenomenon reported	Images can be affected by hysteresis of the piezoelectric material and cross talk between the *x*, *y*, *z* axes that may require filtering which can lead to flattening of topographical features

method for the visualization of lipid bilayers or single polymer molecules in liquid medium.

Table 3.3 provides a comparison of SEM and AFM.

3.2.4.2 Advantages of AFM

AFM is superior in comparison to other microscopic techniques in the quality and the measure of resolution. These are provided below:

1. Biomaterials can be imaged with minimal damage to the sample surface.
2. No specific requirement of complex sample preparation that could irreversibly change or prevent reuse of the specimen.
3. Imaging of a wide variety of materials (polymers, composites, clay, thick film coatings, biomolecules, and even living organisms) can be carried out with high resolution at the nanoscale in ambient air, ultra-high vacuum (UHV), controlled environment, and liquid medium.
4. A true three-dimensional surface profile of the specimen is obtained.
5. Qualitative as well as quantitative information can be obtained with respect to size, morphology, and surface texture.
6. High-resolution AFM offers resolution that is at par to STM and TEM but with lesser costs.
7. Requires substantially less laboratory space and is much simpler to operate than an SEM/TEM.

3.2.4.3 Disadvantages/Limitations of AFM

The limitations of AFM are as below:

1. Unlike fluorescence techniques, detection or location of specific molecules cannot be carried out. However, this limitation has been somewhat removed with an AFM cantilever tip carrying a ligand, cell adhesion molecule, or chemical entities, which have been found to be effective in detecting functional molecules present on the cell surfaces.
2. AFM can only image maximum scanning area of around 150 × 150 μm and maximum height in the order of few micrometers which is quite less as compared to SEM.
3. The size of the cantilever tip is sometimes larger than the dimensions of the nanostructures that frequently leads to unfavorable overestimation of the sample dimensions.
4. Lack of chemical information limits its widespread uses.
5. Artifacts are sometimes obtained due to non-perfect probe, operating environment, or due to the sample itself.

3.3 Spectroscopic Techniques

Spectroscopy is the technique that involves exploration of spectra produced when materials interact with or emit electromagnetic radiation. This exploratory tool facilitates detailed investigation and analysis of matter at atomic, molecular, and macroscale. Spectroscopy has been utilized in various forms for characterization of nanoparticles (Fig. 3.3).

3.3.1 Ultraviolet-Visible Spectroscopy

UV-visible spectroscopy is a facile, low-cost characterization method for characterization of nanomaterials in the ultraviolet-visible spectral region. This technique, routinely used in analytical chemistry, involves the measurement of light intensity reflected from a sample and its comparison to the intensity of light reflected from a reference material. This task is performed using a specialized instrument known as UV/Vis spectrophotometer. The absorbance is based on their ratio termed as *transmittance* which is usually expressed in percentage (%). Figure 3.4 depicts the structure of a double beam spectrophotometer.

Nanoparticles have a much-reduced particle size than their bulk counterparts and optical properties that are sensitive to several factors like size, shape, concentration, agglomeration, and refractive index. These features are utilized in UV-Vis spectroscopy for identification and characterization of nanomaterials along with evaluation of the stability of their colloidal solutions. The reduced particle size of nanoparticles induces a shift in the absorption wavelength toward a shorter wavelength due to plasmon resonance properties. It is measured by the UV-spectrophotometer which

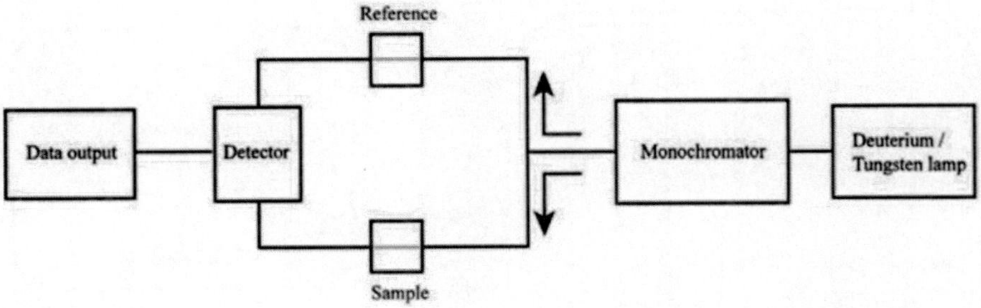

Fig. 3.4 UV. Double beam UV—visible spectrophotometer (Source: Wikipedia)

operates in the range of 200–800 nm and provides information about numerous physical aspects of the nanoparticles. Metal nanoparticles exhibit specific absorbance bands in characteristic spectra like 380–470 nm for silver, 500–550 nm for gold, 540–590 nm for copper, and 370–385 nm for zinc oxide.

3.3.1.1 Advantages of UV-Spectroscopy

1. No requirement of skilled personnel to operate the instrument.
2. Data analysis generally requires minimum processing.
3. Spectrophotometers are inexpensive and accessible for most of the laboratories.
4. Nondestructive method that allows the sample to be reused or used for further processing and analyses.

3.3.1.2 Limitations of UV-Spectroscopy

1. Cannot accurately quantify the extent of metal nanoparticle functionalization: The peak wavelength often shifts due to changes in the medium in which the metal nanoparticles exist along with changes in morphology. Light scattering often caused by suspended solids in liquid samples like unconjugated polymers, protein, reaction intermediates, reactants, impurities, and even bubbles can lead to change in the plasmon resonance resulting in serious measurement errors.
2. Misaligned positioning especially of the cuvette may lead to irreproducible and inaccurate results.
3. Cannot unambiguously account for nanoparticle aggregation.

3.3.2 Fourier Transform Infrared (FTIR) Spectroscopy

FTIR spectroscopy or FTIR analysis is a convenient analytical technique that uses infrared light (within the mid-infrared region of 4000–400 cm^{-1} or 25–2.5 μm wavelength) to scan test samples for identification of different functional groups present in the nanomaterials. Both organic and inorganic compounds in the sample can be easily identified using this technique. It works on the principle that when IR radiation is passed through a sample, some of it is absorbed, while the rest is

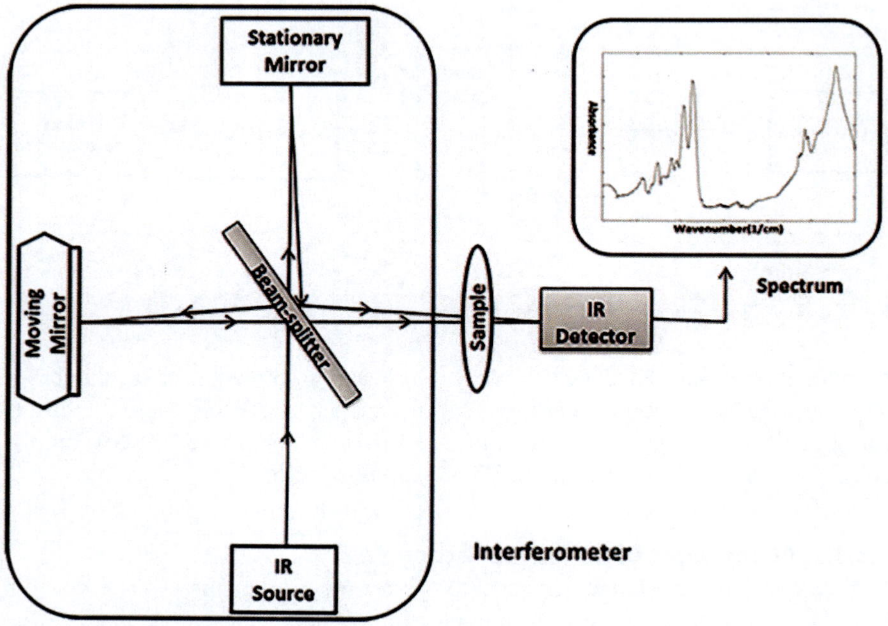

Fig. 3.5 Schematic components of FTIR spectroscopy (Reprinted with permission from Faghihzadeh et al., 2016)

transmitted. The resulting signal represents the molecular "fingerprint" of the sample.

Figure 3.5 depicts the different components of FTIR spectroscopy that are as below:

- Infrared source: typically silicon carbide element, tungsten-halogen lamp, or mercury lamp.
- Interferometer: for spectral encoding.
- Beam splitter for transmitting and reflecting 50% of the incident radiation.
- Fixed and moving mirrors.
- IR detector: for measuring the special interferogram signal.

3.3.2.1 Advantages of FTIR Spectroscopy

1. Sample preparation is quite simple and results are obtained in a short span of time.
2. Both solid and liquid samples can be analyzed with equal efficiency.
3. Provides information about the different functional groups present on the nanoparticle by measurement of the vibrational frequency of the chemical bonds.
4. The molecular data obtained through FTIR allows researchers to establish structural and conformational changes in the functional groups present on the surface of nanoparticles.
5. Small quantity of the sample is required, viz., in μg for solids and μL for liquids.

6. Nondestructive method.
7. High signal-to-noise ratio known as "Fellgett advantage."
8. High-energy throughput known as "Jacquinot advantage."
9. High accuracy and stability known as "Connes advantage."

3.3.3 X-Ray Photoelectron Spectroscopy (XPS)

XPS is a photoemission spectroscopic technique, which is used to analyze the elements existing within a specimen, empirical formulas, surface chemistry, and the density of the electronic states in the specimen. This analytical technique involves irradiation of the sample with X-rays which leads to knocking of electrons from the surface atoms. The enumeration of electrons and measurement of their energy provides valuable information about the surface elements, bonding, and electronic state of these atoms which leads to ascertaining the accurate chemical composition of the sample. A typical XPS setup includes the following:

- Ultra-high vacuum system with mu-metal magnetic shielding.
- X-ray source: an anode usually made up of aluminum that is bombarded with high-energy electrons from a heated filament.
- Electron collection lens.
- Electron detector system: measures the number of electrons arriving at a particular kinetic energy.
- Electron energy analyzer.

XPS offers the advantage in providing information to a depth of about 10 nm below the surface without damaging the samples. Another advantage is the range of materials that can be analyzed which include ceramics, polymers, metals, and composite materials. However, XPS analysis is marred by limitations such as size of sample as the limiting factor, tedious preparation of samples in dry solid form devoid of impurities, and complex interpretation of data. Another important limitation is that XPS is unable detect hydrogen or helium which sometimes leads to inaccurate results since a lot of samples have at least some amount of hydrogen.

3.3.4 Energy Dispersive X-Ray Analysis (EDAX)

This technique, also known as energy-dispersive X-ray microanalysis, is an analytical method used for elemental analysis and chemical characterization of a sample. It involves bombarding the sample with high-energy electron beams (10–20 keV), protons, or X-rays and subsequent compositional analysis of a given sample. The high-energy beams excite an electron of the inner shell and eject it. This leads to creation of an electron hole having a positive charge and subsequent filling of the gap from electron released from a higher-energy shell. The difference in energy between the higher and the lower-energy shells is captured using an energy-dispersive spectrometer, thus allowing a researcher to ascertain the composition of the sample. The EDX setup consists of the following components:

- Electron beam source
- X-ray detector required for converting X-ray energy into voltage signals
- Pulse processor: converts the charge pulse to a voltage pulse
- Analyzer

3.4 Nuclear Magnetic Resonance (NMR)

NMR is a noninvasive technique which utilizes the intrinsic spin properties of atomic nuclei for ascertaining the molecular/chemical environments. It involves putting nuclei possessing a non-zero spin into a magnetic field which produces NMR signal by excitation of the nuclei using radio waves. This signal is then detected using sensitive radio receivers. NMR spectroscopy has been harnessed for determining the 3-D structure, content, and purity of the sample. For known compounds, NMR can characterize the molecular structure along with physical properties like structural and phase changes, solubility, and diffusibility. If the compound is unknown, its molecular structure can be ascertained and matched against available libraries.

In the case of nanomaterials, NMR provides novel information regarding their structure and interactions at the atomic level. The technique finds importance in determination of physiochemical characteristics of nanomaterials like structure, purity, and functionality in dendrimers, polymers, and fullerene derivatives, conformational changes occurring in the interactions between the ligand and the surface of diamagnetic or antiferromagnetic nanoparticles. NMR spectroscopy aids in the detailed investigation of nanoparticle synthesis and their morphology at molecular level in both solid and liquid phases. It is of immense utility in studying the formation as well as the final architecture of noble metal nanoparticles.

3.4.1 Advantages of NMR

NMR affords the following advantages in the characterization of nanoparticles:

1. Capping ligands can also be studied, and information can be gathered on the ligand density, arrangement, and dynamics.
2. Information on particle core like structure, concentration, and compositional architecture can be easily obtained.
3. Nondestructive.
4. Although the spatial resolution of conventional NMR is relatively low (2 to several μm), higher resolution at nanometer scale can be obtained using high-resolution solid-state NMR spectroscopy.

3.4.2 Disadvantages of NMR

NMR has some inherent disadvantages due to which its applications are affected. These are as below:

1. Low detection sensitivity of NMR requires a hefty amount of the sample for measurement.
2. In some instances, NMR takes a lot of time especially when a specific signal-to-noise ratio is required for spectral analysis.

3.5 X-Ray Diffraction (XRD)

XRD is a powerful nondestructive technique for characterizing whether the material is crystalline or amorphous. This analytical method is based on Bragg's law that is given by the following equation:

$$n = 2d\sin$$

where θ = Bragg angle, n = integer, λ = wavelength of X-ray, d = spacing of the crystal layers.

The technique involves projection of monochromatic beam of X-rays on a crystal surface which results in production of constructive interference along with a diffracted ray if the conditions meet Bragg's law. This technique provides valuable information on the identification and characterization of unknown crystalline/semi-crystalline structures, ratio of crystalline to non-crystalline regions, as well as structural parameters like particle size (through Debye Scherer formula), crystal defects and thickness, roughness, and density of thin films.

3.5.1 Advantages of X-Ray Diffraction

XRD offers the following advantages in nanoparticle characterization:

1. A fast, powerful, and well-established technique for identification and characterization of nanomaterials.
2. Provides accurate information on orientation of atoms in space.
3. Minimal sample preparation is required.
4. Results in statistically representative data that is easy to analyze.
5. Composition of the particles can be easily determined.
6. Data interpretation is relatively straightforward.

3.5.2 Disadvantages of X-Ray Diffraction

The following are the major disadvantages of XRD technique:

1. The technique is not suitable for non-crystalline materials.
2. Use of the conventional single peak approach can give erroneous results.
3. Large-sized particles lead to broadening of peaks. Thus, nanoparticles below 3 nm are either not detected or give erroneous results.
4. Low intensity of diffracted X-rays is obtained for materials having low atomic numbers.
5. Anisotropic peak broadening due to defects and microstrain occurring frequently in the material lead to complicated analysis.
6. Requires prior availability of the standard reference file in case of inorganic materials.

3.6 Dynamic Light Scattering (DLS)

DLS or quasi-elastic light scattering (QELS) is an advanced method for measuring the diameter and size distribution of different types of particles dispersed in a liquid medium. It is best suited for submicron particles but has also been efficiently used for measurement of particles in the nanometer range by relating size of the particles with the Brownian motion which is generally faster for small particles as compared to larger particles. The translational diffusion coefficients are calculated by irradiation of the samples using a laser beam and analysis of the intensity fluctuations of the light scattered by the particles using the Stokes-Einstein equation. The large particles induce small fluctuation rates in the scattered light, while the active small particles have high fluctuation rates (Fig. 3.6). These fluctuations are later converted to numerical values for calculation of the coefficient as well as particle radius.

3.6.1 Advantages of DLS Technique

The DLS technique provides the following advantages:

1. Noninvasive nature.
2. Results are obtained in a few minutes.
3. Shows good accuracy in determination of the hydrodynamic size of monodisperse sample.
4. Diluted samples can be measured.
5. Analysis of samples can be carried out in a wide range of concentrations.
6. High-molecular-weight samples can be efficiently detected.
7. Low cost.
8. High reproducibility when compared to other methods.

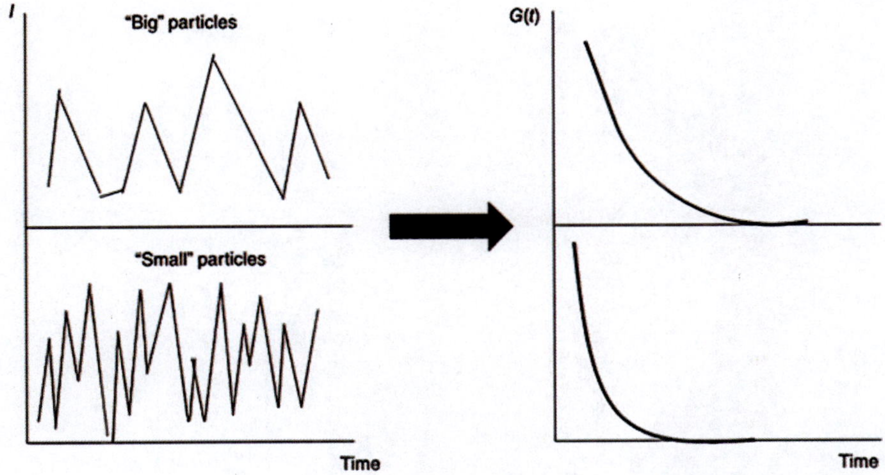

Fig. 3.6 Fluctuations of intensity of scattered light as a function of time and their respective autocorrelation functions [G(t)] (Reprinted with permission from Ramos, 2017)

3.6.2 Disadvantages of Dynamic Light Scattering

The disadvantages and/or limitations of DLS are as below:

1. Difficulty in correlation of size fractions with a particular composition since large-sized particles even when present in small measure are accounted for in this technique.
2. Considerable interference by dust particles in the scattering intensity.
3. Use of dispersants having viscosity >100 mPa.s leads to erroneous estimates.
4. Analysis of particle size is carried out within a narrow range (1 nm–3 μm).
5. Closely spaced particles may sometimes lead to low resolution.
6. Samples with varying size distribution may lead to inaccurate results.
7. Not suitable for accurate measurement of the sizes of non-spherical nanomaterials.
8. Prevalence of multiple light scattering gives unreliable estimates, especially in concentrated samples.

3.7 Zeta Potential

Zeta potential, or ζ potential, conceptualized by Marian Smoluchowski in 1903, refers to the electrokinetic potential between the dispersion medium (aqueous or organic environment) and the stationary layer of fluid attached to the particle referred to as the electric double layer (EDL) (Fig. 3.7). It is analyzed by a zeta potential analyzer which applies an electric charge across the sample and measures the velocity of the charged particles toward the electrode. Zeta potential values typically

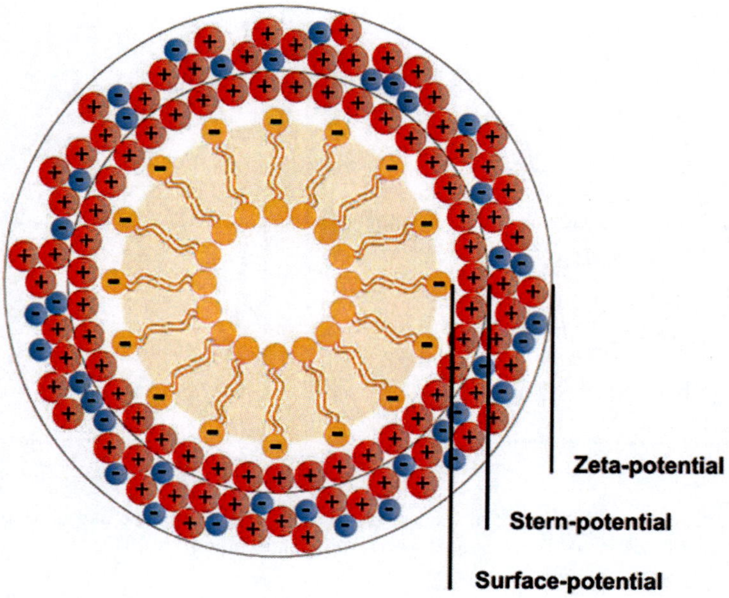

Fig. 3.7 Schematic representation of the double layer that surrounds the nanoparticle in aqueous medium, considering that it has negative charge (Reprinted from Carvalho et al., 2018)

occur in the range of +100 to −100 mV with the values between −10 and + 10 mV considered as close to neutral. The ζ potential measures the magnitude of electrostatic repulsion/attraction between particles and provides information about the stability of colloidal dispersions. A zeta potential of greater than +30 mV indicates stability, while a value more negative than −30 mV is considered stable. The calculation of zeta potential is an important characterization technique of nanoparticles and is increasingly being used for estimating the surface charge, which helps in the understanding the physical stability of nanosuspensions. However, zeta potential is extremely sensitive to pH, ionic strength, nature of the dispersant, and the surface. This often leads to calculation of different zeta potential values for dilute and concentrated samples.

References

Agarwal, N., Nair, M. S., Mazumder, A., & Poluri, K. M. (2018). Characterization of nanomaterials using nuclear magnetic resonance spectroscopy. In: Bhagyaraj SM, Oluwafemi OS, Kalarikkal N, Thomas S. (Eds) Characterization of nanomaterials. Elsevier, Amsterdam pp. 61–10

Akhtar K, Khan SA, Khan SB, Asiri AM (2018) Scanning electron microscopy: principle and applications in nanomaterials characterization. In: Sharma S (ed) Handbook of materials characterization. Springer, Cham, pp 113–145

Behzadi S, Ghasemi F, Ghalkhani M, Ashkarran AA, Akbari SM, Pakpour S, Hormozi-Nezhad MR, Jamshidi Z, Mirsadeghi S, Dinarvand R, Atyabi F, Mahmoudi F (2015) Determination of nanoparticles using UV-vis spectra. Nanoscale 7(12):5134–5139

Binnig G (1990) Atomic Force Microscope and Method for Imaging Surfaces with Atomic Resolution. 4724318A. U.S. Patent

Binnig G, Quate CF, Gerber C (1986) Atomic force microscope. Phys Rev Lett 56:930–933

Binnig G, Rohrer H (1986) Scanning tunneling microscopy. IBM J Res Dev 30(4):355–369

Binnig G, Rohrer H, Gerber C, Weibel E (1982a) Surface studies by scanning tunneling microscopy. Phys Rev Lett 49:57–61

Binnig G, Rohrer H, Gerber C, Weibel E (1982b) Tunneling through a controllable vacuum gap. Appl Phys Lett 40:178

Binnig G, Rohrer H, Gerber C, Weibel E (1983) 7×7 reconstruction on Si(111) resolved in real space. Phys Rev Lett 50:120–123

Bootz A, Vogel V, Schubert D, Kreuter J (2004) Comparison of scanning electron microscopy, dynamic light scattering and analytical ultracentrifugation for the sizing of poly(butyl cyanoacrylate) nanoparticles. Eur J Pharm Biopharm 57:369–375

Brabazon D, Raffer A (2010) 3 – advanced characterization techniques for nanostructures. In: Ahmed W, Jackson MJ (eds) Emerging nanotechnologies for manufacturing. William Andrew Publishing, Boston, pp 59–91

Brar SK, Verma M (2011) Measurement of nanoparticles by light-scattering techniques. TrAC Trends Anal Chem 30:4–17

Carvalho PM, Felício MR, Santos NC, Gonçalves S, Domingues MM (2018) Application of light scattering techniques to nanoparticle characterization and development. Front Chem 6:237

Clogston J, Patri A. Zeta potential measurement. In: McNeil SE. Characterization of nanoparticles intended for drug delivery. Humana Press, New York; 2011. pp. 63–70

Clogston JD, Patri AK (2011) Zeta potential measurement. Methods Mol Biol 697:63–70

Eigler DM, Schweizer EK (1990) Positioning single atoms with a scanning tunneling microscope. Nature 344:524–526

Everett DH (1994) Basic principles of colloid science. The Royal Society of Chemistry, London

Farré M, Barceló D (2012) Introduction to the analysis and risk of nanomaterials in environmental and food samples. In: Farré M, Barceló D (eds) Comprehensive analytical chemistry. Elsevier, Amsterdam, pp 1–32

Faghihzadeh F, Anaya NM, Schifman LA, Oyanedel-Craver V (2016) Fourier transform infrared spectroscopy to assess molecular-level changes in microorganisms exposed to nanoparticles. Nanotechnol Environ Eng 1:1

Filipe V, Hawe A, Jiskoot W (2010) Critical evaluation of nanoparticle tracking analysis (NTA) by nanosight for the measurement of nanoparticles and protein aggregates. Pharm Res 27:796–810

Gmoshinski IV, Khotimchenko SA, Popov VO, Dzantiev BB, Zherdev AV, Demin VF et al (2013) Nanomaterials and nanotechnologies: methods of analysis and control. Russ Chem Rev 82:48

Goldstein JK, Yakowitz H (1975) Practical scanning electron microscopy: electron and ion microprobe analysis. Plenum Press, New York

Hunter RJ (2013) Zeta potential in colloid science: principles and applications. Academic Press, New York

Joseph E, Singhvi G (2019) Multifunctional nanocrystals for cancer therapy: a potential nanocarrier. In: Grumezescu AM (ed) Nanomaterials for drug delivery and therapy. Elsevier, Amsterdam, pp 91–116

Kaliva M, Vamvakaki M (2020) Nanomaterials characterization. In: Narain R (ed) . Polymer science and nanotechnology: fundamentals and applications, Elsevier, Amsterdam, pp 401–433

Khare T, Oak U, Shriram V, Verma SK, Kumar V (2019) Biologically synthesized nanomaterials and their antimicrobial potentials. In: Verma SK, Das AK (eds) Comprehensive analytical chemistry. Elsevier, Amsterdam, pp 263–289

Lin PC, Lin S, Wang PC, Sridhar R (2014) Techniques for physicochemical characterization of nanomaterials. Biotechnol Adv 32(4):711–726

Magonov SN (1993) Surface characterization of materials at ambient conditions by scanning tunneling microscopy and atomic force microscopy. Appl Spect Rev 28:1

Magonov SN, Whangbo M-H (1996) Surface analysis with STM and AFM. VCH, New York

Maraghechi S, Hoefnagels J, Peerlings R, Geers M (2018) Correction of scan line shift artifacts in scanning electron microscopy: an extended digital image correlation framework. Ultramicroscopy 187:144–163

Maraghechi S, Hoefnagels JPM, Peerlings RHJ, Rokoš O, Geers MGD (2019) Correction of scanning electron microscope imaging artifacts in a novel digital image correlation framework. Exp Mech 59:489–516

Mourdikoudis S, Pallares RM, Thanh NTK (2018) Characterization techniques for nanoparticles comparison and complementarity upon studying nanoparticle properties. Nanoscale 10:12871–12934

Neikov OD, Yefimov NA (2019) Powder characterization and testing. In: Neikov OD, Naboychenko SS, Yefimov NA (eds) Handbook of non-ferrous metal powders: technologies and applications. Elsevier, Amsterdam, pp 3–62

Parot P, Dufrêne YF, Hinterdorfer P, Le Grimellec C, Navajas D, Pellequer J-L et al (2007) Past, present and future of atomic force microscopy in life sciences and medicine. J Mol Recognit 20: 418–431

Pı̇collo M, Aceto M, Vitorino T (2019) UV-Vis spectroscopy. Phys Sci Rev 4:4

Ramos AP (2017) Dynamic light scattering applied to nanoparticle characterization. In: Da Róz AL, Ferreira M, Leite FL, Oliveira ON Jr (eds) Nanocharacterization techniques. Elsevier, Amsterdam, pp 99–110

Ringe E (2014) Nanocrystalline materials: recent advances in crystallographic characterization techniques. IUCrJ 1(Pt 6):530–539

Sapsford KE, Tyner KM, Dair BJ, Deschamps JR, Medintz IL (2011) Analyzing nanomaterial bioconjugates: a review of current and emerging purification and characterization techniques. Anal Chem 83(12):4453–4488

Scimeca M, Bischetti S, Lamsira HK, Bonfiglio R, Bonanno E (2018) Energy dispersive X-ray (EDX) microanalysis: a powerful tool in biomedical research and diagnosis. Eur J Histochem 62(1):2841

Strausser YE, Heaton MG (1991) Scanning probe microscopy technology and recent innovations. Am Lab 5:1–7

Tanaka N (2017) Structure and imaging of a transmission electron microscope (TEM). In: Electron nano-imaging. Springer, Tokyo. https://doi.org/10.1007/978-4-431-56502-4_2

Tanzi MC, Farè S, Candiani G (2019) Foundations of biomaterials engineering. Elsevier, Amsterdam, pp 393–469

Uskokovic V (2012) Dynamic light scattering based microelectrophoresis: main prospects and limitations. J Dispers Sci Technol 33:1762–1786

Wang HW, Chu PK (2013) Surface characterization of nanomaterials. In: Bandopadhyay A, Bose S (eds) Characterization of biomaterials. Elsevier, Amsterdam, pp 105–174

Xu R (2008) Progress in nanoparticles characterization: sizing and zeta potential measurement. Particuology 6(2):112–115

Zhou W, Apkarian RP, Wang ZL, Joy D (2006) Fundamentals of scanning electron microscopy. In: Zhoo W, Wang ZL (eds) Scanning microscopy for nanotechnology. Springer, New York, pp 1–40

Green Synthesis of Nanoparticles

4

Abstract

Nanoparticle synthesis is usually accomplished using three methods, viz., physical, chemical, and biological, each having its own unique features. The physical and chemical methods are considered superior for obtaining stable nanostructures having uniform size, but do not fill into our objective of long-term sustainability goals. Considering the limitations of these methods in nanomaterial synthesis, environmentally friendly methods need to be developed using techniques that are clean, nontoxic, and environment-friendly. This has ushered in the evolution of "green nanotechnology" as a new discipline. Bionanofabrication is its most important component since it uses living organisms or their biomass instead of the expensive physical or chemical methods and provides an environmentally benign, less-toxic, and cheap method for nanoparticle synthesis. Biological synthesis has been used for the green synthesis of diverse range of nanoparticles comprising of metals, semiconductors, quantum dots, and alloys.

4.1 Introduction

Nanoparticles are in great demand by different industries due to their unique properties. Conventionally, the production of nanoparticles has been accomplished using two contrasting approaches, viz., the top-down and the bottom-up (Fig. 4.1).

4.1.1 Top-Down Approach

Using this approach, the bulk material is fragmented into smaller particles of desired size by scaling down a complex entity into its component parts. This approach

S. Srivastava, A. Bhargava, *Green Nanoparticles: The Future of Nanobiotechnology*,
https://doi.org/10.1007/978-981-16-7106-7_4

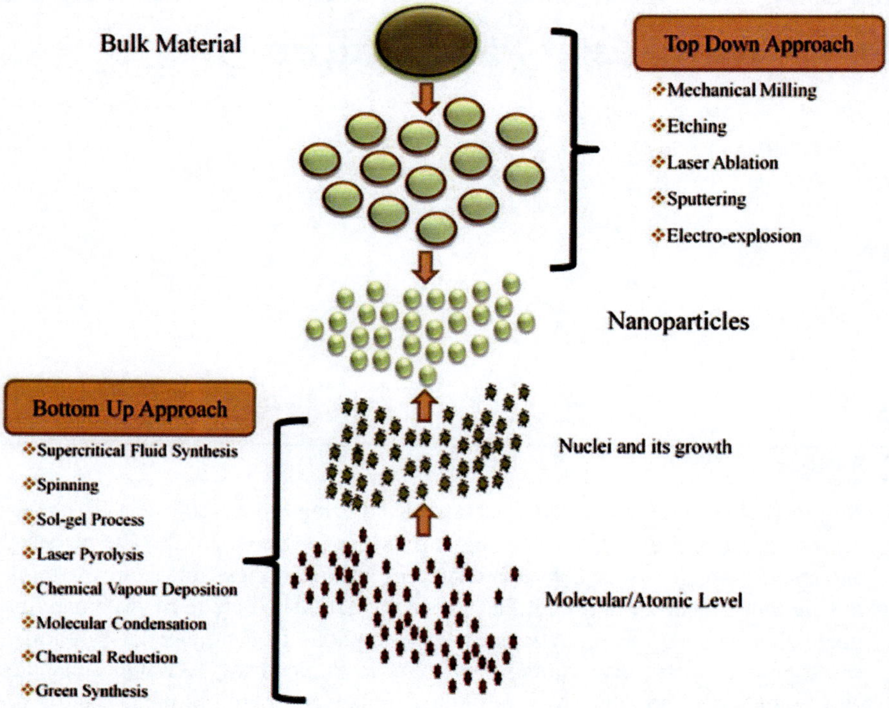

Fig. 4.1 Synthesis of nanoparticles: the top-down and bottom-up approach (Reprinted with permission from Khanna et al., 2019)

carries with it the advantages of simplicity, universality, and low cost. However, the top-down approach has the following limitations:

1. Polydispersity in the resultant materials.
2. Contamination by components of the grinding machinery.
3. Difficulty in attaining nanoscale comminution.

4.1.2 Bottom-Up Approach

The bottom-up approach relies on the molecular self-assembly phenomenon and is primarily a buildup method that envisages building a material from bottom. Thus, in this approach the synthesis of nanostructures is carried out by means of chemical reactions among the atoms/ions/molecules. This impromptu organization of molecules involves different molecular interactions like non-covalent, van der Waals forces and hydrogen bonds. It starts from atoms and gradually builds up a material via atom-by-atom and molecule-by-molecule approach. The bottom-up approach provides spatial control over features, and composition of individual blocks yields well-organized nanostructures as compared to the top-down approach.

4.2 Methods of Nanoparticle Synthesis

Nanoparticle synthesis is accomplished using different methods (Fig. 4.2), each having its own unique features and limitations. Although both the physical and chemical methods are considered superior for obtaining stable nanostructures having uniform size, these do not fill into our objective of long-term sustainability goals.

4.3 Physical Methods

The different physical methods of nanomaterial synthesis have been provided below:

4.3.1 High-Energy Ball Milling

High-energy ball milling, better known as mechanical alloying or mechanical attrition, was developed by John Benjamin in 1970 at the International Nickel Company. It is a novel and efficient technique for grinding powders into fine particles and synthesizing nanoparticles with varying shapes and dimensions. This ultrafine grinding technique has been extensively utilized to grind powders into fine

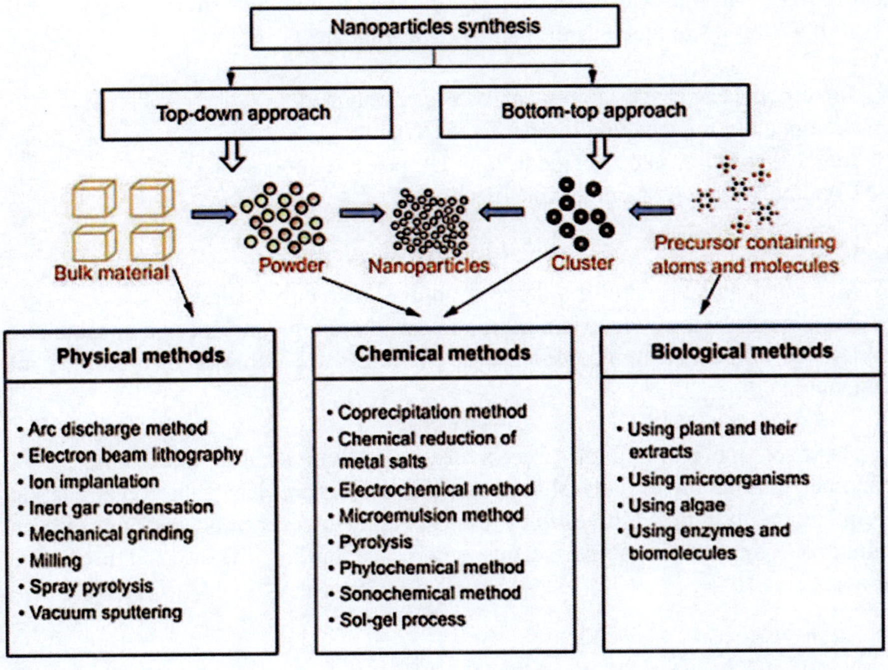

Fig. 4.2 Different methods of nanoparticle synthesis (Reprinted from Synthesis of Inorganic Nanomaterials, Edition 1, Devatha C.P. and Thalla A.K., Green Synthesis of Nanomaterials, Pages 169–184, Copyright (2018), with permission from Elsevier)

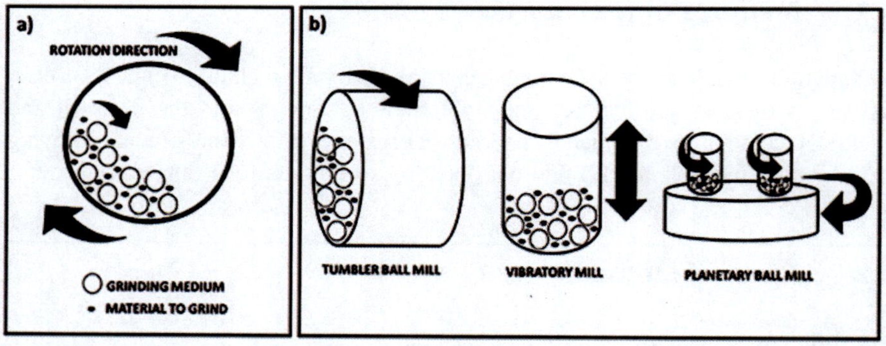

Fig. 4.3 Different types of ball mills (Reprinted with permission from Piras et al., 2019)

particles and obtain nanostructured materials. In this mechanical deformation process, a powdered coarse-grained mixture positioned in a ball mill undergoes dissociation due to high-energy collisions during milling with stiff balls. The simplest ball mill comprises of a hollow cylindrical shell partially filled with balls made of different materials like steel, ceramic, or rubber that rotates on its axis. The technique reduces the grain size of materials from micrometers to nanometers. However, different types of ball mill modifications are available like vibratory, planetary, and tumbler ball mill which are based on the specific requirement (Fig. 4.3).

Ball milling technique has the following advantages:

1. Nanoparticle of size 2–20 nm can be easily obtained.
2. Significantly reduced environmental disposal.
3. Less degradation as compared to high temperature processes.
4. Compatibilization of immiscible blends.
5. Cost-effectiveness.
6. Ease of operation.
7. Materials can be produced in large quantity due to easy scale-up.
8. Results can be easily reproduced due to orderly energy and speed control.
9. Can be efficiently applicable in both wet and dry conditions on a number of materials.

However, the ball milling process also has disadvantages like formation of nanomaterials having irregular shapes. Another serious problem is the contamination from ball and milling additives to the atmosphere. However, this problem has been minimized by the use of surfactants, choice of milling media, and protective atmospheres.

4.3.2 Inert Gas Condensation (IGC)

This technique also known as "gas-phase condensation" (GPC) is the most popular one for nanomaterial synthesis since it produce nanostructures having well-defined grain size with narrow size distribution. After initial experiments in 1930 AD by Pfund and subsequent studies by H.C. Burger and P.H. van Cittert, it was observed that a small sample was obtained and that too with varied particle size. These shortcomings were removed by C.G. Granqvist and R.A. Buhrman in 1976 who modified the method and produced ultrafine particles by IGC method. However, it was H. Gleiter who first synthesized metal (iron) nanoparticles using IGC having properties different than their bulk counterpart. Metals are evaporated in a low-pressure atmosphere in the presence of an inert gas for production of nanoparticles and the particle size manipulated by varying parameters like type of gas, gas pressure, and evaporation rate.

IGC involves two basic steps:

Step 1: Evaporation of the material.
Step 2: Rapid condensation for producing the requisite particle size in a controlled manner.

IGC process involves vaporization of starting materials by evaporation especially in the presence of inert gas (usually helium) atoms. For this the set-up has the following components:

1. Ultrahigh-vacuum (UHV) system: This is usually equipped with evaporation sources.
2. Cluster collection device: For collection of the deposited powder.
3. Scraper: For removal of nanoparticles present in the powder.
4. Compaction device: For consolidation of the collected nanoparticles.

Numerous techniques are employed for vaporization, namely, thermal evaporation, laser vaporization, sputtering, laser arc discharge, and plasma heating. The evaporated atoms lose kinetic energy as a result of collision with the inert gas that leads to formation of clusters by homogenous nucleation near the evaporation source. At higher supersaturation levels, abrupt nucleation takes place that gives rise to a large number of extremely small particles. The ultrafine particles nucleate, develop in the gaseous phase, and finally accumulate on a low temperature surface. Inert gas condensation provides the following advantages:

- Production of particles having small sizes (<10 nm).
- Nanoparticles can be generated within a narrow size range.
- High purity level.
- Good thermal stability.
- Electrically insulated individual particles.

4.3.3 Physical Vapor Deposition

Physical vapor deposition (PVD) refers to set of processes like sputtering, electron beam evaporation, pulsed laser deposition, and vacuum arc. In the physical vapor deposition process, the material transforms from condensed phase to the vapor phase and is later deposited atom-by-atom in the condensed phase on a suitable substrate. The procedure for PVD is as follows:

1. Vaporization of target material from a solid source using a high-power pulsed laser beam.
2. Transfer and condensation of the vapor phase as a thin film in ultra-high vacuum or in an appropriate gas (oxygen, nitrogen, or methane)-rich atmosphere.
3. Nucleation and growth of the nanomaterial in an inert atmosphere.

The following are the main advantages of PVD:

- Simple procedure: The target material is vaporized and transformed into a thin film having similar composition.
- Versatility: A diverse range of materials can be deposited using different gases over varied gas pressures.
- Cost-effectiveness.
- Rapid: Results are obtained within minutes.
- Scalability.

4.3.4 Laser Pyrolysis

It is a gas-phase process for the synthesis of a wide range of nanomaterials in which an infrared (IR) carbon dioxide (CO_2) laser radiation is utilized for heating flowing reactant gases, which results in molecular decomposition to form vapors. This is followed by initiation of nucleation and the growth of nanostructures. Figure 4.4a presents the typical apparatus used for laser pyrolysis. In this process, the gaseous-phase precursors are introduced into a chamber by a carrier gas (usually argon) where they make contact with the laser beam. The high-power (2400 W) beam generates very high temperatures that lead to formation of nanoparticles. Laser pyrolysis facilitates generation of high temperatures in a localized area which leads to rapid nucleation in a small space/volume. The newly formed nanoparticles are later recovered through a catcher equipped with a filter that recovers and separates smaller particles from larger aggregates. Though used to a lesser extent than other methods, laser pyrolysis has been successfully utilized for the synthesis of metals, oxide, non-oxide, carbides, nitrides, and ceramic nanoparticles.

The main advantages of laser pyrolysis are the following:

- Provides highly restricted and quicker heating in a narrow region leading to speedy nucleation and rapid quenching of particle growth.

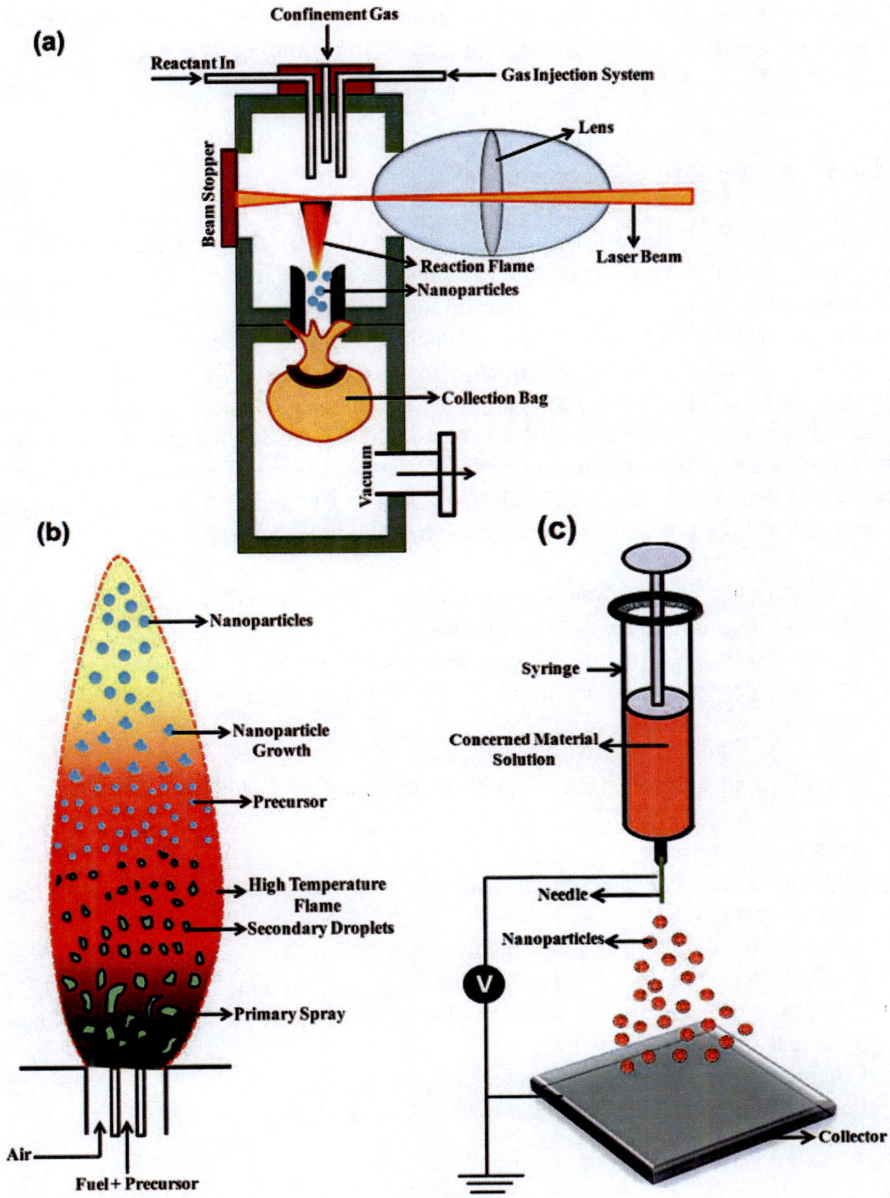

Fig. 4.4 Schematic representation of (**a**) laser pyrolysis, (**b**) flame spray pyrolysis, and (**c**) electrospraying technique for the synthesis of nanoparticles (Reprinted with permission from Dhand et al., 2015)

- Provides greater possibilities for synthesis of nanoparticles having size of 5–60 nm having narrow distribution.
- Formation of high quality end products.

However, the need for a specific reagent/laser resonance and the necessity of complex installation setup leads to higher costs, thus limiting its use.

4.3.5 Flame Spray Pyrolysis

Flame spray pyrolysis is a well-developed, highly versatile, and most exploited technique for the rapid synthesis of complex nanoparticles with engineered functionalities. This one-step combustion process involves the use of a liquid precursor having high combustion enthalpy, normally in an organic solvent (Fig. 4.4b). In this process, an aqueous metal salt solution is sprayed as a fine mist, through a capillary into a high temperature flame. This leads to formation of small droplets, while the solvent burns inside the flame. Salt is converted into the metal oxide through the pyrolysis reaction and the metal oxide atoms aggregate into nanoparticles, which are later collected on a substrate. For the formation of nanoparticles, the precursor spray follows the following sequential steps:

1. Evaporation/decomposition of the precursors to form metal vapors.
2. Nucleation as a result of super saturation.
3. Growth by the twin processes of coalescence and sintering.
4. Particle aggregation by chemical bonds and agglomeration through physical interactions.

The major advantages of flame spray pyrolysis are the following:

- Single-step process.
- Materials are mixed in solution; hence, they are homogeneously mixed on the atomic level at the start.
- Only subsintering temperatures are necessary to form crystallized particles.
- Creation of particle sizes typically in the range of 5–500 nm with high purity.
- Production of nanoparticles with a high specific surface area having low production costs.
- Controllability of primary particle size and film porosity.
- Possibility of in situ deposition of noble metal catalysts.

4.3.6 Electrospraying

Electrospraying, also known as electrodynamic spraying, is a well-established technique based on "electrohydrodynamic atomization" that applies electric field to a precursor solution which forces it to form diminutive droplets with submicron sizes finally culminating in the formation of nanomaterials (Fig. 4.4c). The process of electrospraying is accomplished using the following components:

- High-voltage power supply.
- Syringe made up of metal, plastic, or glass that holds the solution.
- A precise flow syringe pump for controlling the flow rate of the sample solution.
- Collector (solid or liquid-bath collector).

Thus, the presence of high electric field during electrospraying initiates the breakup or dispersion of the liquid stream into fine droplets. The shape and size of the electrosprayed particles can be regulated by alteration in the solvent, polymer solution parameters, polymer molecular weight, and the electrospraying process. Polymer concentration is of immense importance since any alteration in the polymer concentration has a great bearing on the viscosity of the sprayed solution which influences its behavior in the high-voltage medium.

The advantages of electrospraying are as below:

- Process can be carried out at ambient conditions.
- Increased scalable synthesis.
- High encapsulation efficiency due to absence of an external medium.
- Controlled morphology or monodispersity of designed particles.
- Attractive for the production of polymer nanoparticles by control of operational parameters.
- Efficient process for production of natural polymer nanoparticles; stable nanoparticles are produced without alteration in the biological activity of either the drug or the biomolecules.
- Reproducible and ideal process for the synthesis of drug-loaded nanoparticles.

4.3.7 Melt Mixing

This oldest, most common, and adaptable method of nanomaterial production has been primarily used to design clay/polymer nanocomposites of a thermoplastics and elastomeric polymeric matrix. Melt mixing has also been used for synthesizing carbon-based rubber nanocomposites for the different carbon-based nanofillers (carbon nanotubes, graphene, fullerene, nanodiamonds, nanofibers). This process involves melting the polymer and combining it with specific amount of the intercalated clay in the presence of a medium rich in inert gas. Argon, nitrogen, and neon are the gases used for this purpose. Apart from this, the polymer can also be blended with the intercalant and thereafter subjected to heating and subsequent shearing to form the desired polymer nanocomposites.

Melt mixing offers the following advantages:

- Environment friendly since the process does not make use of organic solvents.
- Toxin-free.
- Well-suited for industrial practices.

However, melt mixing has also some disadvantages that have been provided below:

- Poor dispersity of graphene in the matrix due to increased viscosity of the composites.
- May result in buckling of graphene due to intense shearing that are unsuitable for effective conductivity of the composites.

4.4 Chemical Methods

4.4.1 Sol-Gel Synthesis

It is one of the most preferred and versatile bottom-up methods of nanoparticle production. The sol-gel process involves the formation of a diphasic system containing a liquid phase (sol) and a solid phase (gel). Five different types of gels, namely, colloidal, metal oxane polymer, metal complex, polymer complex I in situ polymerizable complex, and polymer complex II coordinating and crosslinking polymers have been found to be useful in the synthesis of nanomaterials through this process. Metal oxides and chlorides, the commonly used precursors, are shaken, stirred, or sonicated that results in the formation of a unique system containing both liquid and solid phases. Thereafter, recovery of nanoparticles is performed by phase separation using different methods like sedimentation, filtration, and centrifugation. Finally, removal of moisture is accomplished by a suitable drying method. The sol-gel method of nanoparticle fabrication is useful in the fabrication of diverse nanoparticles that can be grown within the nanopores, resulting in the synthesis of nanomaterials that are difficult to fabricate.

The procedure of synthesis of nanoparticles using the sol-gel method is as follows:

- Uniform mixing of the precursor materials in a liquid phase.
- Formation of a stable and transparent sol system using techniques like hydrolysis and polycondensation.
- Formation of a 3-D gel having structure formed by the aggregation of colloidal particles after the aging of sols.
- Preparation of nanomaterials after processes like drying and sintering.

The sol-gel method confers several advantages over other chemical synthesis methods. These are as follows:

- The final products are obtained in a homogeneous state which is in sharp contrast to the form of the starting material.
- High-purity products are prepared.
- Control of size, distribution, and arrangement of nanopores.
- Ease in preparation.

- No requirement of high temperature.
- The product's chemical composition can be easily controlled, especially during the preparation of materials having multiple components.
- Dopants, i.e., intentionally added impurities, can be made of which uniformly disperses in the end product.

The disadvantages of the sol-gel method have been provided below:

- Relatively long time required for completing the process.
- The use of some organic solvents can sometimes be damaging to human body.

4.4.2 Microemulsion Method

Microemulsions are thermodynamically stable, homogenous, transparent (or translucent), and optically isotropic mixture that comprises of three components:

1. Polar phase (usually water)—about 45 component.
2. Nonpolar phase (usually oil)—about 45% component.
3. Surfactant—about 10%.

The amphiphilic surfactant molecules (non-ionic, anionic, cationic, and zwitter-ion) act as the emulsifier and stabilize the microemulsion droplets. Cetyl trimethyl ammonium bromide (CTAB), dioctylsulfosuccinate sodium (AOT), phosphatidyl-choline (PC), and Triton X-100 are some of the surfactants that have been utilized for nanofabrication using the microemulsion method. Microemulsion systems comprise of monodispersed spherical droplets (600–8000 nm) of three types, viz., water-in-oil (w/o), oil-in-water (o/w), and bi-continuous microemulsions based on the type of surfactant used. Of these the w/o reverse micellar system has been quite popular for the fabrication of nanoparticles (Fig. 4.5). There are two preferred routes to nano-particle synthesis using the microemulsion technique, viz., one microemulsion and the two microemulsion method (Fig. 4.5). Nanofabrication through the microemulsion method is dependent on the amount of water present, amount of surfactant, presence of cosurfactants, and the concentration of reagents. Microemulsion technique has been largely utilized for the production of core nanoparticles (metals, metal oxide, and semiconductors) as well as core-shell/composite nanoparticles (Table 4.1).

Microemulsion technique provides the following advantages:

- Requirement of mild reaction conditions.
- Easy to prepare: Microemulsions tend to form spontaneously at ambient temperatures and are easy to synthesize as compared to liposomes and macroemulsions, which require high pressure and homogenization.

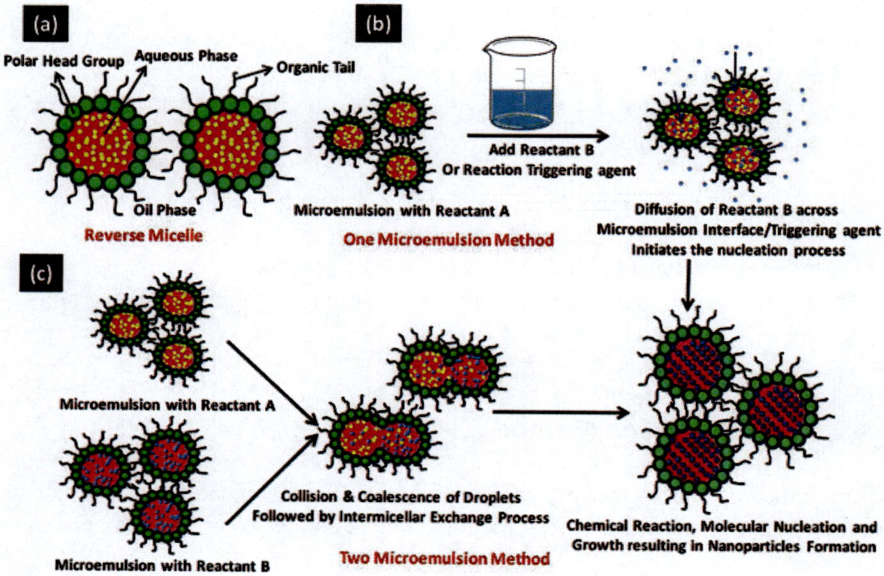

Fig. 4.5 Microemulsions (**a**) typical reverse micelle system, (**b**) different steps in one microemulsion process, and (**c**) reaction sequence involved in the two microemulsion process (Reprinted with permission from Dhand et al., 2015)

- Thermodynamic stability: It is due to the fact that the Gibbs-free energy caused by some effects becomes negative which leads to improved stability of the microemulsion.
- Inorganic nanoparticles can be easily obtained with minimal aggregation.
- Crystalline and monodisperse nanoparticles having high specific surface area can be easily synthesized.
- Size and composition of the nanoparticle can be controlled by adjusting the size of the microemulsion droplet radius and of the precursor concentrations.
- Can be scaled up economically.

Below are given some of the disadvantages of the microemulsion method:

- Often requires a complex setup that is time consuming.
- Requires enormous quantities of surfactant and/or cosurfactants.
- Stability of the microemulsion is affected by external agents like temperature and pH.
- Problems arise while using substances having high melting points which exhibit limited solubilizing capacity.

Table 4.1 Microemulsions in the synthesis of nanomaterials

Category	Type	Reference
Metal nanoparticles	Copper Gold Nickel Platinum Rhenium	Pileni and Lisiecki (1993) Mandal et al. (2014) Chen et al. (2000) Sanchez-Dominguez et al. (2011) Bedia et al. (2015)
Metal sulfide nanoparticles	Cadmium sulfide Chromium sulfide Copper sulfate Copper sulfide Vanadium sulfide	Dhanabalan and Gurunathan (2015) Loukanov and Emin (2016) Haram et al. (1996) Solanki et al. (2010) Loukanov and Emin (2016)
Metal salt nanoparticles	Silver bromide Silver chloride	Monnoyer et al. (1995) Bagwe and Khilar (1997)
Metal oxide nanoparticles	Cobalt hexacyanoferrate Cobalt oxide Copper/cesium Iron oxide Manganese ferrite	Vaucher et al. (2002) Dogra et al. (2019) Pemartin-Biernath et al. (2016) Okoli et al. (2012) Carpenter et al. (1999)
Nanowires	Barium carbonate nanowires Barium sulfate nanowires Cadmium sulfide nanowires Chromium tungstate	Qi et al. (1997) Hopwood and Mann (1997) Routkevitch et al. (1996) Wang et al. (2020)
Nanocomposites	Ag_2S/CdS nanoparticles Au-ag bimetallic nanoparticles Au/Pd bimetallic nanoparticles Pd/Pt bimetallic nanoparticles Pt/Ru bimetallic nanoparticles	Han et al. (1998) Chen and Chen (2002) Wu et al. (2001a) Wu et al. (2001b) Zhang and Chan (2003)

4.4.3 Hydrothermal Synthesis

It refers to the fabrication of crystalline substances/single crystals from high temperature solutions at high pressure and high temperature, normally in specialized equipment known as “hydrothermal autoclave reactor.” The hydrothermal process involves heating of homogeneous solution of the precursors at varying temperatures and high pressures which enables the synthesis of crystalline nanomaterials. It differs from the solvothermal method in having water as the primary solvent instead of an organic one.

The hydrothermal synthesis can be carried out in two types of systems:

1. Batch hydrothermal process carries out a system with the desired ratio phases.
2. Continuous hydrothermal process allows a higher rate of reaction to be achieved within short time duration.

The hydrothermal system has been extensively utilized for synthesizing nanoparticles of various metal oxides especially powdery nanostructures. This method confers the following advantages:

- Experimental setup is easy.
- Production of crystalline phases which are otherwise unstable at higher temperatures.
- Efficiently creates materials/crystalline phases that are not stable at the melting point.
- Creates large-sized, high-quality crystals and nanoparticles with controllable shape, size, and composition.

The hydrothermal method is marred by disadvantages like the need for autoclaves that are costly, difficulty of observing the crystal/nanoparticle as it grows, and hindrance in obtaining the high-quality two-dimensional nanomaterials with controllable thickness and size.

4.4.4 Polyol Synthesis

The polyol method was originally devised by Fievet et al. (1989). This method involves the synthesis of metal-containing compounds using members of the polyol family such as 1,2-ethanediol (ethylene glycol), 1,2-propane diol (propylene glycol), 1,2-butanediol, and 1,5-pentanediol as the reaction medium. These liquid solvents act as reducing and complexing agents and exhibit high boiling point and high polarity allowing the dissolution and reduction of metal salts into nanoparticles with minimal agglomeration. Polyvinylpyrrolidone (PVP) (molecular weight: $M_w \geq 55{,}000$) is the most preferred capping agent which facilitates nanostructure formation via adsorption on the metal surface through the carbonyl group of the pyrrolidone ring. Nanostructures are synthesized in the polyol method through three sequential steps, viz., nucleation, seed evolution, and growth of the nanocrystal. The nanoparticle dimensions depend on the type of polyol used, capping agent, reducing agent, metal precursor concentration, atomic species, reaction temperature, reactant concentration, injection rate, and adsorbates. Polyol method has been effectively utilized in the synthesis of a wide range of metals, metal oxides, and magnetic and metal hybrid nanoparticles.

This technique provides the following advantages:

- The stabilizing and chelating effect of polyols enable synthesis of nanoparticles having uniform size distribution and less agglomeration.
- The abundance of diverse polyols permits for enormous adaptability and flexibility in the synthesis.
- Since polyols show solubility characteristics similar to water, metal salts can be used as the starting material.

- High boiling points of polyols enable the synthesis of crystalline nanostructures at extremely high temperatures without the need of high pressure conditions.

However, the polyol method has some limitations like restricted reducing capacity of polyols and instability of the nonpolar metallic surfaces.

4.4.5 Chemical Vapor Deposition

Chemical vapor deposition (CVD), also known as thin-film deposition, is a material-processing technology that involves condensation of materials from vapor phase for generating solid phase substances through thermally induced chemical reactions. CVD was used to fabricate multi-walled carbon nanotubes (MWNTs) by Endo et al. (1993) and single-walled carbon nanotubes (SWNTs) by Dai et al. (1996). The synthesis of planer nano-graphene was carried out for the first time by Somani et al. (2006) and of single-layer graphene by Pollard et al. (2009). CVD technique can be achieved by exposing the heated target material such as metal or carbon (like methane, carbon monoxide, or acetylene) to one or more volatile precursors (usually a precursor gas or gases) that react and deposit on the substrate surface in a form of a thin film. The main advantages of CVD have been provided below:

- Wide variety of substances can be deposited with high purity.
- Production of high-quality, extremely pure, uniform, and strong nanoparticles with fine structural regularity.
- Highly successful in the synthesis of two-dimensional nanomaterials.

However, CVD has the following drawbacks:

- Requirement of high temperatures (depending on the precursor) which minimizes the list of materials that can be used in this process.
- Difficulty in masking the surface.
- Precursors can sometimes be toxic like nickel tetracarbonyl ($Ni(CO)_4$), explosive like diborane (B_2H_6), or corrosive like silicon tetrachloride ($SiCl_4$).
- Higher production costs due to some costly precursors.
- Prerequisite of specialized equipment.
- Production of toxic by-products.
- In some instances, mechanical instabilities have been observed in the deposited films.

4.4.6 Chemical Vapor Synthesis/Chemical Vapor Precipitation

This bottom-up technique for nanoparticle synthesis, also known as chemical vapor reaction (CVR) and chemical vapor condensation (CVC), involves bringing together the vapor phase precursors (solid, liquid, or gas) into a reactor where nucleation of particles occurs. This is in sharp contrast to chemical vapor deposition where nucleation takes place as deposition of a film on the wall. This method is quite popular for the production of nanoparticles using materials like metal oxides, oxides of group 14 elements, aluminum-doped zinc oxide (AZO), fluorine-doped tin oxide, and chromium-doped ZnO.

4.4.7 Wet-Chemical Method

This versatile and powerful method, also known as solution processing, has been conventionally used for fabrication of nanoparticles having well-defined structural and chemical properties. Solvothermal/hydrothermal synthesis, template synthesis, self-assembly, hot-injection, and interface-mediated synthesis are some of the common processes that come under the purview of the wet-chemical processes. In the solvothermal method, water/organic solvent acts as the reaction medium within a sealed vessel in which the temperature is higher than its boiling point. This results in autogeneration of the solvent at elevated pressure which initiates precursor decomposition and crystal nucleation. If water is used as a solvent, then the process is termed hydrothermal synthesis. The main advantages of solvothermal method are its cost, simplicity, and dissolution of a large number of materials by manipulating temperature and pressure to its critical point. Template synthesis utilizes natural substances or synthetic materials including nanoparticles as templates for the growth of nanorods, nanowire, nanotubes, and nanobelts. The result is formation of aligned nanostructures with desired size, especially the diameter. Template-based synthesis methods are further classified into hard and soft templating methods. Self-assembly process involves spontaneous reorganization of presynthesized low-dimensional nanostructures like nanoparticles and nanowires through noncovalent interactions, such as van der Waals or electrostatic or hydrogen bonds for the fabrication of two-dimensional nanostructures. Two-dimensional-oriented attachment involves association and bonding of adjacent nanocrystals to form two-dimensional nanosheets with a single-crystalline nature with well-defined morphology. Typical examples of two-dimensional nanomaterials include nanodisks, nanoflakes, nanowires, nanosheets, and nanotube arrays. The hot-injection method involves production of homogeneous nuclei through rapid injection of highly reactive reagents into a hot solvent leading to production of monodisperse, highly luminescent semiconductor nanocrystals having uniform shape and purity.

4.4.8 Advantages and Disadvantages of Chemical Methods

Nanomaterials produced through the chemical methods facilitate synthesis of structures with precise dimensions and composition which find application in frontier research areas like electronics, catalysis, data storage, biology, and medicine. However, although chemically produced nanomaterials find applications in diverse areas, their usage has taken a heavy toll on the environment by increasing pollution, a phenomenon termed as nanopollution. The wet chemical methods have certain disadvantages like being capital extensive and costly; using noxious chemicals like Tetrakis(hydroxymethyl)phosphonium chloride, $[(HOCH_2)_4PCl]$, poly–N–vinyl pyrrolidone (PVP) $(C_6H_9NO)_n$, and hydroxylamine (H_3NO); and releasing hazardous by-products. The chemical methods for nanomaterial synthesis usually enhance the particle reactivity and toxicity and can be detrimental to various living forms as well as the environment due to the meager information on their composition and possible interaction with the environmental components. Chemical methods, apart from being expensive and labor intensive, sometimes lead to deposition of noxious substances on the surface of nanoparticles, thus limiting their use in medical biology. They can be a potential hazard for the environment as well as to the living organisms.

4.5 Biological Methods

The initiation of the concept of "green chemistry" and "nanotechnology" has been among the recent approaches that follow the objectives of environmental sustainability. The amalgamation of these doctrines has paved the way to a novel technology known as "green nanotechnology." Considering the negative effects of both the physical and chemical methods of nanomaterial fabrication, efforts began in earnest to develop environmentally friendly methods using techniques that are clean, less-toxic, and environment-friendly. Within the domain of green nanotechnology, biological synthesis of nanoparticles is the most important component and is becoming increasingly popular considering the theme of sustainable development. Green synthesis is a bottom-up approach that uses living organisms or their biomass instead of the expensive physical or chemical methods and provides an eco-friendly cost-effective method for nanoparticle synthesis. Biological synthesis of nanoparticles is cheap and eco-friendly and allows for easy isolation of nanoparticles from the media. The living forms serve as "biological nanofactories" or "biological laboratories" and have immense potential in the production of nanoparticles owing to their cosmopolitan distribution, accessibility, safe handling, and occurrence of a wide range of metabolites. The dimensions and properties of nanoparticles obtained biologically are dependent on the type of biological entity used, its concentration, type of reducing agent, pH, temperature and concentration of the medium, as well as exposure time. Various capping and stabilizing agents required for reducing aggregation and regulating the growth of the nanoparticle also affects its characteristics. These biologically produced nanoparticles are biocompatible owing to the coating of biomolecules on their surface making them ideal candidates for wide applications in biomedicine and agriculture. Dhand et al. (2015) have divided the biological synthesis of nanomaterials into three types:

1. Biosynthesis using microorganisms.
2. Biosynthesis using biomolecules.
3. Biosynthesis using plant extracts.

Biological synthesis using microorganisms involves harnessing different microorganisms like bacteria, actinomycetes, fungi, algae, and yeast for green synthesis of nanostructures having different dimensions. With reference to formation of metal nanoparticles, it has been observed that microorganisms utilize their metabolic enzymes to convert ions present in the environment into element metal, thus forming nanoparticles. The nanofabrication is either intracellular (ions are transported into the microbial cells leading to fabrication of nanoparticles) or extracellular (involves trapping and subsequent reduction of metal ions on the surface of the microbial cells) depending upon the location of nanoparticle synthesis, both requiring the presence of enzymes. Many biomolecules function both as reducing and capping agents for the biosynthesis of nanoparticles with unique structure and properties. Among the different biomolecules, DNA has been the molecule of choice for synthesis of nanoparticles. DNA, first used to program nanomaterials in 1996, has been utilized as an excellent biomolecular template or organic scaffold to produce nanoparticles as well as nanoparticle assemblies like self-assembled $NiWO_4$, $ZnWO_4$, and $MnWO_4$ nanoparticles, metal nanowires, and nanoparticle superlattices. Likewise, amino acids like histidine, lysine, and arginine have been explored in the production of multifunctional metal and metal oxide nanoparticles especially ZnO, FeO, Au, and Ag nanoparticles. Amino acids have the ability to direct and assemble superstructures by acting as surfactant, catalyst, and capping agents. Plant extracts or plant biomass is increasingly gaining popularity for eco-friendly synthesis of a wide range of nanoparticles, primarily of noble metals, metal oxides, and bi-metallic alloys.

References

Anastas PT, Kirchhoff MM (2002) Origins, current status, and future challenges of green chemistry. Acc Chem Res 35:686–694

Anisimov S, Inogamov N, Oparin A, Rethfeld B, Yabe T, Ogawa M, Fortov VE (1999) Pulsed laser evaporation: equation-of-state effects. Appl Phys A Mater Sci Process 69:617–620

Aromal SA, Vidhu VK, Philip D (2012) Green synthesis of well dispersed gold nanoparticles using *Macrotyloma uniflorum*. Spectrochim Acta Part A 85:99–104

Bagwe RP, Khilar KC (1997) Effects of the intermicellar exchange rate and cations on the size of silver chloride nanoparticles formed in reverse micelles of AOT. Langmuir 13:6432–6438

Bedia J, Calvo L, Lemus J, Quintanilla A, Casas J, Mohedano A, Zazo J, Rodriguez J, Gilarranz M (2015) Colloidal and microemulsion synthesis of rhenium nanoparticles in aqueous medium. Colloid Surf A 469:202–210

Bhattacharya D, Rajinder G (2008) Nanotechnology and potential of microorganisms. Crit Rev Biotechnol 25:199–204

Biswas A, Bayer IS, Biris AS, Wang T, Dervishi E, Faupel F (2012) Advances in top–down and bottom–up surface nanofabrication: techniques, applications & future prospects. Adv Colloid Interface Sci 170(1):2–27

Bock N, Woodruff MA, Hutmacher DW, Dargaville TR (2011) Electrospraying, a reproducible method for production of polymeric microspheres for biomedical applications. Polymers 3:131–149

Carpenter EE, Connor CJO, Harris VG (1999) Atomic structure and magnetic properties of $MnFe_2O_4$ nanoparticles produced by reverse micelle synthesis. J Appl Phys 85:5175–5177

Chen DH, Chen CJ (2002) Formation and characterization of au-ag bimetallic nanoparticles in water-in-oil microemulsions. J Mater Chem 12:1557–1562
Chen DH, Wu SH (2000) Synthesis of nickel nanoparticles in water-in-oil microemulsions. Chem Mater 12:1354–1360
Dai H, Rinzler AG, Nikolaev P, Thess A, Colbert DT, Smalley RE (1996) Single-wall nanotubes produced by metal-catalyzed disproportionation of carbon monoxide. Chem Phys Lett 260:471–475
Danks AE, Hall SR, Schnepp Z (2016) The evolution of 'sol–gel' chemistry as a technique for materials synthesis. Materials Horizon 2:91–112
Devatha CP, Thalla AK (2018) Green synthesis of nanomaterials. In: Bhagyaraj SM, Oluwafemi OS, Kalarikkal N, Thomas S (eds) Synthesis of inorganic nanomaterials. Elsevier, Amsterdam, pp 169–184
Dhanabalan K, Gurunathan K (2015) Microemulsion mediated synthesis and characterization of CdS nanoparticles and its anti-biofilm efficacy against *Escherichia coli* ATCC 25922. J Nanosci Nanotechnol 15(6):4200–4204
Dhand C, Dwivedi N, Loh XJ, Jie Ying AN, Verma NK, Beuerman RW, Lakshminarayan R, Ramakrishna S (2015) Methods and strategies for the synthesis of diverse nanoparticles and their applications: a comprehensive overview. RSC Adv 5(127):105003–105037
Dogra V, Kaur G, Jindal S, Kumar R, Kumar S, Singhal NK (2019) Bactericidal effects of metallosurfactants based cobalt oxide/hydroxide nanoparticles against *Staphylococcus aureus*. Sci Total Environ 681:350–364
Dong H, Chen Y-C, Feldmann C (2015) Polyol synthesis of nanoparticles: status and options regarding metals, oxides, chalcogenides, and non-metal elements. Green Chem 17(8):4107–4132
Dwivedi AD, Gopal K (2010) Biosynthesis of silver and gold nanoparticles using Chenopodium album leaf extract. Colloids Surf A Physicochem Eng Asp 369:27–33
Eid K, Wang H, Wang L (2017) Nanoarchitectonic metals. In: Ariga K, Aono M (eds) Supra-materials nanoarchitectonics. Elsevier, Amsterdam, pp 135–171
Endo M, Takeuchi K, Igarashi S, Kobori K, Shiraishi M, Kroto HW (1993) The production and structure of pyrolytic carbon nanotubes. J Phys Chem Solid 54:1841–1848
Eslamian M, Heine MC (2008) Characteristics of spray flames and the effect of group combustion on the morphology of flame-made nanoparticles. Nanotechnology 19:45712
Fievet F, Lagier JP, Blin B, Beaudoin B, Figlarz M (1989) Homogeneous and heterogeneous nucleations in the polyol process for the preparation of micron and submicron size metal particles. Solid State Ion 32(33):198–205
Gericke M, Pinches A (2006) Biological synthesis of metal nanoparticles. Hydrometallurgy 83: 132–140
Gorrasi G, Sorrentino A (2015) Mechanical milling as a technology to produce structural and functional bio-nanocomposites. Green Chem 17(5):2610–2625
Granqvist CG, Buhrman RA (1976) Ultrafine metal particles. J Appl Phys 47:2200–2219
Hakim LF, Portman JL, Casper MD, Weimer AW (2005) Aggregation behavior of nanoparticles in fluidized beds. Powder Technol 160:149–160
Han MY, Huang W, Chew CH, Gan LM, Zhang XJ, Ji W (1998) Large nonlinear absorption in coated Ag_2S/CdS nanoparticles by inverse microemulsion. J Phys Chem B 102:1884–1887
Haram SK, Mahadeshwar AR, Dixit SG (1996) Synthesis and characterization of copper sulphate nanoparticles in triton-X 100 water-in-oil microemulsion. J Phys Chem 100:5868–5873
Harvey E, Ghantasala M (2006) Nanofabrication. In: Hannink RHJ, Hill AJ (eds) Nanostructure control of materials. Elsevier, Amsterdam, pp 303–330
Hopwood JD, Mann S (1997) Synthesis of barium sulfate nanoparticles and nanofilaments in reverse micelles and microemulsions. Chem Mater 9:1819–1828
Huang J, Lin L, Sun D, Chen H, Yang D, Li Q (2015) Bioinspired synthesis of metal nanomaterials and applications. Chem Soc Rev 44:6330–6374

Hussain I, Singh NB, Singh A, Singh H, Singh SC (2016) Green synthesis of nanoparticles and its potential applications. Biotechnol Lett 38:545–560

Inoue M, Hayashi Y, Takizawa H, Suganuma K (2018) Nanoparticle fabrication. In: Morris J (ed) Nanopackaging. Springer, Cham, pp 219–242

Ivanoska-Dacikj A, Bogoeva-Gaceva G (2019) Fabrication methods of carbon-based rubber nanocomposites. In: Yaragalla S, Mishra RK, Thomas S, Kalarikkal N, Maria HJ (eds) Carbon-based Nanofillers and their rubber nanocomposites: fundamentals and applications. Elsevier, Amsterdam, pp 27–47

Jaworek A (2007) Micro and nanoparticle production by electrospraying. Powder Technol 176:18–35

Jha AK, Prasad K, Kumar V (2009) Biosynthesis of silver nanoparticles using *Eclipta* leaf. Biotechnol Prog 25:1476–1479

Khan MKI, Nazir A, Maan AA (2017) Electrospraying: a novel technique for efficient coating of foods. Food Eng Rev 9:112–119

Khanna P, Kaur A, Goyal D (2019) Algae-based metallic nanoparticles: synthesis, characterization and applications. J Microbiol Methods 163:105656

Kharissova OV, Dias HV, Kharisov BI, Pérez BO, Pérez VM (2013) The greener synthesis of nanoparticles. Trends Biotechnol 31:240–248

Kulandaisamy AJ, Rayappan JBB (2018) Significance of nanoparticles and the role of amino acids in structuring them-a review. J Nanosci Nanotechnol 18(8):5222–5233

Kumar V, Yadav SK (2009) Plant–mediated synthesis of silver and gold nanoparticles and their applications. J Chem Technol Biotechnol 84:151–157

Li X, Xu H, Chen ZS, Chen G (2011) Biosynthesis of nanoparticles by microorganisms and their applications. J Nanomat 2011:1–16

Loukanov A, Emin S (2016) Biotinylated vanadium and chromium sulfide nanoparticles as probes for co-localization of membrane proteins. Microsc Res Tech 79(9):799–805

Lu F, Gang O (2017) DNA functionalization of nanoparticles. Methods Mol Biol 1500:99–107

Mackenzie JD, Bescher EP (2007) Chemical routes in the synthesis of nanomaterials using the sol-gel process. Acc Chem Res 40(9):810–818

Madkour LH (2019) Processing of nanomaterials (NMs). In: Nanoelectronic materials. Advanced structured materials, vol 116. Springer, Cham, pp 309–353

Madler L, Kammler HK, Mueller R, Pratsinis SE (2002) Controlled synthesis of nanostructured particles by flame spray pyrolysis. J Aerosol Sci 33:369–389

Malik MA, Wani MY, Hashim MA (2012) Microemulsion method: a novel route to synthesize organic and inorganic nanomaterials. First Nano update. Arab J Chem 5:397–417

Mandal D, Ghosh M, Maiti S, Das K, Das PK (2014) Water-in-oil microemulsion doped with gold nanoparticle decorated single walled carbon nanotube: scaffold for enhancing lipase activity. Colloids Surf B Biointerfaces 113:442–449

Mazid RR, Si KJ, Cheng W (2014) DNA based strategy to nanoparticle superlattices. Methods 67(2):215–226

Mijatovic D, Eijkel J, Van D, Berg A (2005) Technologies for nanofluidic systems: top-down versus bottom-up- a review. Lab Chip 5(5):492–500

Mohan Bhagyaraj S, Oluwafemi OS (2018) Nanotechnology: the science of the invisible. Synthesis of inorganic nanomaterials. In: Bhagyaraj SM, Oluwafemi OS, Kalarikkal N, Thomas S (eds) Synthesis of inorganic nanomaterials. Elsevier, Amsterdam, pp 1–18

Mohanpuria P, Rana NK, Yadav SK (2008) Biosynthesis of nanoparticles: technological concepts and future applications. J Nanopart Res 10:507–517

Monnoyer P, Fonseca A, Nagy JB (1995) Preparation of colloidal AgBr particles from microemulsions. Colloid Surf A: Physicochem Eng Asp 100:233–243

Mueller R, Madler L, Pratsinis SE (2003) Nanoparticle synthesis at high production rates by spray flame pyrolysis. Chem Eng Sci 58:1969–1976

Mukherjee P, Ahmad A, Mandal D, Senapati S, Sainkar SR, Khan MI, Parishcha R, Ajaykumar PV, Alam M, Kumar R, Sastry M (2001) Fungus–mediated synthesis of silver nanoparticles and their immobilization in the mycelial matrix: a novel biological approach to nanoparticle synthesis. Nano Lett 1:515–519

Narayanan KB, Sakthivel N (2010) Biological synthesis of metal nanoparticles by microbes. Adv Colloid Interface Sci 156:1–13

Oda M (2002) Metal nano-particles. J Jpn Inst Electron Packag 5:523–528

Okoli C, Sanchez-Dominguez M, Boutonnet M, Järås S, Civera C, Solans C, Kuttuva GR (2012) Comparison and functionalization study of microemulsion-prepared magnetic iron oxide nanoparticles. Langmuir 28(22):8479–8485

Pawar A, Thakkar S, Misra M (2018) A bird's eye view of nanoparticles prepared by electrospraying: advancements in drug delivery field. J Control Release 286:179–200

Pemartin-Biernath K, Vela-González AV, Moreno-Trejo MB, Leyva-Porras C, Castañeda-Reyna IE, Juárez-Ramírez I, Solans C, Sánchez-Domínguez M (2016) Synthesis of mixed cu/Ce oxide nanoparticles by the oil-in-water microemulsion reaction method. Materials (Basel) 9(6):480

Pereira L, Mehboob F, Stams AJM, Mota MM, Rijnaarts HHM, Madalena Alves MM (2015) Metallic nanoparticles: microbial synthesis and unique properties for biotechnological applications, bioavailability and biotransformation. Crit Rev Biotechnol 35:114–128

Pileni M, Lisiecki I (1993) Nanometer copper metallic particles synthesis in reverse micelles. Colloids Surf A Physicochem Eng Asp 80:63–68

Piras CC, Fernández-Prieto S, De Borggraeve WM (2019) Ball milling: a green technology for the preparation and functionalisation of nanocellulose derivatives. Nanoscale Adv 1:937–947

Pollard AJ, Nair RR, Sabki SN, Staddon CR, Perdigao LMA, Hsu CH, Garfitt JM, Gangopadhyay S, Gleeson HF, Geim AK, Beton PH (2009) Formation of monolayer graphene by annealing sacrificial nickel thin films. J Phys Chem C 113:16565–16567

Prasad R, Pandey R, Burman I (2016) Engineering tailored nanoparticles with microbes: *quo vadis*? WIREs Nanomed Nanobiotechnol 8:316–330

Qi LM, Ma J, Chen H, Zhao Z (1997) Reverse micelle based formation of $BaCO_3$ nanowires. J Phys Chem B 101:3460–3463

Raffi M, Rumaiz AK, Hasan MM, Shah SI (2007) Studies of the growth parameters for silver nanoparticle synthesis by inert gas condensation. J Mater Res 22:3378–3384

Rane AV, Kanny K, Abitha VK, Thomas S (2018) Methods for synthesis of nanoparticles and fabrication of nanocomposites. In: Bhagyaraj SM, Oluwafemi OS, Kalarikkal N, Thomas S (eds) Synthesis of inorganic nanomaterials. Elsevier, Amsterdam, pp 121–139

Rao BG, Mukherjee D, Reddy BM (2017) Novel approaches for preparation of nanoparticles. In: Ficai D, Grumezescu AM (eds) Nanostructures for novel therapy. Elsevier, Amsterdam, pp 1–36

Rodríguez-San-Miguel D, Amo-Ochoa P, Zamora F (2016) MasterChem: cooking 2D-polymers. Chem Commun (Camb) 52(22):4113–4127

Routkevitch D, Bigoni T, Moskovits M, Xu JM (1996) Electrochemical fabrication of CdS nanowire arrays in porous anodic aluminum oxide templates. J Phys Chem 100:14037–14047

Sanchez-Dominguez M, Koleilat H, Boutonnet M, Solans C (2011) Synthesis of Pt nanoparticles in oil-in-water microemulsions: phase behavior and effect of formulation parameters on nanoparticle characteristics. J Dispers Sci Technol 32:1765–1770

Sharma VK, Yngard RA, Lin Y (2009) Silver nanoparticles: green synthesis and their antimicrobial activities. Adv Colloid Interf 145:83–96

Shi W, Song S, Zhang H (2013) Hydrothermal synthetic strategies of inorganic semi-conducting nanostructures. Chem Soc Rev 42:5714–5743

Simon P, Bahrig L, Baburin IA, Formanek P, Röder F, Sickmann J, Hickey SG, Eychmüller A, Lichte H, Kniep R, Rosseeva E (2014) Interconnection of nanoparticles within 2D superlattices of PbS/oleic acid thin films. Adv Mater 26(19):3042–3049

Singh A, Singh NB, Hussain I, Singh H, Singh SC (2015) Plant nanoparticle interaction: an approach to improve agricultural practices and plant productivity. Int J Pharm Sci Inven 4: 25–40

Singh H, Du J, Singh P, Yi TH (2018) Ecofriendly synthesis of silver and gold nanoparticles by *Euphrasia officinalis* leaf extract and its biomedical applications. Artif Cells Nanomed Biotechnol 46:1163–1170

Sintubin L, De Gusseme B, Van der Meeren P, Pycke BF, Verstraete W, Boon N (2011) The antibacterial activity of biogenic silver and its mode of action. Appl Microbiol Biotechnol 91: 153–162

Solanki JN, Sengupta R, Murthy Z (2010) Synthesis of copper sulphide and copper nanoparticles with microemulsion method. Solid State Sci 12:1560–1566

Soleimani Zohr Shiri M, Henderson W, Mucalo MR (2019) A review of the lesser-studied microemulsion-based synthesis methodologies used for preparing nanoparticle systems of the noble metals, Os, re, Ir and Rh. Materials (Basel) 12(12):1896

Solero G (2017) Synthesis of nanoparticles through flame spray pyrolysis: experimental apparatus and preliminary results. Nanosci Nanotechnol 7(1):21–25

Somani PR, Somani SP, Umeno M (2006) Planer nano-graphenes from camphor by CVD. Chem Phys Lett 430:56–59

Sonawane GH, Patil SP, Sonawane SH (2018) Nanocomposites and its applications. In: Bhagyaraj SM, Oluwafemi OS, Kalarikkal M, Thomas S (eds) Applications of nanomaterials. Elsevier, Amsterdam, pp 1–22

Sosnik A (2014) Production of drug-loaded polymeric nanoparticles by electrospraying technology. J Biomed Nanotechnol 10:2200–2217

Sridhar R, Ramakrishna S (2013) Electrosprayed nanoparticles for drug delivery and pharmaceutical applications. Biomatter 3(3):e24281

Srivastava S, Bhargava A (2016) Green nanotechnology. J Nanotech Material Sci 3:1–7

Srivastava S, Usmani Z, Atanasov AG, Singh VK, Singh NP, Abdel-Azeem AM, Prasad R, Gupta G, Sharma M, Bhargava A (2021) Biological nanofactories: using living forms for metal nanoparticle synthesis. Mini Rev Med Chem 21:245–265

Suryanarayana C, Prabhu B (2007) Synthesis of nanostructured materials by inert-gas condensation methods. In: Nanostructured materials. Elsevier, Amsterdam, pp 47–90

Thakkar KN, Mhatre SS, Parikh RY (2010) Biological synthesis of metallic nanoparticles. Nanomed Nanotechnol Biol Med 6:257–262

Tricoli A, Nasiri N, Chen H, Wallerand AS, Righettoni M (2016) Ultra-rapid synthesis of highly porous and robust hierarchical ZnO films for dye sensitized solar cells. Sol Energy 136:553–559

Tripp SL, Pusztay SV, Ribbe AE, Wei A (2002) Self–assembly of cobalt nanoparticle rings. J Am Chem Soc 124:7914–7915

Vaucher S, Fielden J, Li M, Dujardin E, Mann S (2002) Molecule-based magnetic nanoparticles: synthesis of cobalt hexacyanoferrate, cobalt pentacyanonitrosyl ferrate and chromium hexacyanochromate coordination polymers in water-in-oil microemulsions. Nano Lett 2:225–229

Verma D, Goh KL (2019) Functionalized graphene-based nanocomposites for energy applications. In: Jawaid M, Bouhfid R, Qaiss AK (eds) Functionalized graphene nanocomposites and their derivatives: synthesis, processing and applications. Elsevier, Amsterdam, pp 219–243

Wang H, Tang M, Shi F, Ding R, Wang L, Wu J, Li X, Liu Z, Lv B (2020) Amorphous Cr_2WO_6-modified WO_3 nanowires with a large specific surface area and rich Lewis acid sites: a highly efficient catalyst for oxidative desulfurization. ACS Appl Mater Interfaces 12(34):38140–38152

Wang J, Jansen JA, Yang F (2019) Electrospraying: possibilities and challenges of engineering carriers for biomedical applications- a mini review. Front Chem 7:258

Wilson MP, Schwarzman MR (2009) Green chemistry: terminology and principles response. Environ Health Perspect 117:A386–A434

Wu ML, Chen DH, Huang TC (2001a) Synthesis of au/Pd bimetallic nanoparticles in reverse micelles. Langmuir 17:3877–3883

Wu ML, Chen DH, Huang TC (2001b) Preparation of Pd/Pt bimetallic nanoparticles in water/AOT/isooctane microemulsions. J Colloid Interface Sci 243:102–108

Zhang X, Chan KY (2003) Water-in-oil microemulsion synthesis of platinum-ruthenium nanoparticles, their characterization and electrocatalytic properties. Chem Mater 15:451–459

Biological Synthesis of Nanoparticles: Bacteria

5

Abstract

Bacteria are unicellular prokaryotic microorganisms that provide numerous advantages over the eukaryotic systems in green synthesis of nanoparticles. These include rapid growth rates and suitable adaptations to tide over unfavorable conditions created in medium accommodating nanoparticles. The unique phenomenon of intracellular and extracellular production of nanostructures by bacteria as a biogenic agent has been exhaustively reported. Biofabrication has been done using different bacterial components like spores, flagellin, exopolysaccharides, rhamnolipids, and bacterial surfactants. The chapter discusses the mechanism and utility of Gram-positive and Gram-negative bacteria in the biosynthesis of nanoparticles.

5.1 Introduction

Bacteria are a large group of single-celled, prokaryotic microorganisms that inhabit every nook and corner of the world ranging from deep sea to snow-covered peaks and from hot water springs to deep sea vents. In fact normal flora of humans contains nonpathogenic "friendly" bacteria residing in different body systems, especially the digestive tract. Bacteria appear singly or in chains and display a wide diversity of shapes and sizes. They can be ovoid or spherical (*Cocci*), rod-shaped (*Bacilli*), curved rods or comma-shaped (*Vibrio*), spiral shaped (*Spirilla*), or tightly coiled (*Spirochaetes*) (Fig. 5.1). Most bacteria are small, with diameters of about 15 μm. However, the rod-shaped *Epulopiscium fishelsoni*, intestinal symbiont of the brown surgeonfish, is one of the largest bacteria having a diameter and length of 50 μm and 500 μm, respectively. *Thiomargarita namibiensis* is visible to the naked eye due to a large diameter of 750 μm. Most bacteria are unicellular, but some clump together in

S. Srivastava, A. Bhargava, *Green Nanoparticles: The Future of Nanobiotechnology*,
https://doi.org/10.1007/978-981-16-7106-7_5

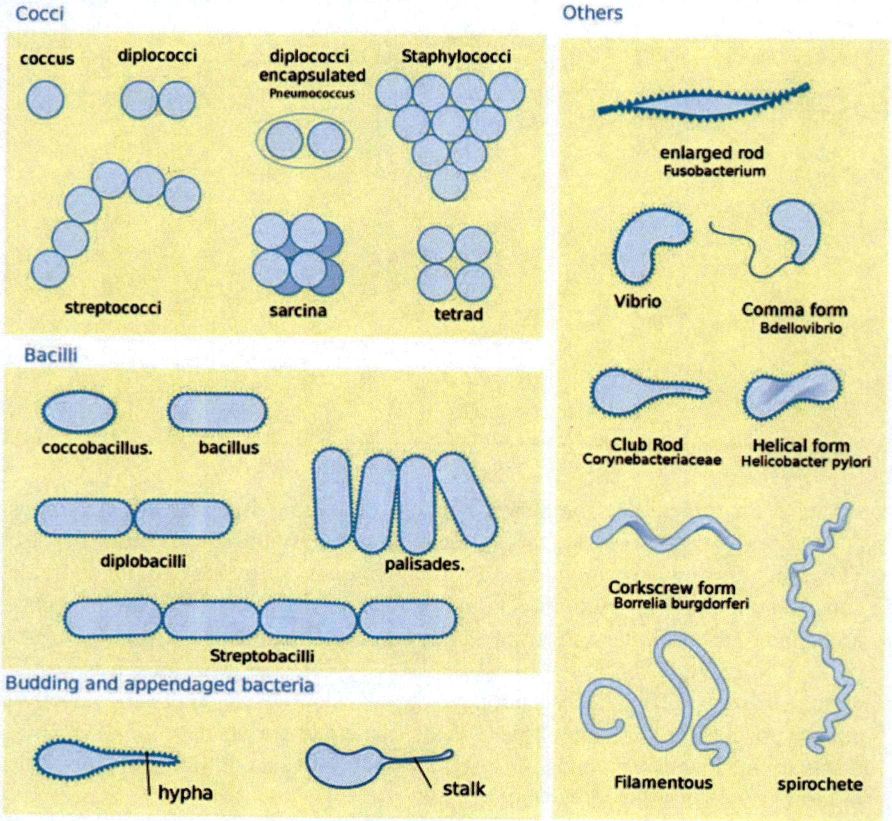

Fig. 5.1 Different shapes of bacterial cells (Courtesy: Wikipedia)

definite pattern, while others can form complex multicellular groups during their life cycles.

Bacteria have the following specific traits:

1. Lack of membrane-bound organelles.
2. A single double-stranded DNA having a single site for replication initiation.
3. Cell walls contain peptidoglycan, a rigid polymer of modified sugars cross-linked by peptides.

Bacteria are often differentiated into two groups on the basis of Gram's staining technique, a bacteriological laboratory technique named after the Danish bacteriologist Hans Christian Gram who introduced the technique in 1884 while identifying microorganisms causing pneumonia. This method is based on differential staining and makes use of two stains, viz., crystal violet and safranin, along with iodine as mordant. After treatment with alcohol, the cell walls of Gram-positive bacteria appear purple since they retain the complex, while Gram-negative organisms

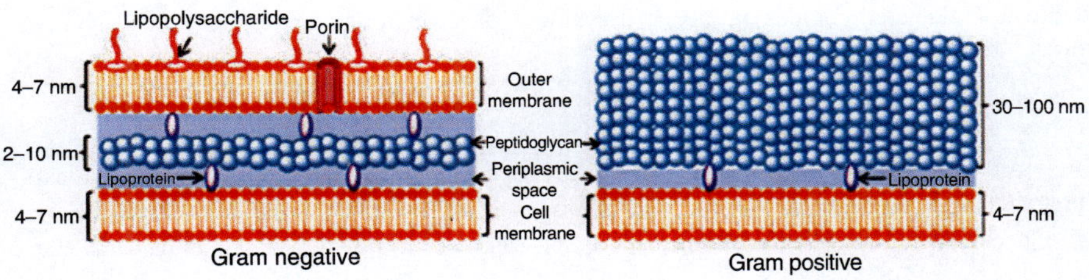

Fig. 5.2 Differences in the cell wall structure and composition of Gram-negative and Gram-positive bacteria (Reprinted from Berezin et al. 2017)

decolorize and appear pink. This differential staining is due to differences in the composition of the bacterial cell surface (Fig. 5.2).

5.2 Bacteria in Green Synthesis of Nanoparticles

The use of prokaryotes like bacteria in nanoparticle synthesis provides numerous advantages over the eukaryotic systems. These include rapid growth rates and suitable adaptations to tide over unfavorable conditions created in medium accommodating nanoparticles. The adaptive mechanisms include metal reduction and/or precipitation and generation of nontoxic or less toxic metal nanoarrays.

The story of utilization of bacteria for fabrication of nanoparticles began in the late 1990s when the research group led by Klaus reported the synthesis of silver-based crystals having size of around 200 nm in *Pseudomonas stutzeri*, a Gram-negative, nonfluorescent denitrifying bacterium. This bacterium isolated from silver mine had the unique ability to survive in metal-rich environments, which aids in the accumulation of nanosilver. Bacteria are popular among the nanobiotechnologists for biofabrication of nanomaterials due to the following reasons:

1. Relative ease of manipulation
2. Abundant presence in the environment
3. Rapid growth
4. Low cost of cultivation
5. Easy adaptation to extreme conditions
6. Easy manipulation of growth conditions

Nanoparticles have been fabricated using different bacterial components like spores, flagellin, exopolysaccharides, rhamnolipids, and bacterial surfactants (Table 5.1).

Several substrate-reducing bacteria have been isolated from diverse habitats. Such is the popularity and ease of using bacteria in nanoparticle synthesis that nanobiotechnologists have successfully carried out biosynthesis of fluorescent nanoparticles from cadmium- and tellurite-resistant bacteria collected from Antarctica. Many polyextremophilic bacteria have been used to biosynthesize

Table 5.1 Bacterial component-mediated synthesis of silver nanoparticles (Reprinted with permission from Singh et al. 2015)

Bacterial component	Bacteria	Shape	Size (nm)	Reference
Actinorhodin pigment	*Streptomyces coelicolor*	Irregular	28–50	Manikprabhu and Lingappa (2013)
Bioflocculant	*Bacillus subtilis* MSBN 17	Spherical	60	Sathiyanarayanan et al. (2013)
Biosurfactant	*Brevibacterium casei* MSA19	–	–	Kiran et al. (2010)
Cellulose	*Gluconacetobacter xylinum*	–	5–40	Liu et al. (2012)
Exopolysaccharide	*Lactobacillus rhamnosus* GG ATCC 53103	Spherical, triangular, rod-shaped, and hexagonal	2–15	Kanmani and Lim (2013)
Extracellular polysaccharide/ matrix	*Nostoc commune*	Spherical	15–54	Morsy et al. (2014)
Flagellin	*Salmonella typhimurium*	–	3–11	Gopinathan et al. (2013)
Rhamnolipids	*Pseudomonas aeruginosa* BS-161R	Spherical	15.1 ± 5.8	Kumar et al. (2010)
Spores	*Bacillus atrophaeus*	Polydispersed	5–30	Hosseini-Abari et al. (2013)
URAK enzyme	*Bacillus cereus NK1*	Spherical	50–80	Deepak et al. (2011)

inorganic nanoparticles. These include production of silver nanoparticles by *Bacillus flexus, Halococcus salifodinae, Deinococcus radiodurans*, and *Nocardiopsis valliformis,* uraninite nanoparticles by *Thermoanaerobacter*, and gold nanoparticles by *Bacillus licheniformis* and *Halomonas salina*.

Both intracellular and extracellular productions of nanoparticles by bacteria have been reported by different researchers. Extracellular synthesis occurs outside the bacterial cell using bacterial cells, culture supernatant, or cell-free extract. Live and dead biomass of bacterial cells produces nanoparticles by two ways. The first one involves liberating the biomolecules of bacteria in the culture media or by secreting the nanoparticles formed inside the bacterial cells into the culture medium. The second method involves collecting supernatant of bacterial cells and incubation at suitable temperature following its exposure to the metal salt. The third method employs collecting a cell-free extract of bacteria after centrifugation or membrane filtration, challenging it with the metal salt and incubation at a particular temperature. The bacterial biomass and components of the growth media are removed by recurrent washings. As a result only the biomolecules released by the bacterial cells remain in solution and are used for nanoparticle synthesis. The recovery of nanoparticles is usually carried out by high-speed centrifugation at 10000–20000 rpm that leads to formation of a pellet containing nanoparticles. This pellet has to be redispersed in the desired solvent for collecting the nanoparticles. Intracellular synthesis of nanoparticles involves addition of bacterial cells to the growth medium carrying the salt and further incubation in conditions that vary according to the experimental design. A number of bacteria like *Idiomarina, Pseudomonas aeruginosa, Bacillus licheniformis, Brevibacterium casei, Pediococcus pentosaceus*, and *Halococcus salifodinae* are known to produce intracellular nanoparticles. Some bacteria like *Lactobacillus* sp., *Aeromonas* sp. SH10, *Vibrio alginolyticus*, and cyanobacterial species like *Plectonema boryanum* UTEX 485 are unique since they produce both extra- and intracellular nanoparticles simultaneously.

5.3 Gram-Positive Bacteria

Gram-positive bacteria, also termed as monoderm bacteria, have thick walls (20–50 nm) composed primarily of peptidoglycans and mucocomplexes. The peptidoglycan layer is essential for osmotic stability, protects the cytoplasmic membrane, and gives shape and structural rigidity to the cell wall. These bacteria contain ample amount of secondary cell wall polymers like teichoic acid, complex of peptides containing d-amino acids, capsular polysaccharides, and/or arabinogalactan, which can account for more than 50% of the total cell wall material. The use of Gram-positive bacteria in nanoparticle production has been provided in Table 5.2. Details of the table reveal that only three bacterial genera, viz., *Bacillus, Lactobacillus*, and *Streptomyces* have been largely utilized for nanoparticle production. In fact *Bacillus anthracis*, the aerobic or facultatively anaerobic endospore-forming anthrax bacteria, has also been explored for the fabrication of lead nanoparticles. Using this group of bacteria, most promising results have been obtained in the biofabrication of metal and metal oxide nanostructures. In one study, a single Gram-positive species, viz., *Streptomyces enissocaesilis*, was maneuvered for the biosynthesis of silver, selenium, and zinc oxide nanostructures using $AgNO_3$, SeO_2, and $ZnSO_4$. Nanoparticles of 5–20 nm size were noticed in the TEM images.

Apart from the abovementioned categories, other types of nanoparticles have also been biosynthesized using Gram-positive bacteria. For example, inulin nanoparticles have been fabricated using inulosucrase from *Lactobacillus reuteri* 121. Inulin, a type of fructans formed of β-2,1 glycosidic linkage, is well known for its numerous health benefits like enhancement of immune response, reduction in the risk of gastrointestinal diseases, and stimulation of mineral absorption, besides being a well-known probiotic. Inulin naturally occurs in several plants like wheat, onion, banana, garlic, asparagus, Jerusalem artichoke, and chicory but can also be synthesized enzymatically from sucrose by inulosucrase (EC2.4.1.9). Inulin nanoparticles were mainly synthesized at pH range of 4–5.5, but their size depended more on reaction temperature rather than pH. The average sizes of inulin nanoparticles synthesized at 50, 40, 30, 20, and 4 °C were 95.9, 115.7, 150.5,

Table 5.2 List of nanoparticles synthesized by Gram-positive bacteria

Microbes	Type/shape	Family	Nanoparticle type	Size (nm)	Reference
Bacillus atrophaeus	Rod	Bacillaceae	Silver	5–30	Hosseini-Abari et al. (2013)
Bacillus brevis	Rod	Bacillaceae	Silver	41–68	Saravanan et al. (2018)
Bacillus cereus	Rod	Bacillaceae	Silver	10–30	Prakash et al. (2011)
Bacillus flexus	Rod	Bacillaceae	Silver	12–65	Priyadarshini et al. (2013)
Bacillus funiculus	Rod	Bacillaceae	Silver	20	Gurunathan et al. (2013)
Bacillus licheniformis	Rod	Bacillaceae	Silver	50	Kalimuthu et al. (2008)
Bacillus megaterium	Rod	Bacillaceae	Gold	1.9	Wen et al. (2009)
Bacillus mycoides	Rod	Bacillaceae	Titanium dioxide	40–60	Aenishanslins et al. (2014)
Bacillus subtilis	Rod	Bacillaceae	Silver	Nr	Kannan et al. (2011)
Bacillus thuringiensis	Rod	Bacillaceae	Cobalt Cobalt oxide Silver	85.3 6.6 43.52–142.97	Marimuthu et al. (2013) Shim et al. (2011) Banu et al. (2014)
Brevibacterium casei	Rod	Brevibacteriaceae	Gold Cobalt oxide Silver	10–50 5–7 10–50	Kalishwaralal et al. (2010) Kumar et al. (2008) Kalishwaralal et al. (2010)
Clostridium pasteurianum	Rod	Clostridiaceae	Molybdenum	5–20	Nordmeier et al. (2018)
Corynebacterium sp.	Rod	Corynebacteriaceae	Silver	10–15	Zhang et al. (2005)
Geobacillus sp.	Rod	Bacillaceae	Gold	5–50	Correa-Llantén et al. (2013)
Halobacillus sp.	Rod	Bacillaceae	Quantum dots	2–5	Bruna et al. (2019)
Lactobacillus sp.	Rod	Lactobacillaceae	Antimony (III) oxide Barium titanate Cadmium sulfide Silver oxide	3–12 20–80 2.5–5.5 2–20	Jha et al. (2009) Jha and Prasad (2010) Prasad and Jha (2010) Dhoondia and Chakraborty (2012)
Lactobacillus acidophilus	Rod	Lactobacillaceae	Silver	30.65	Garmasheva et al. (2016)
Lactobacillus bulgaricus	Rod	Lactobacillaceae	Silver	30.65–100	Naseer et al. (2021)
Lactobacillus casei	Rod	Lactobacillaceae	Gold Selenium	7–56 50–80	Kikuchi et al. (2016) Xu et al. (2018)
Lactobacillus kefiri	Rod	Lactobacillaceae	Platinum	3	Huggias et al. (2020)

Lactobacillus kimchicus	Rod	Lactobacillaceae	Gold	5–30	Markus et al. (2016)
Lactobacillus plantarum	Rod	Lactobacillaceae	Silver Tellurium Zinc oxide	19.92 45.7 191.8–291.1	Garmasheva et al. (2016) Mirjani et al. (2015) Mohd Yusof et al. (2020)
Lactobacillus reuteri	Rod	Lactobacillaceae	Zinc	~100	Yücel et al. (2021)
Lactobacillus ruminis	Rod	Lactobacillaceae	Silver	10–150	Reyes-Escogido et al. (2017)
Listeria innocua	Rod	Listeriaceae	Iron oxide	3	Ceci et al. (2010)
Lysinibacillus sphaericus	Rod	Bacillaceae	Lead sufide	5–10	Wei et al. (2021)
Micrococcus luteus	Coccus	Micrococcaceae	Gold	6–50	Arunkumar et al. (2013)
Micrococcus yunnanensis	Coccus	Micrococcaceae	Gold	53.8	Jafari et al. (2018)
Nocardiopsis alba	Filamentous	Nocardiopsaceae	Gold	32.5	Avilala and Golla (2019)
Nocardiopsis valliformis	Filamentous	Nocardiopsaceae	Silver	5–50	Rathod et al. (2016)
Streptomyces sp.	Filamentous	Streptomycetaceae	Copper oxide Silver	78–80 5	Hassan et al. (2019) Karthik et al. (2014)
Streptomyces atrovirens	Filamentous	Streptomycetaceae	Silver	58	Subbaiya et al. (2017)
Streptomyces coelicolor	Filamentous	Streptomycetaceae	Silver	28–50	Manikprabhu and Lingappa (2014)
Streptomyces griseoruber	Filamentous	Streptomycetaceae	Gold Selenium	5–50 100–250	Ranjitha and Rai (2017) Ranjitha and Ravishankar (2018)
Streptomyces hygroscopicus	Filamentous	Streptomycetaceae	Gold	2–20	Sadhasivam et al. (2012)
Streptomyces laurentii	Filamentous	Streptomycetaceae	Silver	7–15	Eid et al. (2020)
Streptomyces microflavus	Filamentous	Streptomycetaceae	Gold	10–100	Soltani Nejad et al. (2016)
Streptomyces parvus	Filamentous	Streptomycetaceae	Silver	2–5	Al-Dhabi et al. (2019)
Streptomyces xinghaiensis	Filamentous	Streptomycetaceae	Silver	5–20	Wypij et al. (2018)
Thermoactinomyces sp.	Rod	Thermoactinomycetaceae	Silver	20–40	Deepa et al. (2013)
Ureibacillus thermosphaericus	Rod	Planococcaceae	Silver	10–100	Juibari et al. (2011)

nr not reported.

207.3, and 238.5 nm, respectively. Spherical nanostructures of 25–60 nm diameter were observed in the TEM images.

5.4 Gram-Negative Bacteria

The more complex Gram-negative bacteria, also known as diderm bacteria, have a comparatively thin (<10 nm) cell wall. However, they have an additional outer membrane bearing many pores and appendices. They have a comparatively thinner peptidoglycan layer present between an outer lipopolysaccharide-rich membrane and an inner cytoplasmic membrane. They also are adapted to low osmolarity environments. As compared to Gram-positive bacteria, the Gram-negative ones generally show high level of resistance to most classes of antibiotics and are therefore tough to treat.

In contrast to Gram-positive bacteria where only limited genera have been used for green production of nanoparticles, biosynthesis of nanoparticles has been accomplished using more than 20 genera of Gram-negative bacteria. The nanostructures synthesized include gold, silver, and platinum to metal oxide nanoparticles and even quantum dots (Table 5.3).

Intracellular synthesis of cadmium sulfide quantum dots was successfully accomplished for the first time by Gallardo et al. (2014) using *Pseudomonas* spp., a psychrotolerant, oxidative stress-resistant bacteria collected from Antarctic soils. The biosynthesis of cadmium sulfide quantum dots assessed at 15 °C exhibited a time-dependent alteration in the emission of fluorescence after 6 days of treatment which is a typical change associated with nanocrystal growth. TEM studies revealed a large number of vesicles at the cell periphery and numerous electron dense particles at the cell envelope and near the poles. The polydisperse nanoparticles were spherical with sizes ranging 10–40 nm.

The first report of using extremophilic bacteria for the generation of quantum dots was put forth by Ulloa et al. (2016). The group successfully carried out green synthesis of cadmium sulfide (CdS) quantum dots utilizing three species of the genus *Acidithiobacillus*, viz., *A. ferrooxidans, A. thiooxidans*, and *A. caldus*. The bacterial cells were exposed to sublethal cadmium concentrations in the presence of glutathione and the semi-essential amino acid cysteine. The fluorescence of cadmium-exposed cells changed from green to red with rise in incubation time, a typical feature noticed during the production of quantum dots. The biosynthesized nanocrystals displayed an absorption peak showing characteristic features of cadmium sulfide quantum dots and mean size 6 nm for green and 10 nm for red nanoparticles. The quantum dots displayed high tolerance to acid conditions with no alteration in absorbance and fluorescence properties even at pH value of as low as 2.0, which inhibits these properties in the nanoparticles synthesized by mesophilic bacteria.

Twelve Gram-negative bacterial isolates inhabiting Antarctica belonging to *Pseudomonas*, *Psychrobacter* and *Shewanella* were explored by Plaza et al. (2016) for the production of cadmium sulfide and cadmium tellurite quantum dots (QDs). The

Table 5.3 List of nanoparticles synthesized by Gram-negative bacteria

Microbes	Type/shape	Family	Nanoparticle type	Size (nm)	Reference
Acidithiobacillus thiooxidans	Rod	Acidithiobacillaceae	Cadmium sulfide quantum dots	6–10	Ulloa et al. (2018)
Acinetobacter calcoaceticus	Bacilli	Moraxellaceae	Silver	4–40	Gaidhani et al. (2013)
Aeromonas sp.	Rod	Aeromonadaceae	Silver	8–16	Singh et al. (2017)
Aeromonas hydrophila	Rod	Aeromonadaceae	Titanium dioxide Zinc oxide	40.5 57.72	Jayaseelan et al. (2013) Jayaseelan et al. (2012)
Bordetella sp.	Coccobacilli	Alcaligenaceae	Silver	63–90	Thomas et al. (2012)
Citrobacter sedlakii	Rod	Enterobacteriaceae	Silver	4–15	Saeb et al. (2014)
Enterobacter aerogenes	Rod	Enterobacteriaceae	Silver	25–35	Karthik and Radha (2012)
Enterobacter hormaechei	Rod	Enterobacteriaceae	Silver	9.91–92.54	Monowar et al. (2021)
Escherichia coli	Rod	Enterobacteriaceae	Cadmium sulfide Cadmium tellurite quantum dots Cobalt Gold Silver	40–80 ~2–3 473 25 50	Shivashankarappa and Sanjay (2020) Bao et al. (2010) Jang et al. (2015) Du et al. (2007) Gurunathan et al. (2009)
Escherichia hermannii	Rod	Enterobacteriaceae	Silver	4–12	Saeb et al. (2014)
Geobacter sulfurreducens	Rod	Geobacteraceae	Silver	30	Law et al. (2008)
Gluconobacter roseus	Rod	Acetobacteraceae	Silver	10	Krishnaraj and Berchmans (2013)
Halococcus salifodinae	Coccoid	Halobacteriaceae	Selenium	28	Srivastava et al. (2014)
Idiomarina sp.	Rod	Idiomarinaceae	Lead (IV) sulfide Silver	2.38 26	Srivastava and Kowshik (2017) Seshadri et al. (2012)
Klebsiella pneumoniae	Rod	Enterobacteriaceae	Cadmium sulfide Selenium Silver Zinc sulfide	10–25 100–550 15–37 65	Malarkodi et al. (2014) Fesharaki et al. (2010) Kalpana and Lee (2013) Malarkodi et al. (2014)

(continued)

Table 5.3 (continued)

Microbes	Type/shape	Family	Nanoparticle type	Size (nm)	Reference
Magnetospirillum gryphiswaldense	Spirilla	Rhodospirillaceae	Gold	10–40	Cai et al. (2011)
Marinobacter pelagius	Rod	Alteromonadaceae	Gold	2–10	Sharma et al. (2012)
Morganella spp.	Rod	Morganellaceae	Silver	10–50	Parikh et al. (2011)
Pantoea ananatis	Rod	Erwiniaceae	Silver		Monowar et al. (2018)
Pantoea anthophila	Rod	Erwiniaceae	Silver	16	Nirmala and Sridevi (2021)
Paracoccus haeundaensis	Cocci	Rhodobacteraceae	Gold	20	Patil et al. (2019)
Proteus mirabilis	Rod	Morganellaceae	Cobalt Silver	85.3 10–20	Eltarahony et al. (2018) Samadi et al. (2009)
Pseudomonas aeruginosa	Rod	Pseudomonadaceae	Gold Lead sulfide quantum dots Silver	15–30 3.47–11.45 13	Husseiny et al. (2007) Öcal et al. (2020) Kumar and Mamidyala (2011)
Pseudomonas fluorescens	Rod	Pseudomonadaceae	Gold	50–70	Rajasree and Suman (2012)
Pseudomonas fragi	Rod	Pseudomonadaceae	Cadmium sulfide quantum dots	2–16	Gallardo-Benavente et al. (2019)
Pseudomonas hibiscicola	Rod	Pseudomonadaceae	Silver	10–70	Punjabi et al. (2018)
Pseudomonas putida	Rod	Pseudomonadaceae	Cadmium sulfide quantum dots Silver	12.5–30.6 4–30	Oliva-Arancibia et al. (2017) Saeb et al. (2014)
Pseudomonas stutzeri	Rod	Pseudomonadaceae	Silver	200	Klaus et al. (1999)
Rhodopseudomonas capsulata	Rod/ coccobacilli	Bradyrhizobiaceae	Gold	10–20	He et al. (2007)
Rhodopseudomonas palustris	Rod	Bradyrhizobiaceae	Silver	5–20	Chun-Jing and Hong-Juan (2010)
Rhodobacter capsulatus	Rod	Bradyrhizobiaceae	Gold	Nr	Feng et al. (2007)
Rhodobacter sphaeroides	Rod	Rhodobacteraceae	Cadmium chloride Silver	2.3–36.8 9.56	Bai et al. (2009) Bai et al. (2011)

Salmonella typhimurium	Rod	Enterobacteriaceae	Silver	87	Ghorbani (2013)
Shewanella algae	Rod	Shewanellaceae	Platinum	5	Konishi et al. (2007)
Shewanella oneidensis	Rod	Shewanellaceae	Cadmium sulfide	15	Wang et al. (2018)
			Cooper sulfide	~5	Zhou et al. (2016)
			Gold	12	Suresh et al. (2011)
Vibrio alginolyticus	Rod	Vibrionaceae	Silver	50–100	Rajeshkumar et al. (2013)
Xanthomonas oryzae	Rod	Xanthomonadaceae	Silver	14.86	Narayanan and Sakthivel (2013)
Yersinia enterocolitica	Rod	Yersiniaceae	Silver	10–80	Pourali et al. (2012)

nr not reported.

size, spectroscopic characteristics, and time-dependent changes in emission colors exhibited by the nanoparticles produced by different species after exposure to metal salts confirmed the production of nanostructures having fluorescent properties. The bacterial strains demonstrated significant capability for the synthesis of fluorescent nanostructures at room temperatures with no additional requirements. This represented a potential ecofriendly methodology for the synthesis of fluorescent nanoparticles.

The anaerobic, Gram-negative, cadmium-resistant bacteria *Raoultella* sp. X13 (family: Enterobacteriaceae) was utilized by Xu et al. (2019) for the nanofabrication of cadmium sulfide QDs in M9 medium supplemented with 1 mM cadmium nitrate [$Cd(NO_3)_2$] and 4 mM L-cysteine. The nanostructures exhibited an absorption peak at $\lambda = 382$ nm and a fluorescence peak at $\lambda = 468$ nm, and the TEM images showed their size within the range for the cadmium sulfide QDs.

E. coli seemed to be the perfect biomaterial for the synthesis and purification of CdS, CdSAg, and Ag_2S nanocrystals as explored by Órdenes-Aenishanslins et al. (2020). The green method produced the abovementioned nanostructures having different morphologies and minute size (<15 nm). When biosynthesized CdS quantum dots were exposed to $AgNO_3$ concentration, partial replacement of Cd^{+2} by Ag^{+} led to the formation of significantly stable CdSAg ternary quantum dots hetero-structures.

Carrasco et al. (2021) accomplished green synthesis of cadmium sulfide quantum dots using an Antarctic lithobiontic bacterium, *Pedobacter* (family: Sphingobacteriaceae). The bacterial cells were exposed to two experimental sets: 0.0001 M $CdCl_2$ and 0.001 M cysteine and 0.0002 M $CdCl_2$ and 0.002 M cysteine. These were incubated at 28 °C for different time durations. The diameter of the quantum dots were gauged approximately around 2.8 nm and 4.9 nm after 20 and 80 min of incubation, respectively. The HR-TEM images indicated the size of the quantum dots as 3.07 nm. The EDX studies confirmed the presence of nanostructures composed of almost equal quantity of cadmium and sulfide.

5.5 Bacterial Biosurfactants in Nanoparticle Production

Biosurfactants are surface-active molecules of microbial origin released extracellularly or are anchored to the cell surface. These amphiphilic agents contain both hydrophobic (having high affinity for water) and hydrophilic (having low affinity for water) moieties which allow them to lower the surface tension and the interfacial tension of the growth medium. The microbial biosurfactants have been classified as below:

1. Hydroxylated and cross-linked fatty acids: A typical example includes mycolic acids (long chain fatty acids obtained from *Mycobacterium tuberculosis*).
2. Glycolipids: These are carbohydrate compounds having a glycan moiety and a lipid moiety joined by an ester group. Examples of glycolipids include rhamnolipids (produced by *P. aeruginosa* and various species of *Burkholderia*),

Table 5.4 Bacterial biosurfactants used in the synthesis of nanoparticles

Biosurfactant	Source	Nanoparticle	Reference
Rhamnolipid	*Pseudomonas aeruginosa* *Pseudomonas fluorescens*	Silver Zinc oxide Zinc sulfide Silver	Kumar et al. (2010) Singh et al. (2014) Hazra et al. (2013) Maragatham and Govindammal (2014)
Glycolipid	*Brevibacterium. Casei*	Silver	Kiran et al. (2010)
Surfactin	*Bacillus amyloliquefaciens* *Bacillus subtilis*	Cadmium sulfide Gold	Singh et al. (2011) Reddy et al. (2009)
Surfactin and iturin	*Bacillus vallismortis*	Silver	Das et al. (2019)

sophorolipids (produced by several yeast species of the genus *Candida*, *Rhodotorula*, and *Starmerella*); mannosylerythritol lipids (MEL) (produced by the yeast *Pseudozyma*), and trehalolipids (produced by *Mycobacterium*, *Nocardia, Arthrobacter, Rhodococcus*, and *Corynebacterium*).

3. Lipopolysaccharides: These are high molecular weight compounds of lipid and polysaccharide found in the outer cell membrane of Gram-negative bacteria.
4. Lipoproteins-lipopeptides: Cyclic lipopeptides mainly produced by the *Bacillus* spp. These are further categorized into four families, viz., surfactins (cyclic lipoheptapeptide produced by *Bacillus subtilis*), iturins (amphiphilic, cyclic heptapeptides of small molecular mass extracted from *Bacillus subtilis*), fengycins (antifungal lipodecapeptides produced by *Bacillus subtilis* and *Bacillus amyloliquefaciens*), and kurstakins (antifungal lipoheptapeptides extracted from *Bacillus thuringiensis*).
5. Fatty acids, phospholipids, and neutral lipids: Produced by microorganisms that grows on n-alkanes. Typical examples include trehalose lipids produced by *Rhodococcus erythropolis* and *Acinetobacter* sp.
6. Polymeric biosurfactants: Composed of polymers like polysaccharides, proteins, lipopolysaccharides, and lipoproteins. Best known examples include liposan (synthesized by *Candida lipolytica*; contains about 80% carbohydrate and 20% protein), emulsan (polyanionic lipoheteropolysaccharide produced by *Acinetobacter calcoaceticus*), and alasan (synthesized by *Acinetobacter radioresistens*).

Microbial biosurfactants are known to play a dominant role in the aggregation and stabilization of nanomaterials. Table 5.4 provides information on the use of bacterial biosurfactants in the biofabrication of nanoparticles (Table 5.4).

5.6 Mechanistic Aspects of Nanoparticle Production by Bacteria

The following processes are thought to play a pivotal role in the green synthesis of nanoparticles:

1. Alteration of solubility and toxicity via redox reactions
2. Biosorption
3. Bioaccumulation and efflux system
4. Extracellular complexation or precipitation of metals

The bacterial cells need to trap the metallic ions and at the same time protect themselves against their toxicity. This is aided by electrostatic interaction and/or secretion of extracellular polymeric substances which adheres the ions to the bacterial cells. The electrostatic interactions occur between the negatively charged groups present on the microbial cells and positive charge on the metal ions. The functional groups available in the microbial cells include $-NH_2$, –OH, –SH, and –COOH, which play a dual role in reduction as well as the stabilization of nanostructures. This is followed by stabilization of ions in the bacterial cells due to stickiness caused by the secretion of adhesive materials by the cells. The damage to bacterial cells by metals is significantly reduced by the production of electrons and certain enzymes which reduce metal toxicity. The nanoparticles are then formed either extracellularly or intracellularly. The protective action of enzymes is critical in this process since they act as catalysts in furthering the reduction reactions on the cell surface as well as inside the microbial cells.

In extracellular synthesis, enzymes and compounds such as naphthoquinones, anthraquinones, and hydroquinones are involved in the production of metal nanoparticles. Several studies have indicated toward the role of nicotinamide adenine dinucleotide (NADH) and nicotinamide adenine dinucleotide phosphate (NADPH)-dependent enzymes in the bioreduction process through shifting of electron from NADH by NADH-reliant enzymes that function as electron carriers. This is best exemplified in the biofabrication of gold nanomaterials using *Rhodopseudomonas capsulata* (Fig. 5.3). Some lactic acid bacteria produce diverse classes of exopolysaccharides containing different monomers like glucose, galactose, mannose, and fructose which play an important part in the redox reactions culminating in the synthesis of metal nanoparticles. The neutralophilic bacterium *Bacillus megaterium* (family: Bacillaceae) and the facultative anaerobic *Lactobacillus* (family: Lactobacillaceae) produce nanoparticles by reducing metal ions via non-enzymatic process using the functional groups present on the bacterial cell wall. The reduction of ions or the formation of water insoluble complexes is an adaptation to tide over unfavorable conditions, which in this case is the toxicity induced by metal ions.

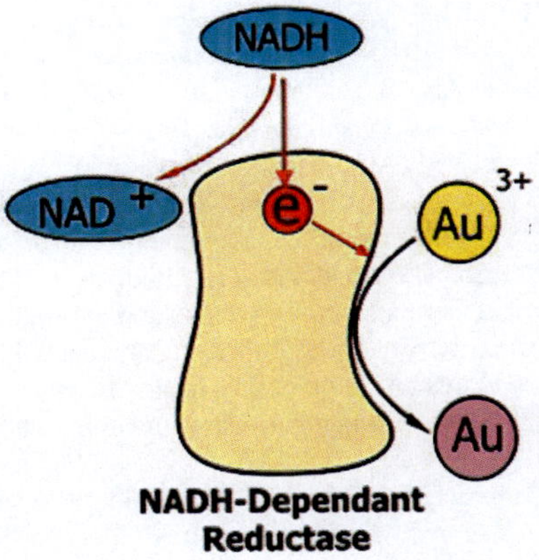

Fig. 5.3 Possible mechanisms of gold ions bioreduction (Reprinted with permission from He et al. 2007)

5.7 Future Prospects

Although bacteria have been successfully utilized in the biosynthesis of a wide range of nanoparticles, there are numerous shortcomings which need to be addressed before they can become economically and technically competitive against physical and chemical methods of nanoparticle production. The following drawbacks need to be looked in depth:

1. Tedious procedure for purification of nanoparticles in pure form.
2. Reliable information on the mechanistic aspects of nanoparticle formation.
3. Effective control over the dimensions of nanoparticles.
4. Attaining uniformity in the shape and size of nanomaterials in solution phase.
5. Scaling up for production level processing.
6. Cost of cultivating bacteria is quite high.

References

Aenishanslins NAO, Saona LA, Duran-Toro VM, Monras JP, Bravo DM, Donoso JMP (2014) Use of titanium dioxide nanoparticles biosynthesized by *Bacillus mycoides* in quantum dot sensitized solar cells. Microb Cell Fact 13:90

Al-Dhabi NA, Ghilan AM, Esmail GA, Arasu MV, Duraipandiyan V, Ponmurugan K (2019) Environmental friendly synthesis of silver nanomaterials from the promising *Streptomyces parvus* strain Al-Dhabi-91 recovered from the Saudi Arabian marine regions for antimicrobial and antioxidant properties. J Photochem Photobiol B 197:111529

Arunkumar P, Thanalakshmi M, Kumar P, Premkumar K (2013) Micrococcus luteus mediated dual mode synthesis of gold nanoparticles: involvement of extracellular α-amylase and cell wall teichuronic acid. Colloids Surf B Biointerfaces 103:517–522

Avilala J, Golla N (2019) Antibacterial and antiviral properties of silver nanoparticles synthesized by marine actinomycetes. Int J Pharm Sci Res 10:1223–1228

Bai H, Zhang Z, Guo Y, Jia W (2009) Biological synthesis of size-controlled cadmium sulfide nanoparticles using immobilized *Rhodobacter sphaeroides*. Nanoscale Res Lett 4:717–723

Bai H-J, Yang B-S, Chai C-J, Yang G-E, Jia W-L, Yi Z-B (2011) Green synthesis of silver nanoparticles using *Rhodobacter sphaeroides*. World J Microbiol Biotechnol 27:2723–2728

Banu AN, Balasubramanian C, Moorthi PV (2014) Biosynthesis of silver nanoparticles using *bacillus thuringiensis* against dengue vector, *Aedes aegypti* (Diptera: Culicidae). Parasitol Res 113:311–316

Bao H, Lu Z, Cui X, Qiao Y, Guo J, Anderson JM, Li CM (2010) Extracellular microbial synthesis of biocompatible CdTe quantum dots. Acta Biomater 6(9):3534–3541

Berezin S, Aviv Y, Aviv H, Goldberg E, Tischler YR (2017) Replacing a century old technique – modern spectroscopy can supplant gram staining. Sci Rep 7:3810

Bruna N, Collao B, Tello A, Caravantes P, Díaz-Silva N, Monrás JP, Órdenes-Aenishanslins N, Flores M, Espinoza-Gonzalez R, Bravo D, Pérez-Donoso JM (2019) Synthesis of salt-stable fluorescent nanoparticles (quantum dots) by polyextremophile halophilic bacteria. Sci Rep 9:1–13

Cai F, Li J, Sun J, Ji Y (2011) Biosynthesis of gold nanoparticles by biosorption using *Magnetospirillum gryphiswaldense* MSR-1. Chem Eng J 175:70–75

Carrasco V, Amarelle V, Lagos-Moraga S, Quezada CP, Espinoza-González R, Faccio R, Fabiano E, Pérez-Donoso JM (2021) Production of cadmium sulfide quantum dots by the lithobiontic Antarctic strain *Pedobacter* sp. UYP1 and their application as photosensitizer in solar cells. Microb Cell Fact 20(1):41

Ceci P, Chiancone E, Kasyutich O, Bellapadrona G, Castelli L, Fittipaldi M, Gatteschi D, Innocenti C, Sangregorio C (2010) Synthesis of iron oxide nanoparticles in *listeria innocua* Dps (DNA-binding protein from starved cells): a study with the wild-type protein and a catalytic centre mutant. Chemistry 16(2):709–717

Charoenwongpaiboon T, Wangpaiboon K, Panpetch P, Field RA, Barclay JE, Pichyangkura R, Kuttiyawong K (2019) Temperature-dependent inulin nanoparticles synthesized by *Lactobacillus reuteri* 121 inulosucrase and complex formation with flavonoids. Carbohydr Polym 223: 115044

Christopher FC, Ponnusamy SK, Ganesan JJ, Ramamurthy R (2019) Investigating the prospects of bacterial biosurfactants for metal nanoparticle synthesis - a comprehensive review. IET Nanobiotechnol 13(3):243–249

Chun-Jing C, Hong-Juan B (2010) Biosynthesis of silver nanoparticles using the phototrophic bacteria *Rhodopseudomonas palustris* and its antimicrobial activity against *Escherichia coli* and *Staphylococcus aureus*. Microbiol China 37:1798–1804

Correa-Llantén DN, Muñoz-Ibacache SA, Castro ME, Muñoz PA, Blamey JM (2013) Gold nanoparticles synthesized by *Geobacillus* sp. strain ID17 a thermophilic bacterium isolated from Deception Island, Antarctica. Microb Cell Fact 12:75

Das M, Borah D, Patowary K, Borah M, Khataniar A, Bhusan KB (2019) Antimicrobial activity of silver nanoparticles synthesised by using microbial biosurfactant produced by a newly isolated *Bacillus vallismortis* MDU6 strain. IET Nanobiotechnol 13(9):967–973

Deepa S, Kanimozhi K, Panneerselvam A (2013) Antimicrobial activity of extracellularly synthesized silver nanoparticles from marine derived actinomycetes. Int J Curr Microbiol App Sci 2:223–230

Deepak V, Umamaheshwaran PS, Guhan K, Nanthini RA, Krithiga B, Jaithoon NMH, Gurunathan S (2011) Synthesis of gold and silver nanoparticles using purified URAK. Colloids Surf B Biointerfaces 86:353–358

Dhoondia ZH, Chakraborty H (2012) Lactobacillus mediated synthesis of silver oxide nanoparticles. Nanomater Nanotechnol 2(2):1

Drakontis CE, Amin S (2020) Biosurfactants: formulations, properties, and applications. Curr Opin Colloid Interface Sci 48:77–90

Du L, Jiang H, Liu X, Wang E (2007) Biosynthesis of gold nanoparticles assisted by *Escherichia coli* DH5a and its application on direct electrochemistry of haemoglobin. Electrochem Commun 9:1165–1170

Eid AM, Fouda A, Niedbała G, Hassan SE, Salem SS, Abdo AM, F Hetta H, Shaheen TI. (2020) Endophytic *Streptomyces laurentii* mediated green synthesis of ag-NPs with antibacterial and anticancer properties for developing functional textile fabric properties. Antibiotics (Basel) 9(10):641

Eltarahony M, Zaki S, ElKady M, Abd-El-Haleem D (2018) Biosynthesis, characterization of some combined nanoparticles, and its biocide potency against a broad spectrum of pathogens. J Nanomater 1:2018

Feng Y, Yu Y, Wang Y, Lin X (2007) Biosorption and bioreduction of trivalent aurum by photosynthetic bacteria *Rhodobacter capsulatus*. Curr Microbiol 55:402–408

Fesharaki PJ, Nazari P, Shakibaie M (2010) Biosynthesis of selenium nanoparticles using *Klebsiella pneumoniae* and their recovery by a simple sterilization process. Braz J Microbiol 41:461–466

Gaidhani S, Singh R, Singh D, Patel U, Shevade K, Yeshvekar R, Chopade BA (2013) Biofilm disruption activity of silver nanoparticles synthesized by *Acinetobacter calcoaceticus* PUCM 1005. Mater Lett 108:324–327

Gallardo C, Monrás JP, Plaza DO, Collao B, Saona LA, Durán-Toro V, Venegas FA, Soto C, Ulloa G, Vásquez CC, Bravo D, Pérez-Donoso JM (2014) Low-temperature biosynthesis of fluorescent semiconductor nanoparticles (CdS) by oxidative stress resistant Antarctic bacteria. J Biotechnol 187:108–115

Gallardo-Benavente C, Carrión O, Todd JD, Pieretti JC, Seabra AB, Durán N, Rubilar O, Pérez-Donoso JM, Quiroz A (2019) Biosynthesis of CdS quantum dots mediated by volatile sulfur compounds released by Antarctic *Pseudomonas fragi*. Front Microbiol 10:1866

Garmasheva I, Kovalenko N, Voychuk S, Ostapchuk A, Livins'ka O, Oleschenko L (2016) Lactobacillus species mediated synthesis of silver nanoparticles and their antibacterial activity against opportunistic pathogens in vitro. Bioimpacts 6(4):219–223

Ghorbani HR (2013) Biosynthesis of silver nanoparticles using *salmonella typhirium*. J Nanostruct Chem 3:29–32

Gopinathan P, Ashok AM, Selvakumar R (2013) Bacterial flagella as biotemplate for the synthesis of silver nanoparticle impregnated bionanomaterial. Appl Surf Sci 276:717–722

Gurunathan S, Han JW, Eppakayala V, Jeyaraj M, Kim J-H (2013) Cytotoxicity of biologically synthesized silver nanoparticles in MDA-MB-231 human breast cancer cells. Biomed Res Int 2013:535796

Gurunathan S, Kalishwaralal K, Vaidyanathan R, Deepak V, Pandian SRK, Muniyandi J, Hariharan N, Eom SH (2009) Biosynthesis, purification and characterization of silver nanoparticles using *Escherichia coli*. Colloids Surf B Biointerfaces 74:328–335

Habibullah G, Viktorova J, Ruml T (2021) Current strategies for noble metal nanoparticle synthesis. Nanoscale Res Lett 16(1):47

Hassan SE, Fouda A, Radwan AA, Salem SS, Barghoth MG, Awad MA, Abdo AM, El-Gamal MS (2019) Endophytic actinomycetes *Streptomyces* spp mediated biosynthesis of copper oxide nanoparticles as a promising tool for biotechnological applications. J Biol Inorg Chem 24(3):377–393

Hazra C, Kundu D, Chaudhari A, Jana T (2013) Biogenic synthesis, characterization, toxicity and photocatalysis of zinc sulfide nanoparticles using rhamnolipids from *Pseudomonas aeruginosa* BS01 as capping and stabilizing agent. J Chem Technol Biotechnol 88:1038–1048

He S, Guo Z, Zhang Y, Zhang S, Wang J, Gu N (2007) Biosynthesis of gold nanoparticles using the bacteria *Rhodopseudomonas capsulata*. Mater Lett 61:3984–3987

Hosseini-Abari A, Emtiazi G, Ghasemi SM (2013) Development of an eco-friendly approach for biogenesis of silver nanoparticles using spores of *bacillus atrophaeus*. World J Microbiol Biotechnol 29:2359–2364

Huggias S, Bolla PA, Serradell MA, Casella M, Peruzzo PJ (2020) Platinum nanoparticles obtained at mild conditions on S-layer protein/polymer particle supports. Langmuir 36(5):1201–1211

Hussain I, Singh NB, Singh A, Singh H, Singh SC (2016) Green synthesis of nanoparticles and its potential applications. Biotech Lett 38:545–560

Husseiny MI, El-Aziz MA, Badr Y, Mahmoud MA (2007) Biosynthesis of gold nanoparticles using *Pseudomonas aeruginosa*. Spectrochim Acta 67:1003–1006

Iravani S (2014) Bacteria in nanoparticle synthesis: current status and future prospects. Int Sch Res Notices 2014:359316

Jafari M, Rokhbakhsh-Zamin F, Shakibaie M, Moshafi MH, Ameri A, Rahimi HR, Forootanfar H (2018) Cytotoxic and antibacterial activities of biologically synthesized gold nanoparticles assisted by *Micrococcus yunnanensis* strain J2. Biocatal Agric Biotechnol 15:245–253

Jang E, Shim HW, Ryu BH, An DR, Yoo WK, Kim KK, Kim DW, Kim TD (2015) Preparation of cobalt nanoparticles from polymorphic bacterial templates: a novel platform for biocatalysis. Int J Biol Macromol 81:747–753

Jayaseelan C, Rahuman AA, Kirthi AV, Marimuthu S, Santhoshkumar T, Bagavan A, Gaurav K, Karthik L, Rao KV (2012) Novel microbial route to synthesize ZnO nanoparticles using *Aeromonas hydrophila* and their activity against pathogenic bacteria and fungi. Spectrochim Acta A 90:78–84

Jayaseelan C, Rahuman AA, Roopan SM, Kirthi AV, Venkatesan J, Kim SK, Iyappan M, Siva C (2013) Biological approach to synthesize TiO_2 nanoparticles using *Aeromonas hydrophila* and its antibacterial activity. Spectrochim Acta A Mol Biomol Spectrosc 107:82–89

Jha AK, Prasad K (2010) Ferroelectric BaTiO3 nanoparticles: biosynthesis and characterization. Colloids Surf B Biointerfaces 75(1):330–334

Jha AK, Prasad K, Prasad K (2009) Biosynthesis of Sb_2O_3 nanoparticles: a low-cost green approach. Biotechnol J 4:1582–1585

Juibari MM, Abbasalizadeh S, Jouzani GS, Noruzi M (2011) Intensified biosynthesis of silver nanoparticles using a native extremophilic *Ureibacillus thermosphaericus* strain. Mater Lett 65: 1014–1017

Kalimuthu K, Suresh Babu R, Venkataraman D, Bilal M, Gurunathan S (2008) Biosynthesis of silver nanocrystals by Bacillus licheniformis. Colloids Surf B Biointerfaces 65:150–153

Kalishwaralal K, Deepak V, Pandian SRK, Kottaisamy M, BarathManiKanth S, Kartikeyan B, Gurunathan S (2010) Biosynthesis of silver and gold nanoparticles using Brevibacterium casei. Colloids Surf B Biointerfaces 77:257–262

Kalpana D, Lee YS (2013) Synthesis and characterization of bactericidal silver nanoparticles using cultural filtrate of simulated microgravity grown Klebsiella pneumoniae. Enzyme Microb Technol 52:151–156

Kanmani P, Lim ST (2013) Synthesis and structural characterization of silver nanoparticles using bacterial exopolysaccharide and its antimicrobial activity against food and multidrug resistant pathogens. Process Biochem 48:1099–1106

Kannan N, Mukunthan KS, Balaji S (2011) A comparative study of morphology, reactivity and stability of synthesized silver nanoparticles using *Bacillus subtilis* and *Catharanthus roseus* (L.) G. Don. Colloids Surf B Biointerfaces 86:378–383

Karthik C, Radha KV (2012) Biosynthesis and characterization of silver nanoparticles using *Enterobacter aerogenes*: a kinetic approach. Dig J Nanomater Biostruct 7:1007–1014

Karthik L, Kumar G, Kirthi AV, Rahuman AA, Rao KVB (2014) Streptomyces sp. LK3 mediated synthesis of silver nanoparticles and its biomedical application. Bioprocess Biosyst Eng 37: 261–267

Kikuchi F, Kato Y, Furihata K, Kogure T, Imura Y, Yoshimura E, Suzuki M (2016) Formation of gold nanoparticles by glycolipids of *lactobacillus casei*. Sci Rep 6:34626

Kiran GS, Sabu A, Selvin J (2010) Synthesis of silver nanoparticles by glicololid biosurfactant produced from marine *Brevibacterium. Casei* MSA 19. J Biotechnol 148:221–225

Klaus T, Joerger R, Olsson E, Granqvist CG (1999) Silver-based crystalline nanoparticles, microbially fabricated. Proc Natl Acad Sci U S A 96(24):13611–13614

Konishi Y, Ohno K, Saitoh N (2007) Bioreductive deposition of platinum nanoparticles on the bacterium *Shewanella algae*. J Biotechnol 128:648–653

Krishnaraj RN, Berchmans S (2013) In vitro antiplatelet activity of silver nanoparticles synthesized using the microorganism Gluconobacter roseus: an AFM-based study. RSC Adv 3:8953–8959

Kumar CG, Mamidyala SK (2011) Extracellular synthesis of silver nanoparticles using culture supernatant of *Pseudomonas aeruginosa*. Colloids Surf B Biointerfaces 84:462–466

Kumar CG, Mamidyala SK, Das B, Sridhar B, Devi GS, Karuna MS (2010) Synthesis of biosurfactant-based silver nanoparticles with purified rhamnolipids isolated from *Pseudomonas aeruginosa* BS-161R. J Microbiol Biotechnol 20:1061–1068

Kumar U, Shete A, Harle AS, Kasyutich O, Schwarzacher W, Pundle A, Poddar P (2008) Extracellular bacterial synthesis of protein-functionalized ferromagnetic Co_3O_4 nanocrystals and imaging of self-organization of bacterial cells under stress after exposure to metal ions. Chem Mater 20(4):1484–1491

Law N, Ansari S, Livens FR, Renshaw JC, Lloyd JR (2008) The formation of nano-scale elemental silver particles via enzymatic reduction by *Geobacter sulfurreducens*. Appl Environ Microbiol 4:7090–7093

Liu C, Yang D, Wang Y, Shi J, Jiang Z (2012) Fabrication of antimicrobial bacterial cellulose–ag/AgCl nanocomposite using bacteria as versatile biofactory. J Nanopart Res 14:1084–1095

Malarkodi C, Rajeshkumar S, Paulkumar K, Vanaja M, Gnanajobitha G, Annadurai G (2014) Biosynthesis and antimicrobial activity of semiconductor nanoparticles against oral pathogens. Bioinorg Chem Appl 2014:347167

Manikprabhu D, Lingappa K (2013) Antibacterial activity of silver nanoparticles against methicillin-resistant *Staphylococcus aureus* synthesized using model Streptomyces sp. pigment by photo-irradiation method. J Pharm Res 6:255–260

Manikprabhu D, Lingappa K (2014) Synthesis of silver nanoparticles using the *Streptomyces coelicolor* klmp33 pigment: an antimicrobial agent against extended-spectrum beta-lactamase (ESBL) producing *Escherichia coli*. Korean J Couns Psychother 45:434–437

Maragatham I, Govindammal M (2014) Biosynthesis ofsilver nanoparticles by rhamnolipid produced by *Pseudomonas fluorescens* MFS-1 from mangrove soil. Int J Pharm Biol Arch 5:135–140

Marimuthu S, Rahuman AA, Kirthi AV, Santhoshkumar T, Jayaseelan C, Rajakumar G (2013) Eco-friendly microbial route to synthesize cobalt nanoparticles using *bacillus thuringiensis* against malaria and dengue vectors. Parasitol Res 112(12):4105–4112

Markus J, Mathiyalagan R, Kim YJ, Abbai R, Singh P, Ahn S, Perez ZEJ, Hurh J, Yang DC (2016) Intracellular synthesis of gold nanoparticles with antioxidant activity by probiotic *lactobacillus kimchicus* DCY51^T isolated from Korean kimchi. Enzyme Microb Technol 95:85–93

Mirjani R, Faramarzi MA, Sharifzadeh M, Setayesh N, Khoshayand MR, Shahverdi AR (2015) Biosynthesis of tellurium nanoparticles by *lactobacillus plantarum* and the effect of nanoparticle-enriched probiotics on the lipid profiles of mice. IET Nanobiotechnol 9(5):300–305

Mohd Yusof H, Abdul Rahman N, Mohamad R, Zaidan UH, Samsudin AA (2020) Biosynthesis of zinc oxide nanoparticles by cell-biomass and supernatant of *lactobacillus plantarum* TA4 and its antibacterial and biocompatibility properties. Sci Rep 10(1):19996

Monowar T, Rahman MS, Bhore SJ, Raju G, Sathasivam KV (2018) Silver nanoparticles synthesized by using the endophytic bacterium *Pantoea ananatis* are promising antimicrobial agents against multidrug resistant bacteria. Molecules 23(12):3220

Monowar T, Rahman MS, Bhore SJ, Sathasivam KV (2021) Endophytic bacteria *Enterobacter hormaechei* fabricated silver nanoparticles and their antimicrobial activity. Pharmaceutics 13(4):511

Morsy FM, Nafady NA, Abd-Alla MH, Elhady DA (2014) Green synthesis of silver nanoparticles by water soluble fraction of the extracellular polysaccharides/matrix of the cyanobacterium *Nostoc commune* and its application as a potent fungal surface sterilizing agent of seed crops. Univers J Microbiol Res 2:36–43

Narayanan KB, Sakthivel N (2010) Biological synthesis of metal nanoparticles by microbes. Adv Colloid Interface Sci 156:1–13

Narayanan KB, Sakthivel N (2013) Biosynthesis of silver nanoparticles by phytopathogen *Xanthomonas oryzae* pv. *Oryzae* strain BXO8. J Microbiol Biotechnol 23:1287–1292

Naseer QA, Xue X, Wang X, Dang S, Din SU, Kalsoom JJ (2021) Synthesis of silver nanoparticles using *lactobacillus bulgaricus* and assessment of their antibacterial potential. Braz J Biol 82: e232434

Nirmala C, Sridevi M (2021) Characterization, antimicrobial and antioxidant evaluation of biofabricated silver nanoparticles from endophytic *Pantoea anthophila*. J Inorg Organomet Polym Mater 29:1–15

Nordmeier A, Merwin A, Roeper DF, Chidambaram D (2018) Microbial synthesis of metallic molybdenum nanoparticles. Chenosphere 203:521–525

Öcal N, Ceylan A, Duman F (2020) Intracellular biosynthesis of PbS quantum dots using *Pseudomonas aeruginosa* ATCC 27853: evaluation of antibacterial effects and DNA cleavage activities. World J Microbiol Biotechnol 36(10):147

Oliva-Arancibia B, Órdenes-Aenishanslins N, Bruna N, Ibarra PS, Zacconi FC, Pérez-Donoso JM, Poblete-Castro I (2017) Co-synthesis of medium-chain-length polyhydroxyalkanoates and CdS quantum dots nanoparticles in *pseudomonas putida* KT2440. J Biotechnol 264:29–37

Órdenes-Aenishanslins N, Anziani-Ostuni G, Monrás JP, Tello A, Bravo D, Toro-Ascuy D, Soto-Rifo R, Prasad PN, Pérez-Donoso JM (2020) Bacterial synthesis of ternary CdSAg quantum dots through cation exchange: tuning the composition and properties of biological nanoparticles for bioimaging and photovoltaic applications. Microorganisms 8(5):631

Parikh RY, Ramanathan R, Coloe PJ, Bhargava SK, Patole MS, Shouche YS, Bansal V (2011) Genus-wide physicochemical evidence of extracellular crystalline silver nanoparticles biosynthesis by *Morganella* spp. PLoS One 6(6):e21401

Patil MP, Kang MJ, Niyonizigiye I, Singh A, Kim JO, Seo YB, Kim GD (2019) Extracellular synthesis of gold nanoparticles using the marine bacterium *Paracoccus haeundaensis* BC74171 (T) and evaluation of their antioxidant activity and antiproliferative effect on normal and cancer cell lines. Colloids Surf B Biointerfaces 183:110455

Plaza DO, Gallardo C, Straub YD, Bravo D, Pérez-Donoso JM (2016) Biological synthesis of fluorescent nanoparticles by cadmium and tellurite resistant Antarctic bacteria: exploring novel natural nanofactories. Microb Cell Fact 15:76

Płaza GA, Chojniak J, Banat IM (2014) Biosurfactant mediated biosynthesis of selected metallic nanoparticles. Int J Mol Sci 15(8):13720–13737

Pourali P, Baserisalehi M, Afsharnezhad S, Behravan J, Alavi H, Bba H (2012) Biological synthesis of silver and gold nanoparticles by bacteria in different temperatures (37°C and 50°C). J Pure Appl Microbiol 6(2):757–763

Pourali P, Baserisalehi M, Afsharnezhad S, Behravan J, Ganjali R, Bahador N, Arabzadeh S (2013) The effect of temperature on antibacterial activity of biosynthesized silver nanoparticles. Biometals 26:189–196

Prakash A, Sharma S, Ahmad N, Ghosh A, Sinha P (2011) Synthesis of AgNPs by *Bacillus cereus* bacteria and their antimicrobial potential. J Biomater Nanobiotechnol 2:155–161

Prasad K, Jha AK (2010) Biosynthesis of CdS nanoparticles: An improved green and rapid procedure. J Colloid Interface Sci 342(1):68–72

Priyadarshini S, Gopinath V, Meera Priyadharsshini N, MubarakAli D, Velusamy P (2013) Synthesis of anisotropic silver nanoparticles using novel strain, *Bacillus flexus* and its biomedical application. Colloids Surf B Biointerfaces 102:232–237

Punjabi K, Mehta S, Chavan R, Chitalia V, Deogharkar D, Deshpande S (2018) Efficiency of biosynthesized silver and zinc nanoparticles against multi-drug resistant pathogens. Front Microbiol 9:2207

Rajasree SR, Suman TY (2012) Extracellular biosynthesis of gold nanoparticles using a gram negative bacterium *Pseudomonas fluorescens*. Asian Pacific J Trop Dis 2:S796–S799

Rajeshkumar S, Malarkodi C, Paulkumar K, Vanaja M, Gnanajobitha G, Annadurai G (2013) Intracellular and extracellular biosynthesis of silver nanoparticles by using marine bacteria *vibrio alginolyticus*. Nanosci Nanotechnol 3:21–25

Ranjitha VR, Rai VR (2017) Actinomycetes mediated synthesis of gold nanoparticles from the culture supernatant of *Streptomyces griseoruber* with special reference to catalytic activity. 3 biotech 7(5):299

Ranjitha VR, Ravishankar VR (2018) Extracellular synthesis of selenium nanoparticles from an actinomycetes *Streptomyces griseoruber* and evaluation of its cytotoxicity on HT-29 cell line. Pharm Nanotechnol 6(1):61–68

Rathod D, Golinska P, Wypij M, Dahm H, Rai M (2016) A new report of *Nocardiopsis valliformis* strain OT1 from alkaline Lonar crater of India and its use in synthesis of silver nanoparticles with special reference to evaluation of antibacterial activity and cytotoxicity. Med Microbiol Immunol 205(5):435–447

Reddy AS, Chen CY, Chen CC, Jean JS, Fan CW, Chen HR, Wang JC, Nimje VR (2009) Synthesis of gold nanoparticles via an environmentally benign route using a biosurfactant. J Nanosci Nanotechnol 9(11):6693–6699

Reyes-Escogido ML, Meneses-Rodrı́guez D, Guardado-Mendoza R (2017) Carbohydrate source affects the synthesis of silver nanoparticles by *lactobacillus plantarum* 1449 and *lactobacillus ruminis* 1313. IET Nanobiotechnol 11(8):1035–1039

Sadhasivam S, Shanmugam P, Veerapandian M, Subbiah R, Yun K (2012) Biogenic synthesis of multidimensional gold nanoparticles assisted by *Streptomyces hygroscopicus* and its electrochemical and antibacterial properties. Biometals 25(2):351–360

Saeb AT, Alshammari AS, Al-Brahim H, Al-Rubeaan KA (2014) Production of silver nanoparticles with strong and stable antimicrobial activity against highly pathogenic and multidrug resistant bacteria. Sci World J 2014:704708

Samadi N, Golkaran D, Eslamifar A, Jamalifar H, Fazeli MR, Mohseni FA (2009) Intra/extracellular biosynthesis of silver nanoparticles by an autochthonous strain of *Proteus mirabilis* isolated from photographic waste. J Biomed Nanotechnol 5:247–253

Santos DK, Rufino RD, Luna JM, Santos VA, Sarubbo LA (2016) Biosurfactants: multifunctional biomolecules of the 21st century. Int J Mol Sci 17(3):401

Saravanan M, Barik SK, MubarakAli D, Prakash P, Pugazhendhi A (2018) Synthesis of silver nanoparticles from *Bacillus brevis* (NCIM 2533) and their antibacterial activity against pathogenic bacteria. Microb Pathog 116:221–226

Sathiyanarayanan G, Kiran GS, Selvin J (2013) Synthesis of silver nanoparticles by polysaccharide bioflocculant produced from marine *Bacillus subtilis* MSBN17. Colloids Surf B Biointerfaces 102:13–20

Seshadri S, Prakash A, Kowshik M (2012) Biosynthesis of silver nanoparticles by marine bacterium, *Idiomarina* sp. p R58–8. Bull Mater Sci 35:1201–1205

Sharma N, Pinnaka AK, Raje M, Fnu A, Bhattacharyya MS, Choudhury AR (2012) Exploitation of marine bacteria for production of gold nanoparticles. Microb Cell Fact 11:86–91

Shim HW, Jin YH, Seo SD, Lee SH, Kim DW (2011) Highly reversible lithium storage in *Bacillus subtilis*-directed porous Co_3O_4 nanostructures. ACS Nano 5(1):443–449

Shivashankarappa A, Sanjay KR (2020) Escherichia coli-based synthesis of cadmium sulfide nanoparticles, characterization, antimicrobial and cytotoxicity studies. Braz J Microbiol 51: 939–948

Singh BN, Rawat AK, Khan W, Naqvi AH, Singh BR (2014) Biosynthesis of stable antioxidant ZnO nanoparticles by *Pseudomonas aeruginosa* rhamnolipids. PLoS One 9:106937

Singh BR, Dwivedi S, Al-Khedhairy AA, Musarrat J (2011) Synthesis of stable cadmium sulfide nanoparticles using surfactin produced by *bacillus amyloliquifaciens* strain KSU-109. Colloids Surf B Biointerfaces 85:207–213

Singh H, Du J, Yi TH (2017) Biosynthesis of silver nanoparticles using *Aeromonas* sp. THG-FG1.2 and its antibacterial activity against pathogenic microbes. Artif Cells Nanomed Biotechnol 45(3):584–590

Singh R, Shedbalkar UU, Wadhwani SA, Chopade BA (2015) Bacteriagenic silver nanoparticles: synthesis, mechanism, and applications. Appl Microbiol Biotechnol 99(11):4579–4593

Slawson RM, Van Dyke MI, Lee H, Trevor JT (1992) Germanium and silver resistance, accumulation and toxicity in microorganisms. Plasmid 27:73–79

Soltani Nejad M, Khatami M, Shahidi Bonjar GH (2016) Extracellular synthesis gold nanotriangles using biomass of *Streptomyces microflavus*. IET Nanobiotechnol 10(1):33–38

Srivastava P, Braganca JM, Kowshik M (2014) In vivo synthesis of selenium nanoparticles by *Halococcus salifodinae* BK18 and their anti-proliferative properties against HeLa cell line. Biotechnol Prog 30(6):1480–1487

Srivastava P, Kowshik M (2017) Fluorescent lead (IV) sulfide nanoparticles synthesized by *Idiomarina* sp. strain PR58-8 for bioimaging applications. Appl Environ Microbiol 83(7): e03091–e03016

Subbaiya R, Saravanan M, Priya AR, Shankar KR, Selvam M, Ovais M, Balajee R, Barabadi H (2017) Biomimetic synthesis of silver nanoparticles from *Streptomyces atrovirens* and their potential anticancer activity against human breast cancer cells. IET Nanobiotechnol 11(8):965–972

Suresh AK, Pelletier DA, Wang W, Broich ML, Moon JW, Gu B, Allison DP, Joy DC, Phelps TJ, Doktycz MJ (2011) Biofabrication of discrete spherical gold nanoparticles using the metal-reducing bacterium *Shewanella oneidensis*. Acta Biomater 7:2148–2152

Thomas R, Jasim B, Mathew J, Radhakrishnan EK (2012) Extracellular synthesis of silver nanoparticles by endophytic *Bordetella* sp. isolated from *Piper nigrum* and its antibacterial activity analysis. Nano Biomed Eng 4:183–187

Ulloa G, Collao B, Araneda M, Escobar B, Álvarez S, Bravo D, Pérez-Donoso JM (2016) Use of acidophilic bacteria of the genus *Acidithiobacillus* to biosynthesize CdS fluorescent nanoparticles (quantum dots) with high tolerance to acidic pH. Enzyme Microb Technol 95: 217–224

Ulloa G, Quezada CP, Araneda M, Escobar B, Fuentes E, Álvarez SA, Castro M, Bruna N, Espinoza-González R, Bravo D, Pérez-Donoso JM (2018) Phosphate favors the biosynthesis of CdS quantum dots in *Acidithiobacillus thiooxidans* ATCC 19703 by improving metal uptake and tolerance. Front Microbiol 9:234

Velusamy P, Kumar GV, Jeyanthi V, Das J, Pachaiappan R (2016) Bio-inspired green nanoparticles: synthesis, mechanism, and antibacterial application. Toxicol Res 32:95–102

Wang L, Chen S, Ding Y, Zhu Q, Zhang N, Yu S (2018) Biofabrication of morphology improved cadmium sulfide nanoparticles using *Shewanella oneidensis* bacterial cells and ionic liquid: for toxicity against brain cancer cell lines. J Photochem Photobiol B 178:424–427

Wei S, Guo C, Wang L, Xu J, Dong H (2021) Bacterial synthesis of PbS nanocrystallites in one-step with l-cysteine serving as both sulfur source and capping ligand. Sci Rep 11:1216

Wen L, Lin Z, Gu P, Zhou J, Yao B, Chen G, Fu J (2009) Extracellular biosynthesis of monodispersed gold nanoparticles by a SAM capping route. J Nanopart Res 11:279–288

Wypij M, Czarnecka J, Świecimska M, Dahm H, Rai M, Golinska P (2018) Synthesis, characterization and evaluation of antimicrobial and cytotoxic activities of biogenic silver nanoparticles synthesized from *Streptomyces xinghaiensis* OF1 strain. World J Microbiol Biotechnol 34(2):23

Xu C, Qiao L, Guo Y, Ma L, Cheng Y (2018) Preparation, characteristics and antioxidant activity of polysaccharides and proteins-capped selenium nanoparticles synthesized by *lactobacillus casei* ATCC 393. Carbohydr Polym 195:576–585

Xu S, Luo X, Xing Y, Liu S, Huang Q, Chen W (2019) Complete genome sequence of *Raoultella* sp. strain X13, a promising cell factory for the synthesis of CdS quantum dots. 3 Biotech 9(4):120

Yan ZY, Du QQ, Qian J, Wan DY, Wu SM (2017) Eco-friendly intracellular biosynthesis of CdS quantum dots without changing *Escherichia coli*'s antibiotic resistance. Enzyme Microb Technol 96:96–102

Yücel N, İspirli H, Mercan E, Erdoğan Ü, Dertli E (2021) Synthesis of alternan-stabilized zinc nanoparticles: morphological, thermal, antioxidant and antimicrobial characterization. Prep Biochem Biotechnol 51(4):331–339

Zhang H, Li Q, Lu Y, Sun D, Lin X, Deng X, He N, Zheng S (2005) Biosorption and bioreduction of diamine silver complex by *Corynebacterium*. J Chem Technol Biotechnol 80:285–290

Zhou NQ, Tian LJ, Wang YC, Li DB, Li PP, Zhang X, Yu HQ (2016) Extracellular biosynthesis of copper sulfide nanoparticles by *Shewanella oneidensis* MR-1 as a photothermal agent. Enzyme Microb Technol 95:230–235

Biological Synthesis of Nanoparticles: Fungi

6

Abstract

Fungi are multicellular, eukaryotic life forms that lack chlorophyll and play a major role in nutrient and carbon cycling in the biosphere as pathogens, mutualistic symbionts, and opportunistic heterotrophs. Fungi are also considered as attractive agents for the biosynthesis of nanoparticles, both intracellularly as well as extracellularly. Myconanotechnology is an upcoming science and technology innovation where fungi are explored for the green synthesis of nanomaterials. As compared to bacteria, the use of fungus is more advantageous since they provide good biomass and easy recovery and have fewer requirements of additional steps to extract the filtrate. Compared to other prokaryotic entities and even plants, the fungal mycelium has proven to be quite sturdy in withstanding adverse conditions during culturing in bioreactors or other chambers during the nanofabrication process. The chapter discusses the role of filamentous fungi, yeast, mushrooms, and lichens in the green synthesis of nanoparticles.

6.1 Introduction

Fungi belong to the kingdom of multicellular, eukaryotic organisms that lack chlorophyll as well as the vascular tissue system. Fungi are heterotrophic and require preformed organic carbon compounds for their well-being. Therefore, they live by decomposition and absorption of organic matter from dead or living sources by secreting enzymes. The members of this group have a cell wall comprising of complex compounds like chitin, glucan, chitosan, mannan, and glycoproteins (but not cellulose) in various combinations. The fungal cellulose or chitin is quite similar in structure to the exoskeleton of insects and arachnids. Fungi can grow in pH ranging from 2 to 8.5 and require free oxygen for growth. Most of the fungi are

S. Srivastava, A. Bhargava, *Green Nanoparticles: The Future of Nanobiotechnology*,
https://doi.org/10.1007/978-981-16-7106-7_6

mesophilic, growing between 25 and 31 °C and require less moisture than yeasts and bacteria for proper growth. Members of this kingdom utilize foods ranging from simple to complex substrates since they possess different hydrolytic enzymes like amylases, pectinase, proteinase, and lipase. Fungi play a pivotal role in nutrient and carbon cycling in the biosphere primarily as disposers and recyclers of organic substrates. They act as pathogens (humans, plants, and animals), mutualistic symbionts, and opportunistic heterotrophs. Fungi are source of a wide range of extracellular enzymes using which they utilize to disintegrate the organic matter and decompose soil components, thereby making the essential elements available for uptake by plants and bacteria.

6.2 Fungal Classification

It is speculated that about 5.1 million fungal species exist in nature. However, only about 100,000 species of fungi have been identified till date. Phylogenetic analyses on the basis of nucleotide sequence of nuclear small subunit ribosomal RNA (18S rDNA) genes have established that fungi constitute a part of the terminal radiation event within the eukaryotic group that occurred approximately 1 billion years ago (Fig. 6.1). The earliest fossil evidence of fungi comprise of several types of spores,

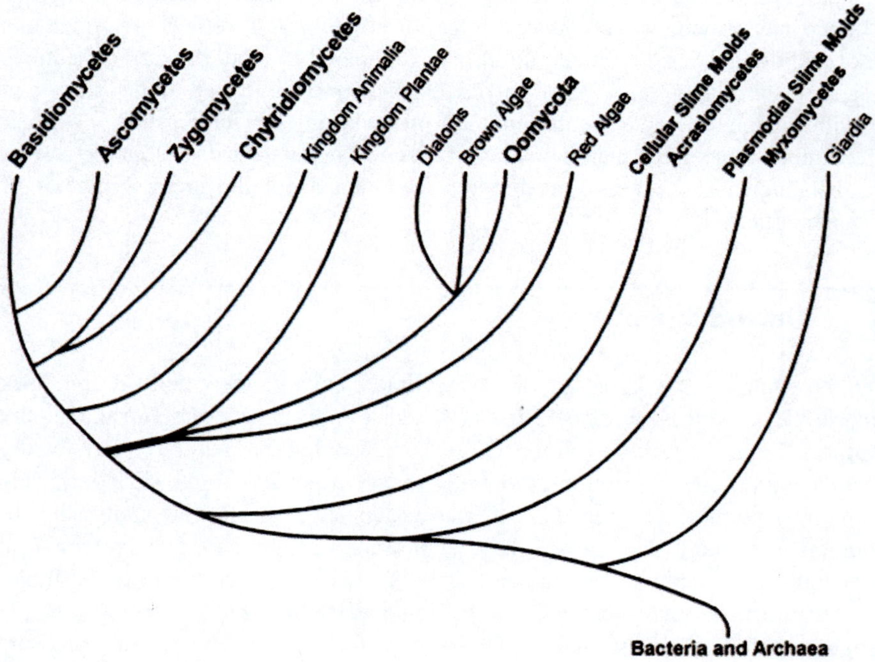

Fig. 6.1 Phylogenetic tree showing relationships of eukaryotes, based on the nucleotide sequence of 18S DNA (Reprinted with permission from Taylor et al. 1994)

zygomycete sporangia, vesicular arbuscular mycorrhizae, *Allomyces*, and perithecial ascomycetous fungus dating back to Lower Devonian Rhynie chert (around 407 million years old). During their course of evolution, the primitive unicellular marine microbes increased in size, complexity, and metabolic functioning. Initial classifications such as those given by Whittaker (1969), based primarily on morphology, sexual spores, and mode of reproduction, divided true fungi or Eumycota into four major phyla, viz., Chytridiomycota, Zygomycota, Ascomycota, and Basidiomycota. However, currently the phylum Zygomycota is no longer accepted due to its polyphyletic nature. Rapid strides in environmental-sequencing-based technologies have brought into light a highly diverse and cosmopolitan clade of fungi-like life forms including *Rozella* and other related genera. The most recent classifications generally divides the true fungi into nine major lineages, viz., Opisthosporidia, Chytridiomycota, Neocallimastigomycota, Blastocladiomycota, Zoopagomycota, Mucoromycota, Glomeromycota, Ascomycota, and Basidiomycota. Chytrids, the unicellular fungi having uniflagellate zoospores and branching hyphae with rhizoids that narrow down into a slender, acute tip, were earlier considered to be the most primitive fungi. However, the presence of flagella in Chytridiomycota warranted its removal from fungi and inclusion in the Protoctista. Later investigations of cell wall polysaccharides and lysine synthesis showed their affinity with the classical phyla of fungi. Their systematic position was settled after analysis of the highly conserved 18S rDNA sequences wherein chytridiomycetes clustered with members of Ascomycota and Basidiomycota in a distinct clade from ciliate protists. The metagenomic studies have shown that some members of this group had the ability to endure the extreme environmental conditions comparable to the icy Precambrian habitats during glacial intervals. On the other hand, Blastocladiomycota, whose first fossil records have also been traced to the Rhynie chert, are different from other chytrids due to differences in zoospore ultrastructure, life cycle, and phylogenetic position. Zoopagomycota and Mucoromycota are thought to have evolved from a common ancestor ~700 million years ago. The hyphal form is considered as an adaptation to invasive, heterotrophic growth wherein the hyphae penetrate, absorb, and digest the food resource as well as helps the fungus by physically and biochemically extending their reach. The phylogeny of Basidiomycota and Ascomycota has been settled leading to several drastic taxonomic revisions in these groups. Basidiomycota, comprising some 35,000 species, are considered to be the most advanced fungi.

6.3 Fungi in Nanoparticle Production

Myconanotechnology is an upcoming science and technology innovation where fungi are explored for the green synthesis of nanomaterials. Fungi are the most promising candidates for the development of new drug candidates, metabolites, and antimicrobials which is primarily due to the production of a large number of bioactive substances (~6400) that are produced by different fungal species. Fungi are also considered as attractive agents for the biosynthesis of nanoparticles, both intracellularly as well as extracellularly. Intracellular synthesis involves the addition of metal salt to the fungal culture that leads to production and internalization of nanoparticles in the biomass. Numerous methods like chemical treatment, centrifugation, and filtration are utilized for disrupting the fungal biomass that leads to extraction of the nanostructures. The more preferred extracellular method involves the mixing of the metal precursor to the fungal filtrate containing the biomolecules leading to production of nanoparticles in non-aggregate form in the medium. Various methods like filtration, membrane filtration, gel filtration, dialysis, and ultracentrifugation are employed for the elimination of fungal residues and other impurities so that pure nanoparticles are obtained.

6.4 Mechanism of Nanoparticle Synthesis by Fungi

Even though the procedure of biosynthesis of nanoparticles seems simple, several experimental parameters like agitation, temperature, light, and reaction time need proper optimization for achieving monodispersion, stability, and biocompatibility of the nanoparticles. Standardization of parameters like temperature, concentration of the inorganic/organic material, pH, type of medium, and quantity of biomass leads to fabrication of nanostructures having desired features. Temperature is a critical factor in some studies since it controls the time required for the synthesis/fabrication of nanoparticles. In general, the rate of nanoparticle formation is increased with rise in temperature and has been reported to increase up to 40 °C in *Trichoderma harzianum*, 90 °C in *Aspergillus oryzae*, and maximum between 60 and 80 °C in *Fusarium oxysporum*. Temperature also has a tremendous effect on the size and stability of the nanoparticles with several reports suggesting an increase in size with rising temperature. AbdelRahim et al. (2017) observed this trend in *Rhizopus stolonifer* with silver nanoparticles of 25.89 and 48.43 nm obtained at 20 °C and 60 °C. Similarly, Shahzad et al. (2019) obtained silver nanoparticles measuring 322.8 nm and 1073.45 nm at 25 °C and 55 °C, respectively, while using *Aspergillus fumigatus*. However, decrease in nanoparticle size with increase in temperatures has been observed in *Fusarium oxysporum* and *Trichoderma viride*.

Another parameter which has a controlling influence on nanoparticle characteristics is pH of the reaction medium. A higher pH is known to enhance the synthesis of metal nanoparticles, especially silver in *Fusarium oxysporum*, *Penicillium oxalicum*, *Epicoccum nigrum*, and *Colletotrichum* sp. This has been attributed to the increased competition between protons and metal ions for initiating bond formation with the negatively charged domains. However, successful synthesis at neutral or slightly alkaline pH has been reported in *Guignardia mangiferae* and *Isaria fumosorosea*. The composition of the medium that contains substrates which induce synthesis and release of enzymes used in metal reduction has modest effect on the formation of nanoparticles. The amount of fungal biomass used has a significant bearing on the characteristics of nanoparticles with several studies reporting rapid rate of nanoparticle synthesis on usage of low biomass concentration, while others are exhibiting higher synthesis rates with higher concentrations. Thus, a proper balance between the amount of fungal biomass and the quantity of metal salt is a prerequisite for optimal synthesis of nanomaterials.

Large quantities of extracellular proteins are secreted by several fungal species which bestow stability to the nanoparticles. The "secretome" consists of all the secreted proteins of fungus into the extracellular space. Fungal biomolecules comprise of polysaccharides, peptides, amino acids, vitamins, enzymes, alkaloids, flavonoids, saponins, steroids, tannins, carboxylic acids, quinones, and other secondary metabolites. Several enzymes like α-NADPH-dependent nitrate reductase, cytochrome b5 reductase, phytochelatins, and glutathione reductase (FAD-dependent) are known to be involved in metal bioreduction. The synthesis of nanoparticles requires an important step involving coating of their surface with polymers and surfactants using the biomolecules derived from the organism itself. Capping of the nanoparticles confers stability and prevents their agglomeration and aggregation. The binding of proteins at the surface of nanoparticles and their consequent stabilization usually involves the following:

1. Free amino groups or cysteine residues
2. Mycelial cell wall enzymes having negative carboxyl groups
3. Nucleophilic hydroxide ions that are adsorbed on the surfaces

Fungi have been extensively researched for the production of nanoparticles (especially metal) of different shapes and size since they exhibit high tolerance toward metals and show easy in handling. As compared to bacteria, the use of fungus is more advantageous since they provide good biomass and easy recovery and do not require additional steps for extracting the filtrate. Fungi are also more efficient than bacteria in biosorption of metal ions at low pH. A comparison between the two groups for immobilization of gold ions (Au^{3+}) has shown that some bacteria have the ability to immobilize up to 0.35 mM/g dry cells, while fungal species like *Aspergillus* sp. can immobilize up to 1 mM/g dry cells, both at low pH (2.5–3.0). When we compare their use in nanoparticles vis-à-vis plants, it has been observed that the mycelial mass is more resistant to turbulence and mechanical stress and therefore more appropriate for large-scale synthesis. Apart from this, the fungal metabolism can be easily manipulated by adjustment of reaction conditions like temperature, pH, amount of biomass, type of culture medium, and reaction time, which results in production of nanoparticles having specific size and morphology.

Among different metals, most of the studies with respect to production have been done with silver followed by gold, titanium, and zinc (Fig. 6.2). However, metal oxide nanoparticles have also been leisurely produced by a large number of fungi. Fig. 6.3 depicts the fungal genera that have been utilized most for nanoparticle production. It is evident from Fig. 6.3 that the filamentous fungi have taken a lead over other fungal forms in nanoparticle production by occupying the first five positions. However, several mushrooms have also been effectively utilized for nanoparticle production which will be discussed in later section.

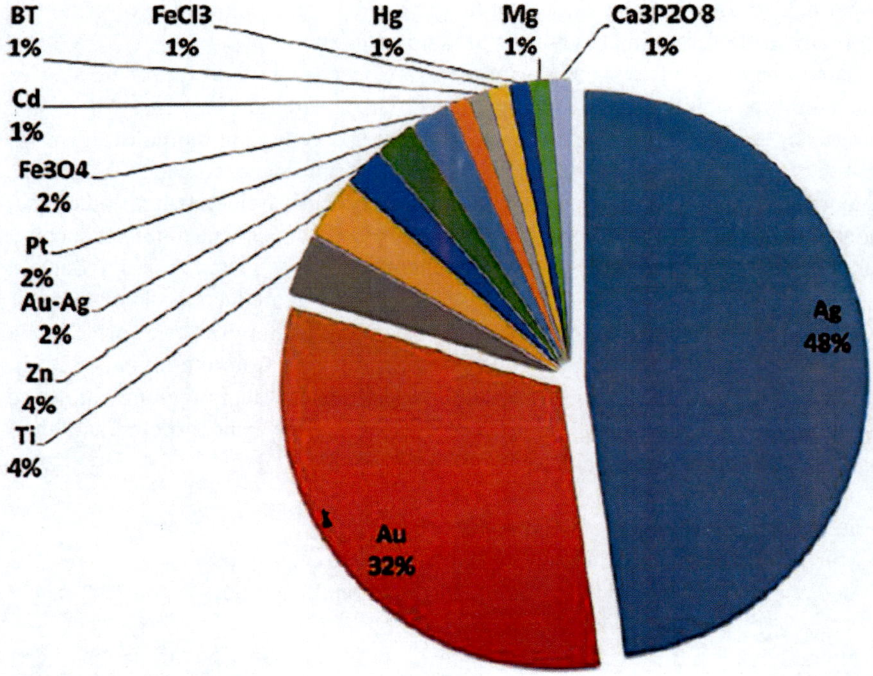

Fig. 6.2 Nanoparticles of different metals synthesized by fungi (Reprinted with permission from Dorcheh and Vahabi 2016)

6.5 Filamentous Fungi in Nanoparticle Production

Filamentous fungi have been at the forefront of biosynthesis of nanoparticles (Tables 6.1, 6.2, and 6.3). *Fusarium* (Hypocreales, Ascomycota) is a ubiquitous, anamorphic filamentous fungi widely distributed in the biosphere which is known for mycotoxins like fumonisins, zearalenone, deoxynivalenol, and trichothecenes. More than 500 species of *Fusarium* have been reported that includes many economically important plant and human pathogens. Together with *Aspergillus*, *Fusarium* has been most researched for the green synthesis of diverse nanostructures (Tables 6.1 and 6.2). Together, these genera account for about 50% of the studies done in fungi with reference to synthesis of metal-derived nanoparticles. However, other filamentous genera have also been used for metal nanoparticle synthesis (Table 6.3). Gaikwad et al. (2013) successfully accomplished green synthesis of silver nanoparticles from six species of *Fusarium*, viz., *F. graminearum, F. solani, F. oxysporum, F. culmorum, F. scirpi*, and *F. tricinctum*. The size of nanoparticles synthesized by different species ranged from 24 to 45 nm with *F. oxysporum* synthesizing the smallest and *F. tricinctum* the largest sized nanoparticles. The zeta potential ranged from −38.3 for *F. tricinctum* to −10.3 for *F. culmorum*, the negative values being probably due to protein capping on nanoparticles. One species

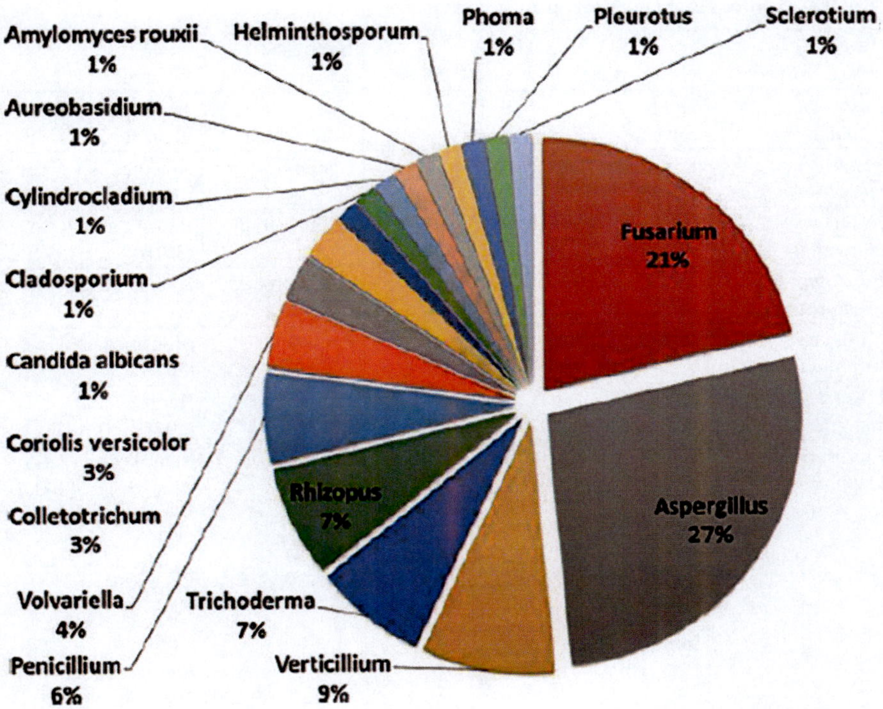

Fig. 6.3 Fungal genera used for synthesis of nanoparticles (Reprinted with permission from Dorcheh and Vahabi 2016)

of *Aspergillus* (*A. aculeatus* strain T6) collected from mining wastes produced semi-quasi spherical iron-containing nanoparticles under ambient conditions with mean size of 15 ± 5 nm. The nanoentities showed potential for enhanced seed germination and increased plant growth (Fig. 6.4).

Successful production of nanoparticles via the intracellular and extracellular routes has been carried out in a single species during a single experiment in several filamentous fungi. Riddin et al. (2006) accomplished formation of both intracellular and extracellular platinum nanoparticles of varying shapes and sizes ranging from 10 to 100 nm using *Fusarium oxysporum* f. sp. *lycopersici*. Gupta and Bector (2013) collected *Aspergillus flavus* and *A. fumigatus* from soil/sludge/effluent samples for production of extracellular (17.76 nm) and intracellular (22.61–26.16 nm) gold nanoparticles of various morphologies (spherical, trigonal, cubic, tetragonal, and hexagonal) when challenged with different concentrations of tetra-chloroauric acid (Fig. 6.5).

Filamentous fungi have also been used in the biosynthesis of carbon-based nanomaterials. *Aspergillus fumigatus* was successfully developed in a catalytic system for the production of Ag nanoparticles/multi-walled carbon nanotubes (Ag/MWf-CNT). Electron microscopy studies revealed well-dispersed spherical silver nanoparticles (≤10 nm) on the carbon nanotubes. The reduction of metal

Table 6.1 Use of Fusarium in nanoparticle production.

Fusarium species	Nanoparticle	Size (nm)	Biological activity	Reference
Fusarium acuminatum	Gold Silver	8–28 5–40	Nr Antibacterial	Tidke et al. (2014) Ingle et al. (2008)
Fusarium chlamydosporum	Silver	6–26	Cytotoxicity	Khalil et al. (2019)
Fusarium culmorum	Silver	5–25	Antibacterial	Bawaskar et al. (2010)
Fusarium graminearum	Silver	Nr	Antimicrobial	Ajah et al. (2018)
Fusarium keratoplasticum	Zinc oxide	10–42	Bactericidal	Mohamed et al. (2019)
Fusarium mangiferae	Silver	25–52	Antibiofilm; anticancer	Hamzah et al. (2018)
Fusarium oxysporum	Barium titanate Cadmium sulfide Cadmium-selenide quantum dots Gold Gold-silver Iron oxide Platinum	4–5 5–20 14–16 22–30 8–14 20–50 5–30	Nr Nr Nr Antibacterial Nr Nr Nr	Bansal et al. (2006) Ahmad et al. (2002) Yamaguchi et al. (2016) Naimi-Shamel et al. (2019) Senapati et al. (2005) Bharde et al. (2006) Syed and Ahmad (2012)
Fusarium scirpi	Silver	0.5–5	Antibiofilm	Rodríguez-Serrano et al. (2020)
Fusarium semitectum	Gold Selenium	25 32.80–130.82	Nr Anticancer; antimicrobial	Sawle et al. (2008) Abbas and Baker (2020)
Fusarium solani	Copper oxide Gold Silver Zinc oxide	9.97–19.49 40–45 5–35 8.55–21.76	Antibacterial; antifungal Anticancer Nr Antibacterial; antifungal	El Sayed and El-Sayed (2020) Clarance et al. (2020) Ingle et al. (2009) El Sayed and El-Sayed (2020)
Fusarium tricinctum	Silver	45	Nr	Gaikwad et al. (2013)
Fusarium verticillioides	Silver	13–19.2	Antibacterial	Mekkawy et al. (2017)

nr not reported.

ions was thought to have been mainly dependent on the biomass weight applied in the biosynthesis process.

Table 6.2 Use of *Aspergillus* in nanoparticle production

Aspergillus species	Nanoparticle	Size (nm)	Biological activity	Reference
Aspergillus aculeatus	Iron	15	Seed germination; plant growth	Bedi et al. (2018)
Aspergillus brasiliensis	Cobalt oxide Silver	20–27 6–21	Antimicrobial Antimicrobial	Omran et al. (2020) Omran et al. (2018)
Aspergillus brunneoviolaceus	Silver	0.72–15.21	Antibacterial; antioxidant	Mistry et al. (2021)
Aspergillus clavatus	Gold	24.4	Nr	Verma et al. (2011)
Aspergillus flavus	Gold Zinc sulfide	17.76–22.61 10–18	Nr Metal detection	Gupta and Bector (2013) Uddandarao et al. (2019)
Aspergillus foetidus	Cadmium sulfide Silver	20–30 35	Nr Bioremediation	Nasrin et al. (2014) Mukherjee et al. (2017)
Aspergillus fumigatus	Gold Silver	17.76–26.16 <84.4	Nr Antimicrobial; antitumor	Gupta and Bector (2013) Othman et al. (2019)
Aspergillus nidulans	Cobalt oxide	20.29	Nr	Vijayanandan and Balakrishnan (2018)
Aspergillus niger	Gold Silver Zinc oxide	12.79 Nr 53–69	Nr Antibacterial Antibacterial; catalytic	Bhambure et al. (2009) Madakka et al. (2018) Kalpana et al. (2018)
Aspergillus oryzae	Selenium Silver	55 7–27	Antibacterial; antifungal Antibacterial	Mosallam et al. (2018) Phanjom and Ahmed (2017)
Aspergillus oryzae var. viridis	Gold	10–60	Nr	Binupriya et al. (2010)
Aspergillus parasiticus	Silica	3–400	Nr	Zielonka et al. (2018)
Aspergillus sydowii	Gold Silver	8.7–15.6 1–24	Nr Antifungal; antiproliferative	Vala (2014) Wang et al. (2021)
Aspergillus terreus	Copper oxide Gold Silver Zinc oxide	15.75 10–50 7–23 10–45	Biomedical Antimicrobial Antimicrobial; antitumor Antibacterial; cytotoxicity	Mousa et al. (2020) Balakumaran et al. (2016) Lotfy et al. (2021) Mohamed et al. (2019)
Aspergillus tubingensis	Phosphorus Silver	28.2 35	Nr Cytotoxicity	Tarafdar et al. (2012) Ottoni et al. (2019)
Aspergillus versicolor	Copper Silver	23–82 3–40	Antifungal Antimicrobial; free radical scavenging	Ammar et al. (2019) Netala et al. (2016a)
Aspergillus welwitschiae	Tellurite	60.80	Antibacterial	Abo Elsoud et al. (2018)

nr not reported.

Table 6.3 Use of other filamentous fungi in nanoparticle production

Fungus	Nanoparticle	Size (nm)	Biological activity	Reference
Alternaria alternata	Gold	12	Nr	Sarkar et al. (2012)
Arthroderma fulvum	Silver	15.5	Antifungal	Xue et al. (2016)
Bjerkandera sp.	Silver	10–100	Nr	Osorio-Echavarría et al. (2021)
Colletotrichum sp.	Gold Silver	8–40 20–50	Nr Antibacterial	Shankar et al. (2003) Azmath et al. (2016)
Cylindrocladium floridanum	Gold	5–35	Dye degradation	Narayanan and Sakthivel (2011a)
Duddingtonia flagrans	Silver	10.29–11.69	Nr	Costa Silva et al. (2017)
Guignardia mangiferae	Silver	5–30	Antibacterial; antifungal	Balakumaran et al. (2015)
Helminthosporium solani	Gold	2–70	Anticancerous	Kumar et al. (2008)
Hormoconis resinae	Gold	3–20	Nr	Mishra et al. (2010)
Humicola sp.	Silver	5–25	Cytotoxicity	Syed et al. (2013)
Neurospora crassa	Gold Platinum	3–100 4–110	Nr Nr	Castro-Longoria et al. (2011) Castro-Longoria et al. (2012)
Neurospora intermedia	Silver	19–84	Antibacterial	Hamedi et al. (2014)
Penicillium sp.	Gold Zirconium	30–50 <100	Nr Bactericidal	Du et al. (2011) Golnaraghi Ghomi et al. (2019)
Penicillium brevicompactum	Gold	10–60	Anticancerous	Mishra et al. (2011a, 2011b)
Penicillium chrysogenum	Platinum Silver	5–40 9–17.5	Cytotoxicity Cytotoxicity	Subramaniyan et al. (2018) Khalil et al. (2019)
Penicillium expansum	Silver	14–25	Antifungal	Ammar and El-Desouky (2016)
Penicillium italicum	Silver	39.5	Antioxidant; anticancer	Taha et al. (2019)
Penicillium oxalicum	Silver	36.53	Catalytic	Du et al. (2015)
Penicillium polonicum	Silver	10–15	Bactericidal	Neethu et al. (2018)
Penicillium purpurogenum	Silver	8–10	Antibacterial	Nayak et al. (2011)
Penicillium rugulosum	Gold	20–80	Nr	Mishra et al. (2012)
Penicillium verrucosum	Silver	10–12	Antifungal	Yassin et al. (2021)
Pestalotiopsis microspora	Silver	2–10	Antioxidant; anticancer	Netala et al. (2016b)

(continued)

Table 6.3 (continued)

Fungus	Nanoparticle	Size (nm)	Biological activity	Reference
Phenerochaete chrysosporium	Silver	34–90	Antibacterial	Saravanan et al. (2018)
Rhizoctonia solani	Silver	2–22	Nr	Ashrafi et al. (2013)
Rhizopus oryzae	Gold	5–65	Catalytic	Das et al. (2012)
Rhizopus stolonifer	Silver	9.47	Nr	AbdelRahim et al. (2017)
Sclerotinia sclerotiorum	Silver	10–15	Antibacterial	Saxena et al. (2016)
Sclerotium rolfsii	Gold	25.2	Nr	Narayanan and Sakthivel (2011b)
Talaromyces purpureogenus	Silver	25	Antibacterial; anticancer	Hu et al. (2019)
Trichoderma hamatum	Gold	5–30	Antimicrobial	Abdel-Kareem and Zohri (2018)
Trichoderma harzianum	Silver	51.10	Antibacterial	Ahluwalia et al. (2014)
Trichoderma longibrachiatum	Silver	5–11	Biocidal	Omran et al. (2019)
Trichoderma reesei	Silver	5–50	Nr	Vahabi et al. (2011)
Trichoderma viride	Silver	2–4	Nr	Fayaz et al. (2009)
Trichothecium sp.	Gold	5–200	Nr	Ahmad et al. (2005)
Verticillium sp.	Gold	20	Nr	Mukherjee et al. (2001)

nr not reported.

6.6 Yeast in Nanoparticle Production

Yeasts are eukaryotic, single-celled fungi which constitute about 1% of the known fungal species. They can grow in foods having low pH, metabolize the food components, and induce physical, chemical, and sensible changes that lead to food being spoiled. However, some yeasts like *Cryptococcus neoformans*, *Cryptococcus gattii*, and several species of *Candida* live as opportunistic pathogens and are the cause of several human diseases.

The diverse metabolic profile of yeast comprising of terpenoid, phenols, amines, amides, proteins, and alkaloids make it an ideal candidate for the production of nanoparticles. Yeasts are preferred over bacteria in mass synthesis of nanoparticles due to several factors like rapid growth, less requirement of complex nutrients, relative ease in culturing, and secretion of numerous enzymes. The presence of membrane-bound (cytosolic) oxidoreductase and quinones plays an important role in the biosynthesis of nanostructures. Of these the oxidoreductases are pH sensitive since increase in pH activates the reductases and reduces the metal ions in the

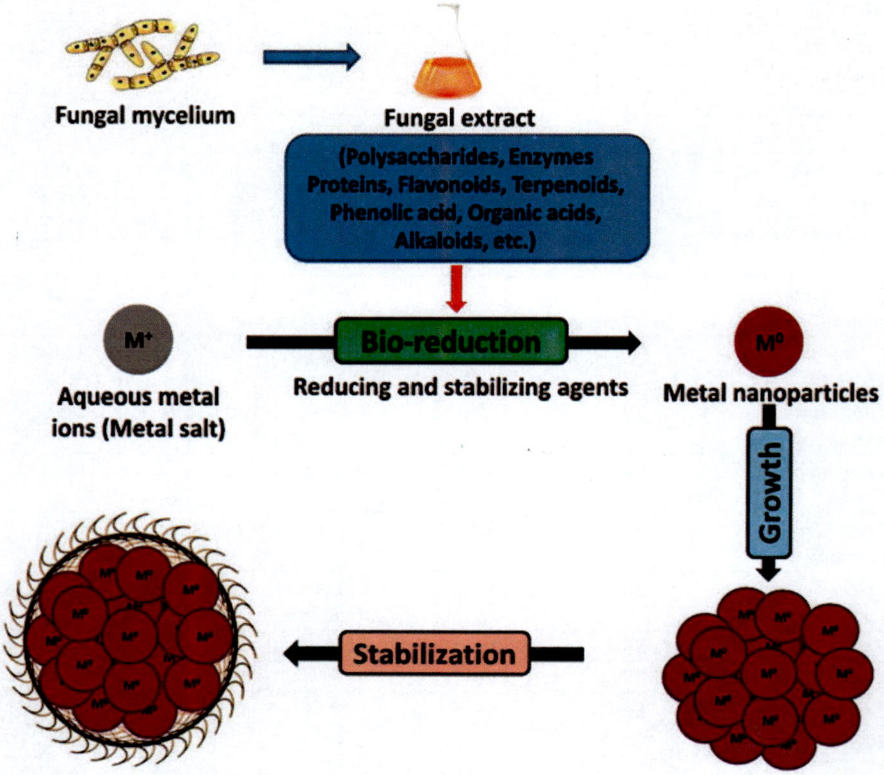

Fig. 6.4 Mechanism of nanoparticle production by fungus (Reprinted from Rai et al. 2021)

medium leading to reduced ability of the yeast cells to form nanoparticles. Several genera of yeasts are known to employ different mechanisms for nanoparticle synthesis which enables collection of nanoparticles of diverse size, particle position, and monodispersity. Some studies have also hinted toward a strong correlation between the size of nanoparticles and the quantity of cells as well as the salt concentrations. Among the yeasts, *Saccharomyces* and *Candida* have been the genera of popular choice for the biosynthesis of nanostructures. *Saccharomyces cerevisiae*, the Baker's yeast, has been used for the green synthesis of gold, silver, palladium, selenium, cadmium sulfide, selenium sulfide, and several metal oxide nanoparticles (Table 6.4). Green synthesis of gold and silver nanoparticles having size ranging from 2 to 80 nm has been successfully carried out using *Candida* sp. However, this area is under-researched, and more studies are needed for successful utilization of yeasts in nanobiotechnology realm.

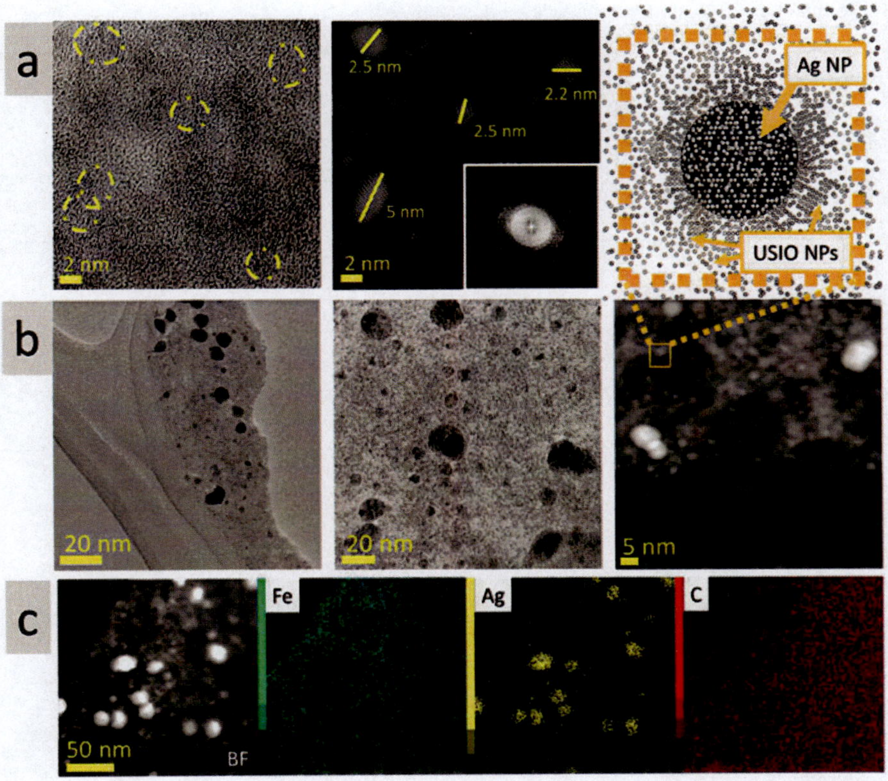

Fig. 6.5 TEM images of (**a**) USIO NPs (left), FFT analysis (inset), and the result of FFT-filtered image using the principal crystalline reflections; (**b**) silver and ultrasmall iron oxide nanoparticles (Mag NPs); (**c**) elements mapping of MAg NPs (Reprinted from Ivashchenko et al. 2018)

6.7 Mushrooms in Nanoparticle Production

Several wild and cultivated mushrooms have been successfully harnessed for the green production of a diverse range of nanoparticles, namely, silver, gold, iron, selenium, palladium, zinc sulfide, and cadmium sulfide (Table 6.5). The first report utilizing mushroom in green synthesis of nanoparticles was published in 2004 when Numata et al. developed nanofibers from the polysaccharides/β-1,3-glucan (Schizophyllan) obtained from *Schizophyllum commune*, commonly known as the split-gill mushroom. Three years later, another research group utilized the protein of spent mushroom substrates (SMS) of oyster mushroom to synthesize silver-protein nanostructures having a mean size of 30.5 nm. This was followed by numerous studies that utilized mushrooms for eco-friendly nanoparticles. Apart from cadmium sulfide and zinc sulfide nanoparticles, biosynthesis of several metallic nanoparticles like silver, gold, iron, and selenium has been successfully achieved using mushrooms. Extracellular synthesis of Au, Ag, and Au-Ag nanoparticles was carried

Table 6.4 Use of yeast in nanoparticle production

Fungus	Nanoparticle	Size (nm)	Biological activity	Reference
Candida albicans	Cadmium sulfide Gold	50–60 20–80	Bactericidal Cancer detection	Kumar et al. (2019) Chauhan et al. (2011)
Candida glabrata	Silver	2–15	Antibacterial; antifungal	Jalal et al. (2018)
Candida guilliermondii	Gold Silver	50–70 10–20	Antibacterial Antibacterial	Mishra et al. (2011b) Mishra et al. (2011b)
Candida utilis	Silver	20–80	Antibacterial	Waghmare et al. (2015)
Cryptococcus humicola	Iron	8–9	Nr	Vainshtein et al. (2014)
Cryptococcus laurentii	Silver	35–400	Antifungal	Fernández et al. (2016)
Kluyveromyces marxianus	Silver	3–12	Antimicrobial	Ashour (2014)
Magnusiomyces ingens	Gold Selenium	10–80 70–90	Catalytic Antibacterial	Zhang et al. (2016) Lian et al. (2019)
Nematospora coryli	Selenium	50–250	Antifungal; antioxidant	Rasouli (2019)
Phaffia rhodozyma	Gold Silver	4–7 5–9	Antifungal Antifungal	Rónavári et al. (2018) Rónavári et al. (2018)
Pichia kudriavzevii	Zinc oxide	10–61	Antimicrobial; antioxidant	Moghaddam et al. (2017)
Saccharomyces boulardii	Silver	3–10	Anticancer	Kaler et al. (2013)
Saccharomyces cerevisiae	Cadmium sulfide Gold Iron oxide Manganese dioxide Palladium Selenium Selenium sulfide Silver Titanium dioxide	3.75 15–20 2–10 34.4 32 75–709 6–153 2–20 6.7	Nr Nr Biocatalysis; biosensing Nr Catalytic Antioxidant Antifungal Nr Antibacterial; photocatalytic	Prasad and Jha (2010) Sen et al. (2011) Mishra et al. (2015) Salunke et al. (2015) Sriramulu and Sumathi (2018) Faramarzi et al. (2020) Asghari-Paskiabi et al. (2019) Korbekandi et al. (2016) Peiris et al. (2018)
Saccharomyces uvarum	Silver	12.4	Anti-inflammatory; antimicrobial	Ammar et al. (2021)
Pichia kudriavzevii	Silver	20.65	Anti-inflammatory; antimicrobial	Ammar et al. (2021)
Rhodotorula glutinis	Silver	15.45	Antifungal; catalytic; cytotoxicity	Cunha et al. (2018)

(continued)

Table 6.4 (continued)

Fungus	Nanoparticle	Size (nm)	Biological activity	Reference
Rhodotorula mucilaginosa	Silver	13.70	Antifungal; catalytic; cytotoxicity	Cunha et al. (2018)
Williopsis saturnus	Gold	5–30	Biocatalysis; biosensing	Mohite et al. (2017)
Yarrowia lipolytica	Gold Silver	7.5–23 12.4	Nr Antibacterial	Pimprikar et al. (2009) Bolbanabad et al. (2020)

nr not reported.

out by Philip (2009) using *Volvariella volvacea*, one of the most widely cultivated edible mushrooms. This fungus is also commonly known as paddy straw mushroom or straw mushroom widely grown in Southeast Asia as a high-quality human food source. Triangular nanoprism and spherical and hexagonal-shaped gold nanoparticles of varying sizes (20–150 nm) were obtained. On the other hand, the silver nanostructures were mainly spherical having dimensions of ~15 nm. Photoluminescent property was found in the nanostructures which showed crystalline structure as shown by SAED and XRD patterns. It was speculated that the amino and the carboxylate groups played an active role in the synthesis of gold and silver nanoparticles, respectively.

Mushrooms have also been explored for the fabrication of gold palladium (Au/Pd) bimetallic and fluorescent carbon dots. It is evident from Table 6.5 that some genera like *Pleurotus* and *Ganoderma* have been utilized the most in this respect. There are also some reports of the use of poisonous mushrooms in the green synthesis of nanoparticles. An extremely toxic mushroom *Amanita muscaria* (Common name: "fly agaric" or "fly amanita") was used for the biofabrication of ultrasmall iron oxide nanoparticles by Ivashchenko et al. (2018). In this study, the fungal extract was used as capping agent and hyaluronic acid as the gel-forming agent. FFT and HRTEM measurements showed the presence of crystalline ultrasmall iron oxide nanoparticles of about 2 nm dimensions having narrow size distribution. The silver nanoparticles of 5–25 nm size visible in the TEM images were distributed throughout the ultrasmall iron oxide matrix.

6.8 Lichens in Nanoparticle Production

Lichens are complex, dual organisms that consist of a mutualistic relationship between two separate organisms, a fungus and an alga; neither of which can survive on its own. Besides having high proportions of phenolic compounds, polysaccharides (lichenan and isolichenan), and various secondary metabolites that impart them with numerous biological activities, lichens are also known as efficient biological indicators. The use of lichens in biosynthesis of nanoparticles has been

Table 6.5 Use of mushrooms in nanoparticle production

Fungus	Nanoparticle	Size (nm)	Biological activity	Reference
Agaricus bisporus	Gold Selenium Silica Silver	20.0–73.3 10–250 30–65 80–100	Antibacterial Nr Nr Antibacterial	Eskandari-Nojehdehi et al. (2016) Loshchinina et al. (2018) Loshchinina et al. (2018) Mirunalini et al. (2012)
Agaricus arvensis	Gold Selenium Silica	2–10 15–550 50–200	Nr Nr Nr	Loshchinina et al. (2018) Loshchinina et al. (2018) Loshchinina et al. (2018)
Amanita muscaria	Iron oxide Silver	2 5–25	Anticancer Anticancer	Ivashchenko et al. (2018) Ivashchenko et al. (2018)
Calocybe indica	Silver	100	Antibacterial	Mirunalini et al. (2012)
Catathelasma ventricosum	Selenium	~50	Antidiabetic	Liu et al. (2018)
Cordyceps militaris	Gold	15–20	Antiproliferative	Ji et al. (2019)
Coriolus versicolor	Cadmium sulfide	10	Nr	Sanghi and Verma (2009)
Flammulina velutipes	Gold Silver	⩽20 21.4	Catalytic Anti-Alzheimer; antidiabetic	Narayanan et al. (2015) Faisal et al. (2021)
Fomes fomentarius	Silver Titanium oxide	10–20 100–120	Antibacterial; anticancer Antibacterial; anticancer	Rehman et al. (2020) Rehman et al. (2020)
Ganoderma applanatum	Silver	133	Antibacterial	Mohanta et al. (2016)
Ganoderma enigmaticum	Silver	15–25	Antibacterial	Gudikandula et al. (2017)
Ganoderma lucidum	Silver	15–22	Antibacterial; antifungal	Aygün et al. (2020)
Ganoderma neo-japonicum	Silver	<6	Anticancer	Gurunathan et al. (2015)
Ganoderma sessiliforme	Silver	~45	Antibacterial	Mohanta et al. (2018)
Hericium erinaceus	Gold	20–40	Anticancer	Raman et al. (2015a, 2015b)

(continued)

Table 6.5 (continued)

Fungus	Nanoparticle	Size (nm)	Biological activity	Reference
Inonotus obliquus	Palladium Silver	77–93 14–35	Anticancer Antioxidant	Gil et al. (2018) Nagajyothi et al. (2013)
Lentinula edodes	Silver	Nr	Antibacterial	Sujatha et al. (2015)
Microporus xanthopus	Silver	40	Antibacterial	Balashanmugam et al. (2013)
Phellinus igniarius	Silver	<100	Antibacterial	Paul et al. (2015)
Pleurotus sp.	Iron	Nr	Nr	Mazumdar and Haloi (2011)
Pleurotus citrinopileatus	Silver	6–10	Bactericidal	Maurya et al. (2016)
Pleurotus cornucopiae var. citrinopileatus	Silver	20–30	Antifungal	Owaid et al. (2015)
Pleurotus cystidiosus	Silver	2–100	Antioxidant	Madhanraj et al. (2017)
Pleurotus djamor	Silver Zinc oxide	5–50 70–80	Cytotoxicity Anticancer; larvicidal	Raman et al. (2015a, 2015b) Manimaran et al. (2020)
Pleurotus eous	Silver	Nr	Antibacterial	Latha (2010)
Pleurotus eryngii	Silver	18.45	Antimicrobial	Acay and Baran (2019)
Pleurotus florida	Gold Gold-platinum Gold-platinum-silver Silver	10–50 16 4–10 20	Anticancer Cytotoxicity Anticancer Antibacterial	Bhat et al. (2013) Chaturvedi et al. (2021) Chaturvedi et al. (2020a) Bhat et al. (2011)
Pleurotus giganteus	Silver	5–25	Antibacterial	Debnath et al. (2019)
Pleurotus ostreatus	Cadmium sulfide Gold Silver Zinc sulfide	4–5 22–39 8–50 2.30–4.04	Nr Catalytic Antibacterial Nr	Borovaya et al. (2015) El-Batal et al. (2015) Devika et al. (2012) Senapati and Sarkar (2014)
Pleurotus pulmonarius	Gold	Nr	Antioxidant	Madhanraj et al. (2017)
Pleurotus sajor-caju	Gold Silver	4–22 30.5	Anticancer Antibacterial	Chaturvedi et al. (2020b) Vigneshwaran et al. (2007)

(continued)

Table 6.5 (continued)

Fungus	Nanoparticle	Size (nm)	Biological activity	Reference
Polyporus rhinocerus	Selenium	50	Anticancer	Wu et al. (2013)
Pycnoporus sanguineus	Silver	1–20	Antimicrobial	Chan and Don (2013a)
Schizophyllum commune	Silver	32.08	Antibacterial; antifungal	Chan and Don (2013b)
Trametes ljubarskyi	Silver	15–25	Antibacterial	Gudikandula et al. (2017)
Trametes trogii	Silver	5–65	Nr	Kobashigawa et al. (2019)
Tricholoma crissum	Silver	5–50	Antibacterial	Ray et al. (2011)
Tricholoma matsutake	Silver	10–20	Antibacterial	Anthony et al. (2014)
Xylaria acuta	Zinc oxide	34–55	Antibacterial	Sumanth et al. (2020)

nr not reported.

limited till date (Table 6.6). A close look at the table reveals that most of the species have been used to fabricate silver nanoparticles, which are incidentally also the most researched ones with respect to production by other organisms. However, there have been some recent reports about production of other metallic and metal oxide nanoparticles as well. Among other metals, gold nanoparticles have been produced using *Acroscyphus sphaerophoroides* (size: 5–35 nm), *Cetraria islandica* (size: 19 nm), *Parmelia sulcata* (size: 54 nm), and *Sticta nylanderiana* (size: 20–50). Devasena et al. (2014) carried out green synthesis of magnesium nanoparticles using magnesium sulfate and extract of *Cladonia rangiferina* (family: Cladoniaceae), commonly known as the "reindeer lichen" or "caribou moss." The nanoparticles measured 30–70 nm and had mean size of 23 nm. Among the metal oxides, zinc oxide (size: 21 nm) and iron oxide (size: 20–40 nm) nanoparticles have been synthesized by *Ramalina fraxinea* and *Ramalina sinensis*. Biosynthesis of $ZnO/TiO_2/SiO_2$ and Fe_3O_4/SiO_2 nanocomposites having size of 50–85 nm and 55–90 nm, respectively, has also been accomplished using *Lecanora muralis* (family: Lecanoraceae), a yellowish-green crustose lichen. An interesting study pertains to the biosynthesis of silver, copper, titanium dioxide, zinc oxide, and iron oxide nanoparticles by Alavi et al. (2019) using the extract of *Protoparmeliopsis muralis*. The average diameter of Ag, Cu, Fe_3O_4, TiO_2, and ZnO nanoparticles were 33.49 ± 22.91, 253.97 ± 57.2, 307 ± 154, 133.32 ± 35.33, and 178.06 ± 49.97 nm, respectively.

Table 6.6 Lichens in biosynthesis of nanoparticles (Reprinted from Rattan et al. 2021).

S. No	Lichen	Type of NPs	Shape of NPs	Size of NPs (nm)	Activity exhibited against	References
1.	*Parmotrema pseudotinctorum* and *Ramalina hossei*	Ag NPs	Circular	100	Gram-negative bacteria 1. *Salmonella typhi* 2. *Escherichia coli*	Kumar et al. (2010)
2.	*Parmotrema praesorediosum*	Ag NPs	Cubical	19	Gram-positive bacteria 1. *Staphylococcus epidermidis* 2. *Staphylococcus aureus* 3. *Bacillus subtilis* 4. *Streptococcus faecalis* Gram-negative bacteria 1. *Proteus vulgaris* 2. *Pseudomonas aeruginosa* 3. *Serratia marcescens* 4. *Salmonella typhi*	Mie et al. (2014)
3.	*Ramalina dumeticola*	Ag NPs	Cubical	13	Gram-positive bacteria	Din et al. (2015)

(continued)

Table 6.6 (continued)

S. No	Lichen	Type of NPs	Shape of NPs	Size of NPs (nm)	Activity exhibited against	References
					1. *Staphylococcus epidermidis* 2. *Bacillus subtilis* 3. *Streptococcus faecalis* Gram-negative bacteria 1. *Proteus vulgaris* 2. *Pseudomonas aeruginosa* 3. *Serratia marcescens* 4. *Salmonella typhi*	
4.	*Parmotrema clavuliferum*	Ag NPs	Spherical	106	Gram-positive bacteria 1. *Bacillus subtilis* 2. *Streptococcus faecalis* 3. *Staphylococcus aureus* Gram-negative bacteria	Alqahtani et al. (2017)

					1. *Pseudomonas aeruginosa*	
5.	*Parmelia perlata*	Ag NPs	Spherical	–	Gram-positive bacteria 1. *Staphylococcus aureus* 2. *Streptococcus spp.* Gram-negative bacteria 1. *Escherichia coli* 2. *Klebsiella pneumoniae* 3. *Salmonella spp.* 4. *Pseudomonas aeruginosa* Fungi 1. *Aspergillus niger* 2. *Candida albicans*	Leela and Anchana Devi (2017)
6.	*Usnea longissima*	Ag NPs	Spherical	9.40–11.23	Gram-positive bacteria 1. *Staphylococcus aureus* 2. *Streptococcus mutans*	Siddiqi et al. (2018)

(continued)

Table 6.6 (continued)

S. No	Lichen	Type of NPs	Shape of NPs	Size of NPs (nm)	Activity exhibited against	References
					3. *Streptococcus pyogenes* 4. *Streptococcus viridans* 5. *Corynebacterium xerosis* 6. *Corynebacterium diphtheriae* Gram-negative bacteria 1. *Escherichia coli* 2. *Klebsiella pneumoniae* 3. *Pseudomonas aeruginosa*	
7.	*Protoparmeliopsis muralis*	Ag NPs cu NPs	Spherical	Ag NPs, 44.87 Cu NPs, 34.38	Gram-positive bacteria 1. *Staphylococcus aureus* Gram-negative bacteria 1. *Escherichia coli* 2. *Pseudomonas aeruginosa*	Alavi et al. (2019)

8.	*Heterodermia boryi Parmotrema stuppeum*	Ag NPs	Cubic	27.91–37.21 27.69–36.00	Gram-positive bacteria 1. *Staphylococcus aureus* 2. *Viridans streptococci* Gram-negative bacteria: 1. *Acinetobacter baumannii* 2. *Escherichia coli* 3. *Klebsiella pneumoniae* 4. *Pseudomonas aeruginosa*	Senthil et al. (2019)
9.	*Flavopunctelia flaventior* and *Xanthoria parietina*	Ag NPs	Spherical	69–145	Gram-positive bacteria 1. *Staphylococcus aureus* Gram-negative bacteria: 1. *Escherichia coli* 2. *Pseudomonas aeruginosa*	Alqahtani et al. (2020)
10.	*Xanthoria elegans, Cetraria islandica, Usnea antarctica*, and *Leptogium puberulum*	Ag NPs	Bimodal	5–100	Gram-positive bacteria	Baláž et al. (2020)

(continued)

Table 6.6 (continued)

S. No	Lichen	Type of NPs	Shape of NPs	Size of NPs (nm)	Activity exhibited against	References
					1. *Staphylococcus aureus* Gram-negative bacteria 1. *Escherichia coli*	
11.	*Pseudevernia furfuracea* and *Lobaria pulmonaria*	Ag NPs	Bimodal	Majority smaller, 10 few reaching, 100	Gram-positive bacteria 1. *Staphylococcus aureus* 2. *Listeria monocytogenes* 3. *Bacillus cereus* Gram-negative bacteria 1. *Escherichia coli* 2. *Pseudomonas aeruginosa* 3. *Salmonella enterica*	Goga et al. (2020)

12.	*Ramalina sinensis*	FeO NPs	Spherical	31.74–53.91	Gram-positive bacteria 1. *Staphylococcus aureus* Gram-negative bacteria 1. *Pseudomonas aeruginosa*	Safarkar et al. (2020)

6.9 Advantages of Using Fungi in Nanoparticle Production

Fungi are the frontrunners in the green fabrication of nanoparticles since their usage is safe and more advantageous as compared to other microorganisms vis-a-vis metal tolerance, high bioaccumulation ability, efficient scale-up mechanisms, easy handling, rapid growth, and higher productivity. Compared to other prokaryotic entities and even plants, the fungal mycelium has proven to be quite sturdy in withstanding adverse conditions during culturing in bioreactors or other chambers during the nanofabrication process. The extracellular secretion of proteins and other metabolites by fungal biomass is appreciably higher which is an added advantage. Another advantage pertains to the precipitation of the biogenic nanoparticles synthesized by fungi outside the cell which enables their utilization in diverse areas.

6.10 Future Prospects

Fungi are quite versatile and produce nanoparticles in a wide range of culture conditions varying in temperature, pH, fungal biomass, and metal concentration leading to production of nanoentities having different physicochemical characteristics. Additionally, capping confers stability as well as biological activity, which is highly required for medicinal uses. Despite several advantages in the use of fungal species for biofabrication of nanostructures, one serious limitation pertains to the use of pathogenic fungi as biological source. It has been observed that many fungal species used for nanoparticle production are either plant or human pathogens. This has been a major impediment in their utilization for biosynthesis of nanoparticles. Thus, the handling of the fungal biomass is an area of concern and ne* to be addressed before proper commercialization of the process. However, recently, several nonpathogenic fungi like *Trichoderma* (*T. reesei*) have come to the rescue and have attracted attention for high-scale production of nanoparticles. Apart from this, the other limitations include the technical know-how regarding the selection of the fungal species, growth criterion, maintenance of aseptic conditions, fungal growth rate, and time needed for completion of nanoparticle formation. There are also problems related to scale-up and insufficient information on the processes involved on the biomolecules and the mechanism involved in formation of capping layers.

References

Abbas H, Baker DA (2020) Biological evaluation of selenium nanoparticles biosynthesized by fusarium semitectum as antimicrobial and anticancer agents. Egypt J Chem 4:18–19

Abdel-Kareem MM, Zohri AA (2018) Extracellular mycosynthesis of gold nanoparticles using Trichoderma hamatum: optimization, characterization and antimicrobial activity. Lett Appl Microbiol 67(5):465–475

AbdelRahim K, Mahmoud SY, Ali AM, Almaary KS, Mustafa AE, Husseiny SM (2017) Extracellular biosynthesis of silver nanoparticles using Rhizopus stolonifer. Saudi J Biol Sci 24:208–216

Abo Elsoud MM, Al-Hagar OEA, Abdelkhalek ES, Sidkey NM (2018) Synthesis and investigations on tellurium myconanoparticles. Biotechnol Rep (Amst) 18:e00247

Acay H, Baran MF (2019) Biosynthesis and characterization of silver nanoparticles using king oyster (Pleurotus eryngii) extract: effect on some microorganisms. Appl Ecol Environ Res 17: 9205–9214

Ahluwalia V, Kumar J, Sisodia R, Shakil NA, Walia S (2014) Green synthesis of silver nanoparticles by Trichoderma harzianum and their bio-efficacy evaluation against Staphylococcus aureus and Klebsiella pneumonia. Ind Crop Prod 55:202–206

Ahmad A, Mukherjee P, Mandal D, Senapati S, Khan MI, Kumar R, Sastry M (2002) Enzyme mediated extracellular synthesis of CdS nanoparticles by the fungus, Fusarium oxysporum. J Am Chem Soc 124(41):12,108–12,109

Ahmad A, Senapati S, Khan MI, Kumar R, Sastry M (2005) Extra−/intracellular biosynthesis of gold nanoparticles by an alkalotolerant fungus, Trichothecium sp. J Biomed Nanotechnol 1(1):47–53

Ajah AH, Hassan AS, Aja HA (2018) Extracellular biosynthesis of silver nanoparticles using fusarium graminearum and their antimicrobial activity. J Global Pharma Technol 10:683–689

Alavi M, Karimi N, Valadbeigi T (2019) Antibacterial, antibiofilm, antiquorum sensing, antimotility, and antioxidant activities of green fabricated Ag, Cu, TiO_2, ZnO, and Fe_3O_4 NPs via Protoparmeliopsis muralis lichen aqueous extract against multi-drug-resistant bacteria. ACS Biomater Sci Eng 5:4228–4243

Alqahtani MA, Al Othman MR, Mohammed AE (2020) Bio fabrication of silver nanoparticles with antibacterial and cytotoxic abilities using lichens. Sci Rep 10:16781

Alqahtani MA, Mohammed AE, Daoud SI, Alkhalifah DHM, Albrahim JS (2017) Lichens (Parmotrema clavuliferum) extracts: bio-mediator in silver nanoparticles formation and antibacterial potential. J Bionanosci 11:410–415

Ammar HA, El Aty AAA, El Awdan SA (2021 Apr) Extracellular myco-synthesis of nano-silver using the fermentable yeasts Pichia kudriavzevii HA-NY2 and Saccharomyces uvarum HA-NY3, and their effective biomedical applications. Bioprocess Biosyst Eng 44(4):841–854

Ammar HA, El-Desouky TA (2016) Green synthesis of nanosilver particles by aspergillus terreus HA1N and Penicillium expansum HA2N and its antifungal activity against mycotoxigenic fungi. J Appl Microbiol 121(1):89–100

Ammar HA, Rabie GH, Mohamed E (2019) Novel fabrication of gelatin-encapsulated copper nanoparticles using Aspergillus versicolor and their application in controlling of rotting plant pathogens. Bioprocess Biosyst Eng 42(12):1947–1961

Anthony K, Murugan M, Jeyaraj M, Rathinam NK, Sangiliyandi G (2014) Synthesis of silver nanoparticles using pine mushroom extract: a potential antimicrobial agent against E. coli and B. subtilis. J Ind Eng Chem 20:2325–2331

Asghari-Paskiabi F, Imani M, Rafii-Tabar H, Razzaghi-Abyaneh M (2019) Physicochemical properties, antifungal activity and cytotoxicity of selenium sulfide nanoparticles green synthesized by Saccharomyces cerevisiae. Biochem Biophys Res Commun 516(4):1078–1084

Ashour SM (2014) Silver nanoparticles as antimicrobial agent from Kluyveromyces marxianus and Candida utilis. Int J Curr Microbiol App Sci 3:384–396

Ashrafi SJ, Rastegar MF, Ashrafi M, Yazdian F, Pourrahim R, Suresh AK (2013) Influence of external factors on the production and morphology of biogenic silver nanocrystallites. J Nanosci Nanotechnol 13:2295–2301

Aygün A, Özdemir S, Gülcan M, Cellat K, Şen F (2020) Synthesis and characterization of Reishi mushroom-mediated green synthesis of silver nanoparticles for the biochemical applications. J Pharm Biomed Anal 178:112970

Azmath P, Baker S, Rakshith D, Satish S (2016) Mycosynthesis of silver nanoparticles bearing antibacterial activity. Saudi Pharm J 24(2):140–146

Balakumaran MD, Ramachandran R, Balashanmugam P, Mukeshkumar DJ, Kalaichelvan PT (2016) Mycosynthesis of silver and gold nanoparticles: optimization, characterization and antimicrobial activity against human pathogens. Microbiol Res 182:8–20

Balakumaran MD, Ramachandran R, Kalaichelvan PT (2015) Exploitation of endophytic fungus, Guignardia mangiferae for extracellular synthesis of silver nanoparticles and their in vitro biological activities. Microbiol Res 178:9–17

Balashanmugam P, Santhosh S, Giyaullah H, Balakumaran MD (2013) Mycosynthesis, characterization and antibacterial activity of silver nanoparticles from Microporus xanthopus: a macro mushroom. Int J Innov Res Sci Eng Technol 2:6262–6270

Baláž M, Goga M, Hegedüs M, Daneu N, Kováčová M, Tkáčiková L, Balážová L, Bačkor M (2020) Biomechanochemical solid-state synthesis of silver nanoparticles with antibacterial activity using lichens. ACS Sustain Chem Engine 8(13):945–13,955

Bansal V, Poddar P, Ahmad A, Sastry M (2006) Room-temperature biosynthesis of ferroelectric barium titanate nanoparticles. J Am Chem Soc 128:11,958–11,963

Bawaskar M, Gaikwad S, Ingle A, Rathod D, Gade A, Durán N, Marcato PD, Rai M (2010) A new report on mycosynthesis of silver nanoparticles by Fusarium culmorum. Curr Nanosci 6:376–380

Bedi A, Singh BR, Deshmukh SK, Adholeya A, Barrow CJ (2018) An Aspergillus aculeatus strain was capable of producing agriculturally useful nanoparticles via bioremediation of iron ore tailings. J Environ Manage 215:100–107

Berbee ML, James TY, Strullu-Derrien C (2017) Early diverging fungi: diversity and impact at the dawn of terrestrial life. Annu Rev Microbiol 71:41–60

Berman JJ (2019) Fungi. In: Berman JJ (ed) Taxonomic guide to infectious diseases. Elsevier, Amsterdam, pp 229–262

Bhambure R, Bule M, Shaligram N, Kamat M, Singhal R (2009) Extracellular biosynthesis of gold nanoparticles using Aspergillus niger—its characterisation and stability. Chem Eng Technol 32: 1036–1041

Bharde A, Rautaray D, Bansal V, Ahmad A, Sarkar I, Yusuf SM, Sanyal M, Sastry M (2006) Extracellular biosynthesis of magnetite using fungi. Small 2:135–141

Bhardwaj K, Sharma A, Tejwan N, Bhardwaj S, Bhardwaj P, Nepovimova E, Shami A, Kalia A, Kumar A, Abd-Elsalam KA, Kuča K (2020) Pleurotus macrofungi-assisted nanoparticle synthesis and its potential applications: a review. J Fungi 6(4):351

Bhat R, Deshpande R, Ganachari SV, Huh DS, Venkataraman A (2011) Photo-irradiated biosynthesis of silver nanoparticles using edible mushroom Pleurotus florida and their antibacterial activity studies. Bioinorg Chem Appl 2011:5

Bhat R, Sharanabasava VG, Deshpande R, Shetty U, Sanjeev G, Venkataraman A (2013) Photo-bio-synthesis of irregular shaped functionalized gold nanoparticles using edible mushroom Pleurotus florida and its anti-cancer evaluation. J Photochem Photobiol B Biol 125:63–69

Binupriya AR, Sathishkumar M, Vijayaraghavan K, Yun SI (2010) Bioreduction of trivalent aurum to nano-crystalline gold particles by active and inactive cells and cell free extract of aspergillus oryzae var. viridis. J Hazard Mater 177:539–545

Bolbanabad EM, Ashengroph M, Darvishi F (2020) Development and evaluation of different strategies for the clean synthesis of silver nanoparticles using Yarrowia lipolytica and their antibacterial activity. Process Biochem 94:319–328

Borovaya M, Pirko Y, Krupodorova T, Naumenko A, Blume Y, Yemets A (2015) Biosynthesis of cadmium sulphide quantum dots by using Pleurotus ostreatus (Jacq.) P. Kumm. Biotechnol Biotechnol Equip 29:1156–1163

Castro-Longoria E, Velasquez SDM, Nestor ARV, Berumen EA, Borja MA (2012) Production of platinum nanoparticles and nanoaggregates using Neurospora crassa. J Microbiol Biotechnol 22:1000–1004

Castro-Longoria E, Vilchis-Nestor AR, Avalos-Borja M (2011) Biosynthesis of silver, gold and bimetallic nanoparticles using the filamentous fungus Neurospora crassa. Colloids Surf B Biointerfaces 83:42–48

Chan Y, Don MM (2012) Characterization of Ag nanoparticles produced by white-rot fungi and its in vitro antimicrobial activities. Int Arab J Antimicrob Agents 2:12

Chan Y, Don MM (2013a) Biosynthesis and structural characterization of ag nanoparticles from white rot fungi. Mater Sci Eng C 33(2013):282–288

Chan YS, Don MM (2013b) Biosynthesis of silver nanoparticles from Schizophyllum commune and in-vitro antibacterial and antifungal activity studies. J Physiol Sci 24(2013):83–96

Chaturvedi VK, Rai SN, Tabassum N, Yadav N, Singh V, Bohara RA, Singh MP (2020a) Rapid eco-friendly synthesis, characterization, and cytotoxic study of trimetallic stable nanomedicine: A potential material for biomedical applications. Biochem Biophys Rep 24:100812

Chaturvedi VK, Yadav N, Rai NK, Bohara RA, Rai SN, Aleya L, Singh MP (2021) Two birds with one stone: oyster mushroom mediated bimetallic Au-Pt nanoparticles for agro-waste management and anticancer activity. Environ Sci Pollut Res Int 28(11):13761–13,775

Chaturvedi VK, Yadav N, Rai NK, Ellah NHA, Bohara RA, Rehan IF, Marraiki N, Batiha GE, Hetta HF, Singh MP (2020b) Pleurotus sajor-caju-mediated synthesis of silver and gold nanoparticles active against colon cancer cell lines: a new era of herbonanoceutics. Molecules 25(13):3091

Chauhan A, Zubair S, Tufail S, Sherwani A, Sajid M, Raman SC, Azam A, Owais M (2011) Fungus-mediated biological synthesis of gold nanoparticles: potential in detection of liver cancer. Int J Nanomedicine 6:2305–2319

Clarance P, Luvankar B, Sales J, Khusro A, Agastian P, Tack JC, Al Khulaifi MM, Al-Shwaiman HA, Elgorban AM, Syed A, Kim HJ (2020) Green synthesis and characterization of gold nanoparticles using endophytic fungi Fusarium solani and its in-vitro anticancer and biomedical applications. Saudi J Biol Sci 27(2):706–712

Costa Silva LP, Oliveira JP, Keijok WJ, da Silva AR, Aguiar AR, Guimarães MCC, Ferraz CM, Araújo JV, Tobias FL, Braga FR (2017) Extracellular biosynthesis of silver nanoparticles using the cell-free filtrate of nematophagous fungus Duddingtonia flagrans. Int J Nanomedicine 12: 6373–6381

Cunha FA, Da Cso Cunha M, Da Frota SM, Mallmann EJJ, Freire TM, Costa LS, Paula AJ, Menezes EA, Fechine PBA (2018) Biogenic synthesis of multifunctional silver nanoparticles from Rhodotorula glutinis and Rhodotorula mucilaginosa: antifungal, catalytic and cytotoxicity activities. World J Microbiol Biotechnol 34:127

Das SK, Dickinson C, Laffir F, Brougham DF, Marsili E (2012) Synthesis, characterization and catalytic activity of gold nanoparticles biosynthesized with Rhizopus oryzae protein extract. Green Chem 14:1322–1344

Debnath G, Das P, Saha AK (2019) Green synthesis of silver nanoparticles using mushroom extract of Pleurotus giganteus: characterization, antimicrobial, and α-amylase inhibitory activity. BioNanoSci 9:611–619

Devasena T, Ashok V, Dey N, Arul Prakash F (2014) Phytosynthesis of magnesium nanoparticles using lichens. World J Pharmaceut Res 3:4625–4632

Devika R, Elumalai S, Manikandan E, Eswaramoorthy D (2012) Biosynthesis of silver nanoparticles using the fungus Pleurotus ostreatus and their antibacterial activity. Open Access Sci Reports 1:10–14

Din LB, Mie R, Samsudin MW, Ahmad A, Ibrahim N (2015) Biomimetic synthesis of silver nanoparticles using the lichen Ramalina dumeticola and the antibacterial activity. Malays J Anal Sci 19:369–376

Dorcheh SK, Vahabi K (2016) Biosynthesis of nanoparticles by fungi: large-scale production. Fungal Metabolites 5:1–20

Du L, Xian L, Feng J-X (2011) Rapid extra−/intracellular biosynthesis of gold nanoparticles by the fungus Penicillium sp. J Nanopart Res 13:921–930

Du L, Xu Q, Huang M, Xian L, Feng J-X (2015) Synthesis of small silver nanoparticles under light radiation by fungus Penicillium oxalicum and its application for the catalytic reduction of methylene blue. Mater Chem Phys 160:40–47

El Sayed MT, El-Sayed ASA (2020) Biocidal activity of metal nanoparticles synthesized by fusarium solani against multidrug-resistant bacteria and mycotoxigenic fungi. J Microbiol Biotechnol 30(2):226–236

El-Batal AI, Elkenawy NM, Yassin AS, Amin MA (2015) Laccase production by Pleurotus ostreatus and its application in synthesis of gold nanoparticles. Biotechnol Rep 5:31–39

Eskandari-Nojehdehi M, Jafarizadeh-Malmiri H, Rahbar-Shahrouzi J (2016) Optimization of processing parameters in green synthesis of gold nanoparticles using microwave and edible mush-room (Agaricus bisporus) extract and evaluation of their antibacterial activity. Nanotechnol Rev 5:537–548

Faisal S, Khan MA, Jan H, Shah SA, Abdullah, Shah S, Rizwan M, Ullah W, Akbar MT, Redaina (2021) Edible mushroom (Flammulina velutipes) as biosource for silver nanoparticles: from synthesis to diverse biomedical and environmental applications. Nanotechnology 32(6):065101

Faramarzi S, Anzabi Y, Jafarizadeh-Malmiri H (2020) Nanobiotechnology approach in intracellular selenium nanoparticle synthesis using Saccharomyces cerevisiae-fabrication and characterization. Arch Microbiol 202(5):1203–1209

Fayaz AM, Balaji K, Kalaichelvan PT, Venkatesan R (2009) Fungal based synthesis of silver nanoparticles- an effect of temperature on the size of particles. Colloids Surf B Biointerfaces 74: 123–126

Fernández JG, Fernández-Baldo MA, Berni E, Camí G, Durán N, Raba J, Sanz MI (2016) Production of silver nanoparticles using yeasts and evaluation of their antifungal activity against phyto-pathogenic fungi. Process Biochem 51(9):1306–1313

Frąc M, Hannula SE, Bełka M, Jędryczka M (2018) Fungal biodiversity and their role in soil health. Front Microbiol 9:707

Gaikwad SC, Birla SS, Ingle AP, Gade AK, Marcato PD, Rai MK, Durán N (2013) Screening of different fusarium species to select potential species for the synthesis of silver nanoparticles. J Braz Chem Soc 24:1974–1982

Gil YG, Kang S, Chae A, Kim YK, Min DH, Jang H (2018) Synthesis of porous Pd nanoparticles by therapeutic chaga extract for highly efficient tri-modal cancer treatment. Nanoscale 10(42):19810–19,817

Goga M, Baláž M, Daneu N (2020) Biological activity of selected lichens and lichen-based Ag nanoparticles prepared by a green solid-state mechano chemical approach. Mater Sci Eng C 119: 111640

Golnaraghi Ghomi AR, Mohammadi-Khanaposhti M, Vahidi H, Kobarfard F, Ameri Shah Reza M, Barabadi H (2019) Fungus-mediated extracellular biosynthesis and characterization of zirconium nanoparticles using standard Penicillium species and their preliminary bactericidal potential: a novel biological approach to nanoparticle synthesis. Iran J Pharm Res 18(4):2101–2110

Guarro J, Gené J, Stchigel AM (1999) Developments in fungal taxonomy. Clin Microbiol Rev 12(3):454–500

Gudikandula K, Vadapally P, Charya MAS (2017) Biogenic synthesis of silver nanoparticles from white rot fungi: their characterization and antibacterial studies. OpenNano 2:64–78

Guilger-Casagrande M, Lima R (2019) Synthesis of silver nanoparticles mediated by fungi: a review. Front Bioeng Biotechnol 7:287

Gupta S, Bector S (2013) Biosynthesis of extracellular and intracellular gold nanoparticles by aspergillus fumigatus and A. flavus. Antonie Van Leeuwenhoek 103(5):1113–1123

Gurunathan S, Rraman J, Abd Malek S et al (2015) Green synthesis of silver nanoparticles using Ganoderma neo-japonicum Imazeki: a potential cytotoxic agent against breast cancer cells. Int J Nanomedicine 8:4399–4413

Hamedi S, Shojaosadati SA, Shokrollahzadeh S, Hashemi-Najafabadi S (2014) Extracellular biosynthesis of silver nanoparticles using a novel and non-pathogenic fungus, Neurospora intermedia: controlled synthesis and antibacterial activity. World J Microbiol Biotechnol 30(2):693–704

Hamida RS, Ali MA, Abdelmeguid NE, Al-Zaban MI, Baz L, Bin-Meferij MM (2021) Lichens—a potential source for nanoparticles fabrication: a review on nanoparticles biosynthesis and their prospective applications. J Fungi (Basel) 7(4):291

Hamzah HM, Salah RF, Maroof MN (2018) Fusarium mangiferae as new cell factories for producing silver nanoparticles. J Microbiol Biotechnol 28(10):1654–1663

Hass H, Taylor TN, Remy W (1994) Fungi from the lower Devonian Rhynie chert: Mycoparasitism. Am J Bot 81:29–37

Hu X, Saravanakumar K, Jin T, Wang MH (2019) Mycosynthesis, characterization, anticancer and antibacterial activity of silver nanoparticles from endophytic fungus Talaromyces purpureogenus. Int J Nanomedicine 14:3427–3438

Ingle A, Gade A, Pierrat S, Sonnichsen C, Rai M (2008) Mycosynthesis of silver nanoparticles using the fungus Fusarium acuminatum and its activity against some human pathogenic bacteria. Curr Nanosci 4:141–144

Ingle A, Rai M, Gade A, Bawaskar M (2009) Fusarium solani: A novel biological agent for the extracellular synthesis of silver nanoparticles. J Nanopart Res 11:2079–2085

Ivashchenko O, Przysiecka Ł, Peplińska B, Jarek M, Coy E, Jurga S (2018) Gel with silver and ultrasmall iron oxide nanoparticles produced with Amanita muscaria extract: physicochemical characterization, microstructure analysis and anticancer properties. Sci Rep 8(1):13260

Jalal M, Ansari MA, Alzohairy MA, Ali SG, Khan HM, Almatroudi A et al (2018) Biosynthesis of silver nanoparticles from oropharyngeal Candida glabrata isolates and their antimicrobial activity against clinical strains of bacteria and fungi. Nanomaterials 8:586

Ji Y, Cao Y, Song Y (2019) Green synthesis of gold nanoparticles using a Cordyceps militaris extract and their antiproliferative effect in liver cancer cells (HepG2). Artif Cells Nanomed Biotechnol 47(1):2737–2745

Kaler A, Jain S, Banerjee UC (2013) Green and rapid synthesis of anticancerous silver nanoparticles by saccharomyces boulardii and insight into mechanism of nanoparticle synthesis. Biomed Res Int 2013:872940

Kalpana VN, Kataru BAS, Sravani N, Vigneshwari T, Panneerselvam A, Rajeswari VD (2018) Biosynthesis of zinc oxide nanoparticles using culture filtrates of aspergillus Niger: antimicrobial textiles and dye degradation studies. OpenNano 3:48–55

Khalil NM, Abd El-Ghany MN, Rodríguez-Couto S (2019) Antifungal and anti-mycotoxin efficacy of biogenic silver nanoparticles produced by Fusarium chlamydosporum and Penicillium chrysogenum at non-cytotoxic doses. Chemosphere 218:477–486

Kitching M, Ramani M, Marsili E (2015) Fungal biosynthesis of gold nanoparticles: mechanism and scale up. J Microbial Biotechnol 8(6):904–917

Kobashigawa JM, Robles CA, Martínez Ricci ML, Carmarán CC (2019) Influence of strong bases on the synthesis of silver nanoparticles (AgNPs) using the ligninolytic fungi Trametes trogii. Saudi J Biol Sci 26(7):1331–1337

Korbekandi H, Mohseni S, Mardani Jouneghani R, Pourhossein M, Iravani S (2016) Biosynthesis of silver nanoparticles using Saccharomyces cerevisiae. Artif Cells Nanomed Biotechnol 44(1):235–239

Kumar RMP, Venkatesh A, Moorthy VHS (2019) Nanopits based novel hybrid plasmonic nanosensor fabricated by a facile nanofabrication technique for biosensing. Mater Res Express 6:1150b6

Kumar SA, Peter YA, Nadeau JL (2008) Facile biosynthesis, separation and conjugation of gold nanoparticles to doxorubicin. Nanotechnology 19:495101

Kumar SVP, Kekuda TRP, Vinayaka KS, Yogesh M (2010) Synergistic efficacy of lichen extracts and silver nanoparticles against bacteria causing food poisoning. Asian J Res Chem 3:67–70

Latha S (2010) Extracellular biosynthesis, characterization and in vitro antimicrobial potential of silver nanoparticles using myconanofactoris. Bharathidasan University, Tiruchirappalli

Leela K, Anchana Devi C (2017) A study on the applications of silver nanoparticle synthesized using the aqueous extract and the purified secondary metabolites of lichen Parmelia perlata. Int J Pharm Sci Invent 6:42–59

Lian S, Diko CS, Yan Y, Li Z, Zhang H, Ma Q, Qu Y (2019) Characterization of biogenic selenium nanoparticles derived from cell-free extracts of a novel yeast Magnusiomyces ingens. 3 Biotech 9(6):221

Liu Y, Zeng S, Liu Y, Wu W, Shen Y, Zhang L, Li C, Chen H, Liu A, Shen L, Hu B, Wang C (2018) Synthesis and antidiabetic activity of selenium nanoparticles in the presence of polysaccharides from Catathelasma ventricosum. Int J Biol Macromol 114:632–639

Loshchinina EA, Vetchinkina EP, Kupryashina MA, Kursky VF, Nikitina VE (2018) Nanoparticles synthesis by Agaricus soil basidiomycetes. J Biosci Bioeng 126(1):44–52

Lotfy WA, Alkersh BM, Sabry SA, Ghozlan HA (2021) Biosynthesis of silver nanoparticles by Aspergillus terreus: characterization, optimization, and biological activities. Front Bioeng Biotechnol 9:633468

Madakka M, Jayaraju N, Rajesh N (2018) Mycosynthesis of silver nanoparticles and their characterization. MethodsX 5:20–29

Madhanraj R, Eyini M, Balaji P (2017) Antioxidant assay of gold and silver nanoparticles from edible basidiomycetes mushroom fungi. Free Radic Antioxid 7:137–142

Manimaran K, Balasubramani G, Ragavendran C, Natarajan D, Murugesan S (2020) Biological applications of synthesized ZnO nanoparticles using Pleurotus Djamor against mosquito larvicidal, histopathology, antibacterial, antioxidant and anticancer effect. J Clust Sci 52:1–13

Maurya S, Bhardwaj AK, Gupta KK, Agarwal S, Kushwaha A, Vk C, Pathak RK, Gopal R, Uttam KN, Singh AK (2016) Green synthesis of silver nanoparticles using Pleurotus and its bactericidal activity. Cell Mol Biol 62:131

Mazumdar H, Haloi N (2011) A study on biosynthesis of iron nanoparticles by Pleurotus sp. J Microbiol Biotechnol Res 1:39–49

Mekkawy AI, El-Mokhtar MA, Nafady NA, Yousef N, Hamad MA, El-Shanawany SM, Ibrahim EH, Elsabahy M (2017) In vitro and in vivo evaluation of biologically synthesized silver nanoparticles for topical applications: effect of surface coating and loading into hydrogels. Int J Nanomedicine 12:759–777

Mie R, Samsudin MW, Din LB, Ahmad A, Ibrahim N, Adnan SNA (2014) Synthesis of silver nanoparticles with antibacterial activity using the lichen Parmotrema praesorediosum. Int J Nanomedicine 9:121–127

Mirunalini S, Arulmozhi V, Deepalakshmi K, Krishnaveni M (2012) Intracellular biosynthesis and antibacterial activity of silver nanoparticles using edible mushrooms. Not Sci Biol 4:55–61

Mishra A, Ahmad R, Sardar M (2015) Biosynthesized iron oxide nanoparticles mimicking peroxidase activity: application for biocatalysis and biosensing. J Nanoeng Nanomanufactur 5:37–42

Mishra A, Tripathy SK, Wahab R, Jeong SH, Hwang I, Yang YB, Kim YS, Shin HS, Yun SI (2011a) Microbial synthesis of gold nanoparticles using the fungus Penicillium brevicompactum and their cytotoxic effects against mouse mayo blast cancer C2C12 cells. Appl Microbiol Biotechnol 92:617–630

Mishra A, Tripathy SK, Yun S-I (2011b) Bio-synthesis of gold and silver nanoparticles from Candida guilliermondii and their antimicrobial effect against pathogenic bacteria. J Nanosci Nanotechnol 11:243–248

Mishra A, Tripathy SK, Yuna SI (2012) Fungus mediated synthesis of gold nanoparticles and their conjugation with genomic DNA isolated from Escherichia coli and Staphylococcus aureus. Process Biochem 47:701–711

Mishra AN, Bhadaurla S, Singh Gaur M, Pasricha R (2010) Extracellular microbial synthesis of gold nanoparticles using fungus Hormoconis resinae. JOM 62:45–48

Mistry H, Thakor R, Patil C, Trivedi J, Bariya H (2021) Biogenically proficient synthesis and characterization of silver nanoparticles employing marine procured fungi aspergillus brunneoviolaceus along with their antibacterial and antioxidative potency. Biotechnol Lett 43(1):307–316

Moghaddam A, Namvar F, Moniri M, Tahir P, Azizi S, Mohamad R (2015) Nanoparticles biosynthesized by fungi and yeast: a review of their preparation, properties, and medical applications. Molecules 20:16540–16,565

Moghaddam AB, Moniri M, Azizi S, Rahim RA, Ariff AB, SaadWZ NF, Navaderi M, Mohamad R (2017) Biosynthesis of ZnO nanoparticles by a new Pichia kudriavzevii yeast strain and evaluation of their antimicrobial and antioxidant activities. Molecules 22(6):872

Mohamed AA, Fouda A, Abdel-Rahman MA, Hassan SE-D, El-Gamal MS, Salem SS, Shaheen TI (2019) Fungal strain impacts the shape, bioactivity and multifunctional properties of green synthesized zinc oxide nanoparticles. Biocatal Agric Biotechnol 19:101103

Mohanta YK, Nayak D, Biswas K, Singdevsachan SK, Abd Allah EF, Hashem A, Alqarawi AA, Yadav D, Mohanta TK (2018) Silver nanoparticles synthesized using wild mushroom show potential antimicrobial activities against food borne pathogens. Molecules 23(3):655

Mohanta YK, Singdevsachan SK, Parida UK, Panda SK, Mohanta TK, Bae H (2016) Green synthesis and antimicrobial activity of silver nanoparticles using wild medicinal mushroom Ganoderma applanatum (Pers.) pat. From Similipal biosphere reserve, Odisha, India. IET Nanobiotechnol 10(4):184–189

Mohite P, Kumar AR, Zinjarde S (2017) Relationship between salt tolerance and nanoparticle synthesis by Williopsis saturnus NCIM 3298. World J Microbiol Biotechnol 33(9):163

Money NP (2016) Fungal diversity. In: Watkinson SC, Boddy L, Money NP (eds) The fungi. Elsevier, New York, pp 1–36

Moretti A., Sarrocco S. 2016. Fungi. In: Caballero B., Finglas P.M. and Toldrá F. Encyclopedia of food and health. Elsevier, Amsterdam Pp. 162–168.

Mosallam FM, El-Sayyad GS, Fathy RM, El-Batal AI (2018) Biomolecules-mediated synthesis of selenium nanoparticles using Aspergillus oryzae fermented Lupin extract and gamma radiation for hindering the growth of some multidrug-resistant bacteria and pathogenic fungi. Microb Pathog 122:108–116

Mousa AM, Abdel Aziz OA, Al-Hagar OEA, Gizawy MA, Allan KF, Attallah MF (2020) Biosynthetic new composite material containing CuO nanoparticles produced by Aspergillus terreus for ^{47}Sc separation of cancer theranostics application from irradiated Ca target. Appl Radiat Isot 166:109389

Mukherjee P, Ahmad A, Mandal D, Senapati S, Sainkar SR, Khan MI, Ramani R, Parischa R, Ajayakumar PV, Alam M, Sastry M, Kumar R (2001) Bioreduction of $AuCl_4$ ions by the fungus, Verticillium sp. and surface trapping of the gold nanoparticles formed. Angew Chem Int Ed Engl 40:3585–3588

Mukherjee T, Chakraborty S, Biswas AA, Das TK (2017) Bioremediation potential of arsenic by non-enzymatically biofabricated silver nanoparticles adhered to the mesoporous carbonized fungal cell surface of aspergillus foetidus MTCC8876. J Environ Manage 201:435–446

Nagajyothi PC, Sreekanth TVM, Lee J, Lee KD (2013) Mycosynthesis: antibacterial, antioxi-dant and antiproliferative activities of silver nanoparticles synthesized from Inonotus obliquus (Chaga mushroom) extract. J Photochem Photobiol B Biol 130:299

Naimi-Shamel N, Pourali P, Dolatabadi S (2019) Green synthesis of gold nanoparticles using Fusarium oxysporum and antibacterial activity of its tetracycline conjugant. J de Mycologie Médicale 29(1):7–13

Narayanan KB, Park HH, Han SS (2015) Synthesis and characterization of biomatrixed-gold nanoparticles by the mushroom Flammulina velutipes and its heterogeneous catalytic potential. Chemosphere 141:169–175

Narayanan KB, Sakthivel N (2011a) Synthesis and characterization of nano-gold composite using Cylindrocladium floridanum and its heterogeneous catalysis in the degradation of 4-nitrophenol. J Hazard Mater 189(1–2):519–525

Narayanan KB, Sakthivel N (2011b) Facile green synthesis of gold nanostructures by NADPH – dependent enzyme from the extract of Sclerotium rolfsii. Colloids Surf A 380:156–161

Nasrin T, Roy S, Das TK (2014) Aspergillus foetidus mediated biosynthesis of CdS nano particles and its characterization. Int J Innov Res Sci Eng 2:633–639

Nayak RR, Pradhan N, Behera D, Pradhan KM, Mishra S, Sukla LB et al (2011) Green synthesis of silver nanoparticle by Penicillium purpurogenum NPMF: the process and optimization. J Nanopart Res 13:3129–3137

Neethu S, Midhun SJ, Sunil MA, Soumya S, Radhakrishnan EK, Jyothis M (2018) Efficient visible light induced synthesis of silver nanoparticles by Penicillium polonicum ARA 10 isolated from

Chaetomorpha antennina and its antibacterial efficacy against salmonella enterica serovar typhimurium. J Photochem Photobiol B 180:175–185

Netala VR, Bethu MS, Pushpalatha B, Baki VB, Aishwarya S, Rao JV, Tartte V (2016b) Biogenesis of silver nanoparticles using endophytic fungus Pestalotiopsis microspora and evaluation of their antioxidant and anticancer activities. Int J Nanomedicine 11:5683–5696

Netala VR, Kotakadi VS, Bobbu P, Gaddam SA, Tartte V (2016a) Endophytic fungal isolate mediated biosynthesis of silver nanoparticles and their free radical scavenging activity and anti microbial studies. 3 Biotech 6(2):132

Omran BA, Nassar HN, Fatthallah NA, Hamdy A, El-Shatoury EH, El-Gendy NS (2018) Characterization and antimicrobial activity of silver nanoparticles mycosynthesized by Aspergillus brasiliensis. J Appl Microbiol 125(2):370–382

Omran BA, Nassar HN, Younis SA, El-Salamony RA, Fatthallah NA, Hamdy A, El-Shatoury EH, El-Gendy NS (2020) Novel mycosynthesis of cobalt oxide nanoparticles using Aspergillus brasiliensis ATCC 16404-optimization, characterization and antimicrobial activity. J Appl Microbiol 128(2):438–457

Omran BA, Nassar HN, Younis SA, Fatthallah NA, Hamdy A, El-Shatoury EH, El-Gendy NS (2019) Physiochemical properties of Trichoderma longibrachiatum DSMZ 16517-synthesized silver nanoparticles for the mitigation of halotolerant sulphate-reducing bacteria. J Appl Microbiol 126(1):138–154

Osorio-Echavarría J, Osorio-Echavarría J, Ossa-Orozco CP, Gómez-Vanegas NA (2021) Synthesis of silver nanoparticles using white-rot fungus Anamorphous Bjerkandera sp. R1: influence of silver nitrate concentration and fungus growth time. Sci Rep 11(1):3842

Othman AM, Elsayed MA, Al-Balakocy NG, Hassan MM, Elshafei AM (2019) Biosynthesis and characterization of silver nanoparticles induced by fungal proteins and its application in different biological activities. J Genet Eng Biotechnol 17(1):8

Ottoni CA, Maria DA, Gonçalves PJRO, de Araújo WL, de Souza AO (2019) Biogenic Aspergillus tubingensis silver nanoparticles' in vitro effects on human umbilical vein endothelial cells, normal human fibroblasts, HEPG2, and Galleria mellonella. Toxicol Res (Camb) 8(6):789–801

Owaid MN, Raman J, Lakshmanan H et al (2015) Mycosynthesis of silver nanoparticles by Pleurotus cornucopiae var. citrinopileatus and its inhibitory effects against Candida sp. Mater Lett 153:186–190

Paul S, Singh ARJ, Sasikumar CS (2015) Green synthesis of bio-silver nanoparticles by Parmelia perlata, Ganoderma lucidum and Phellinus igniarius and their fields of application. Indian J Res Pharm Biotechnol 5674:100–110

Peiris M, Gunasekara T, Jayaweera PM, Fernando S (2018) TiO_2 nanoparticles from Baker's yeast: a potent antimicrobial. J Microbiol Biotechnol 28(10):1664–1670

Phanjom P, Ahmed G (2017) Effect of different physicochemical conditions on the synthesis of silver nanoparticles using fungal cell filtrate of Aspergillus oryzae (MTCC no. 1846) and their antibacterial effects. Adv Nat Sci Nanosci Nanotechnol 8:1–13

Philip D (2009) Biosynthesis of Au, Ag and Au-Ag nanoparticles using edible mushroom extract. Spectrochim Acta A Mol Biomol Spectrosc 73:374–381

Pimprikar PS, Joshi SS, Kumar AR, Zinjarde SS, Kulkarni SK (2009) Influence of biomass and gold salt concentration on nanoparticle synthesis by the tropical marine yeast Yarrowia lipolytica NCIM 3589. Colloids Surf B Biointerfaces 74(1):309–316

Prasad K, Jha AK (2010) Biosynthesis of CdS nanoparticles: an improved green and rapid procedure. J Colloid Interface Sci 342:68–72

Rai M, Bonde S, Golinska P, Trzcińska-Wencel J, Gade A, Abd-Elsalam KA, Shende S, Gaikwad S, Ingle AP (2021) Fusarium as a novel fungus for the synthesis of nanoparticles: mechanism and applications. J Fungi (Basel) 7(2):139

Raman J, Lakshmanan H, John PA, Zhijian C, Periasamy V, David P, Naidu M, Sabaratnam V (2015a) Neurite outgrowth stimulatory effects of mycosynthesized AuNPs from Hericium Neurite outgrowth stimulatory effects of myco synthesized AuNPs from Hericium erinaceus (Bull.: Fr.) Pers. on pheochromocytoma (PC-12) cells. Int J Nanomedicine 10:5853–5863

Raman J, Reddy GR, Lakshmanan H, Selvaraj V, Gajendran B, Nanjian R, Chinnasamy A, Sabaratnam V (2015b) Mycosynthesis and characterization of silver nanoparticles from Pleurotus djamor var. roseus and their in vitro cytotoxicity effect on PC3 cells. Process Biochem 50:140–147

Rasouli M (2019) Biosynthesis of selenium nanoparticles using yeast Nematospora coryli and examination of their anti-candida and anti-oxidant activities. IET Nanobiotechnol 13(2):214–218

Rattan R, Shukla S, Sharma B, Bhat M (2021) A mini-review on lichen-based nanoparticles and their applications as antimicrobial agents. Front Microbiol 12:633090

Ray S, Sarkar S, Kundu S (2011) Extracellular biosynthesis of silver nanoparticles using the mycorrhizal mushroom Tricholoma crassum (Berk.) Sacc.: its antimicrobial activity against pathogenic bacteria and fungus, including multidrug resistant plant and human bacteria. Dig J Nanomater Biostruct 6:1289–1299

Rehman S, Farooq R, Jermy R, Asiri SM, Ravinayagam V, Jindan RA, Alsalem Z, Shah MA, Reshi Z, Sabit H, Khan FA (2020) A wild Fomes fomentarius for biomediation of one pot synthesis of titanium oxide and silver nanoparticles for antibacterial and anticancer application. Biomolecules 10(4):622

Rodríguez-Serrano C, Guzmán-Moreno J, Ángeles-Chávez C, Rodríguez-González V, Ortega-Sigala JJ, Ramírez-Santoyo RM, Vidales-Rodríguez LE (2020) Biosynthesis of silver nanoparticles by Fusarium scirpi and its potential as antimicrobial agent against uropathogenic Escherichia coli biofilms. PLoS One 15(3):e0230275

Riddin TL, Gericke M, Whiteley CG (2006) Analysis of the inter- and extracellular formation of platinum nanoparticles by Fusarium oxysporum f.sp. Lycopersici using response surface methodology. Nanotechnology 17:1–8

Rónavári A, Igaz N, Gopisetty MK, Szerencsés B, Kovács D, Papp C, Vágvölgyi C, Boros IM, Kónya Z, Kiricsi M (2018) Biosynthesized silver and gold nanoparticles are potent antimycotics against opportunistic pathogenic yeasts and dermatophytes. Int J Nanomedicine 13:695

Safarkar R, Rajaei GE, Khalili-Arjagi S (2020) The study of antibacterial properties of iron oxide nanoparticles synthesized using the extract of lichen Ramalina sinensis. Asian J Nanosci Mater 3:157–166

Salunke BK, Sawant SS, Lee SI, Kim BS (2015) Comparative study of MnO_2 nanoparticle synthesis by marine bacterium Saccharophagus degradans and yeast Saccharomyces cerevisiae. Appl Microbiol Biotechnol 99(13):5419–5427

Sanghi R, Verma P (2009) Biomimetic synthesis and characterization of protein capped silver nanoparticle. Biomagn Res Technol 100:501–504

Saravanan M, Arokiyaraj S, Lakshmi T, Pugazhendhi A (2018) Synthesis of silver nanoparticles from Phanerochaete chrysosporium (MTCC-787) and their antibacterial activity against human pathogenic bacteria. Microb Pathog 117:68–72

Sarkar J, Ray S, Chattopadhyay D, Laskar A, Acharya K (2012) Mycogenesis of gold nanoparticles using a phytopathogen Alternaria alternata. Bioprocess Biosyst Eng 35:637–643

Sawle BD, Salimath B, Deshpande R, Bedre MD, Prabhakar BK, Venkataraman A (2008) Biosynthesis and stabilization of Au and Au-Ag alloy nanoparticles by fungus. Sci Technol Adv Mater 9:5

Saxena J, Sharma PK, Sharma MM, Singh A (2016) Process optimization for green synthesis of silver nanoparticles by Sclerotinia sclerotiorum MTCC 8785 and evaluation of its antibacterial properties. Springerplus 5:861

Sen K, Sinha P, Lahiri S (2011) Time dependent formation of gold nanoparticles in yeast cells: a comparative study. Biochem Eng J 55:1–6

Senapati S, Ahmad A, Khan MI, Sastry M, Kumar R (2005) Extracellular biosynthesis of bimetallic Au-Ag alloy nanoparticles. Small 1:517–520

Senapati US, Sarkar D (2014) Characterization of biosynthesized zinc sulphide nanoparticles using edible mushroom Pleurotus ostreatus. Indian J Phys 88:557–562

Senthil PS, Ramanujam JR, Sudha SS (2019) Antibacterial activity of silver nanoparticles synthesized by using lichens Heterodermia boryi and Parmotrema stuppeum. Int J Pharm Biol Sci 9:1397–1402

Shahzad A, Saeed H, Iqtedar M, Hussain SZ, Kaleem A, Abdullah R (2019) Size-controlled production of silver nanoparticles by Aspergillus fumigatus BTCB10: likely antibacterial and cytotoxic effects. J Nanomater 2019:5168698

Shankar SS, Ahmad A, Pasricha R, Sastry M (2003) Bioreduction of chloroaurate ions by geranium leaves and its endophytic fungus yields gold nanoparticles of different shapes. J Mater Chem 13(7):1822–1826

Siddiqi KS, Rashid M, Rahman A, Tajuddin H, A., and Rehman, S. (2018) Biogenic fabrication and characterization of silver nanoparticles using aqueous-ethanolic extract of lichen (Usnea longissima) and their antimicrobial activity. Biomater Res 22:23

Sriramulu M, Sumathi S (2018) Biosynthesis of palladium nano-particles using Saccharomyces cerevisiae extract and its photocatalytic degradation behaviour. Adv Nat Sci Nanosci Nanotechnol 9(2):025018

Srivastava S, Bhargava A, Pathak N, Srivastava P (2019) Production, characterization and antibacterial activity of silver nanoparticles produced by Fusarium oxysporum and monitoring of protein-ligand interaction through in-silico approaches. Microb Pathog 129:136–145

Srivastava S, Pathak N, Srivastava P (2011) Identification of limiting factors for the optimum growth of Fusarium oxysporum in liquid media. Toxicol Int 18:111–116

Subramaniyan SA, Sheet S, Vinothkannan M, Yoo DJ, Lee YS, Belal SA, Shim KS (2018) One-pot facile synthesis of Pt nanoparticles using cultural filtrate of microgravity simulated grown P. chrysogenum and their activity on bacteria and cancer cells. J. Nanosci Nanotechnol 18(5):3110–3125

Sujatha S, Kanimozhi G, Panneerselvam A (2015) Synthesis of silver nanoparticles from Lentinula edodes and antibacterial activity. Int J Pharm Sci Rev Res 33:189–191

Sumanth B, Lakshmeesha TR, Ansari MA, Alzohairy MA, Udayashankar AC, Shobha B, Niranjana SR, Srinivas C, Almatroudi A (2020) Mycogenic synthesis of extracellular zinc oxide nanoparticles from Xylaria acuta and its nanoantibiotic potential. Int J Nanomedicine 15:8519–8536

Syed A, Ahmad A (2012) Extracellular biosynthesis of platinum nanoparticles using the fungus Fusarium oxysporum. Colloids Surf B Biointerfaces 97:27–31

Syed A, Saraswati S, Kundu GC, Ahmad A (2013) Biological synthesis of silver nanoparticles using the fungus Humicola sp. and evaluation of their cytoxicity using normal and cancer cell lines. Spectrochim Acta A Mol Biomol Spectrosc 114:144–147

Taha ZK, Hawar SN, Sulaiman GM (2019) Extracellular biosynthesis of silver nanoparticles from Penicillium italicum and its antioxidant, antimicrobial and cytotoxicity activities. Biotechnol Lett 41(8–9):899–914

Tarafdar JC, Raliya R, Rathore I (2012) Microbial synthesis of phosphorous nanoparticle from tri-calcium phosphate using Aspergillus tubingensis TFR-5. J Bionanosci 6:84–89

Taylor JW, Swann FC, Berbee ML (1994) Molecular evolution of ascomycete fungi: phylogeny and conflict. In: Hawksworth DL (ed) Ascomycete systematics: problems and perspectives in the nineties. Plenum Press, New York, pp 201–212

Taylor TN, Hass H, Remy W (1992a) Devonian fungi: interactions with the green alga Palaeonitella. Mycologia 84:901–910

Taylor TN, Remy W, Hass H (1992b) Fungi from the lower Devonian Rhynie chert: Chytridiomycetes. Am J Bot 79:1233–1241

Tidke PR, Gupta I, Gade AK, Rai M (2014) Fungus-mediated synthesis of gold nanoparticles and standardization of parameters for its biosynthesis. IEEE Trans Nanobioscience 13(4):397–402

Uddandarao P, Balakrishnan RM, Ashok A, Swarup S, Sinha P (2019) Bioinspired ZnS:Gd nanoparticles synthesized from an endophytic fungi Aspergillus flavus for fluorescence-based metal detection. Biomimetics (Basel) 4(1):11

Vahabi K, Mansoori GA, Karimi S (2011) Biosynthesis of silver nanoparticles by fungus Trichoderma reesei (A route for large-scale production of AgNPs). Insci J 1(1):65–79

Vainshtein M, Belova N, Kulakovskaya T, Suzina N, Sorokin V (2014) Synthesis of magneto-sensitive iron-containing nanoparticles by yeasts. J Ind Microbiol Biotechnol 41(4):657–663

Vala AK (2014) Exploration on green synthesis of gold nanoparticles by a marine-derived fungus aspergillus sydowii. Environ Prog Sustain Energy 34:194

Verma VC, Singh SK, Solanki R, Prakash S (2011) Biofabrication of anisotropic gold nanotriangles using extract of endophytic Aspergillus clavatus as a dual functional reductant and stabilizer. Nanoscale Res Lett 6:16–22

Vigneshwaran N, Kathe AA, Varadarajan PV, Nachane RP, Balasubramanya RH (2007) Silver-protein (core–shell) nanoparticle production using spent mushroom substrate. Langmuir 23: 7113–7117

Vijayanandan AS, Balakrishnan RM (2018) Biosynthesis of cobalt oxide nanoparticles using endophytic fungus aspergillus nidulans. J Environ Manage 218:442–450

Volk TJ (2013) Fungi. In: Levin SA (ed) Encyclopedia of biodiversity. Elsevier, Amsterdam, pp 624–640

Waghmare SR, Mulla MN, Marathe SR, Sonawane KD (2015) Ecofriendly production of silver nanoparticles using Candida utilis and its mechanistic action against pathogenic microorganisms. 3 Biotech 5:33–38

Wang D, Xue B, Wang L, Zhang Y, Liu L, Zhou Y (2021) Fungus-mediated green synthesis of nano-silver using Aspergillus sydowii and its antifungal/antiproliferative activities. Sci Rep 11(1):10356

Whittaker RH (1969) New concepts of kingdoms or organisms. Evolutionary relations are better represented by new classifications than by the traditional two kingdoms. Science 163:150–160

Wu H, Zhu H, Li X et al (2013) Induction of apoptosis and cell cycle arrest in A549 human lung adenocarcinoma cells by surface-capping selenium nanoparticles: an effect enhanced by polysaccharide–protein complexes from Polyporus rhinocerus. J Agric Food Chem 61:9859–9866

Xue B, He D, Gao S, Wang D, Yokoyama K, Wang L (2016) Biosynthesis of silver nanoparticles by the fungus Arthroderma fulvum and its antifungal activity against genera of Candida, Aspergillus and Fusarium. Int J Nanomed 11:1899–1906

Yamaguchi T, Tsuruda Y, Furukawa T, Negishi L, Imura Y, Sakuda S, Yoshimura E, Suzuki M (2016) Synthesis of CdSe quantum dots using Fusarium oxysporum. Materials (Basel) 9(10):855

Yassin MA, Elgorban AM, El-Samawaty AEMA, Almunqedhi BMA (2021) Biosynthesis of silver nanoparticles using Penicillium verrucosum and analysis of their antifungal activity. Saudi J Biol Sci 28(4):2123–2127

Zhang X, Qu Y, Shen W, Wang J, Li H, Zhang Z et al (2016) Biogenic synthesis of gold nanoparticles by yeast Magnusiomyces ingens LH-F1 for catalytic reduction of nitrophenols. Colloids Surf 497:280–285

Zielonka A, Żymańczyk-Duda E, Brzezińska-Rodak M, Duda M, Grzesiak J, Klimek-Ochab M (2018) Nanosilica synthesis mediated by Aspergillus parasiticus strain. Fungal Biol 122(5):333–344

Check for updates

Biological Synthesis of Nanoparticles: Algae 7

Abstract

Algae constitute a polyphyletic and non-cohesive grouping of organisms that are capable of performing photosynthesis using oxygen. Although algae have been utilized as biological sources for the green synthesis of metal nanoparticles in the last few years, their usage has been less as compared to other living forms. The different functional groups and enzymes present in the algal cell walls are thought to play an important role in the fabrication of metal and metal oxide nanoparticles. Nanoparticle production has been reported from all algal groups including blue-green algae, green algae, diatoms, brown algae, and red algae.

7.1 Introduction

Algae constitute a polyphyletic and non-cohesive consortium of ancient plants that are capable of performing photosynthesis using oxygen. These are thallophytes which unlike plants are devoid of root, stem, and leaves. Members of this group lack the vascular system, as well as the sterile covering around the reproductive cells with most members having chlorophyll *a* as the principal photosynthetic pigment. Algal diversity ranges from unicellular forms like *Chlamydomonas* and *Acetabularia*, and the diatoms like Pinnularia and Navicula, to complex, multicellular brown algae like *Laminaria* and *Macrocystis* that are hundreds of meters in length. In fact giant kelps, the large brown algae seaweeds, grow as large coastal seaweeds in cold, shallow waters and form extensive underwater kelp forests (Fig. 7.1). Strangely, algae are an assemblage of plant forms that are sometimes not closely related. One nice example of this is Cyanophyceae that is evolutionary closer to bacteria as compared to the rest of the algae. Algae are economically (sources of food, fodder, fertilizers, medicine, and industrially important products)

S. Srivastava, A. Bhargava, *Green Nanoparticles: The Future of Nanobiotechnology*,
https://doi.org/10.1007/978-981-16-7106-7_7

Fig. 7.1 Kelp forests are widely distributed around the Norwegian coastline (Reprinted with permission from Harman et al., 2019)

and ecologically highly significant being the source of more than half of the world's oxygen generated through photosynthesis.

7.2 Algal Classification

Algae are classified on the basis of the following characteristics:

1. Pigments
2. Photosynthetic apparatus
3. Reserve food material
4. Cell wall components
5. Flagellar apparatus

The first classification of algae was put forward by the algologist W.H. Harvey. F.E. Fritsch, Father of Phycology, proposed a comprehensive classification of algae in 1935 in his comprehensive work entitled "The Structure and Reproduction of the Algae," which is also one of the most accepted systems of classification. Fritsch (1945) classified algae into 11 classes (Table 7.1).

Another famous classification was put forward by G.M. Smith who based it on Pascher (1914, 1931) and divided algae into seven divisions that are enumerated below:

1. Chlorophyta
2. Euglenophyta
3. Pyrrophyta

Table 7.1 Outline of classification proposed by F.E. Fritsch (1935)

Class	Number of orders	Details of orders
Chlorophyceae (green algae)	9	Volvocales Chlorococcales Ulotrichales Cladophorales Chaetophorales Oedogoniales Conjugales Sipohonales Charales
Xanthophyceae (yellow-green)	4	Heterochloridales Heterococcales Heterotrichales Heterosiphonales
Chrysophyceae (orange)	3	Chrysomonadales Chrysophaerales Chrysotrichales
Bacillariophyceae (diatoms or golden brown algae)	2	Centrales Pennales
Cryptophyceae (nearly brown)	2	Cryptomonadales Cryptococcales
Dinophyceae (dark yellow or brown)	6	Desmomonadales Thecatales Dinophysiales Dinoflagellata Dinococcales Dinotrichales
Chloromonadineae (bright green)	1	Chloromonadales
Euglenophyceae	Divided into 3 families	Euglenaceae Astasiaceae Peranemaceae
Phaeophyceae (Brown algae)	9	Ectocarpales Tilopteridales Cutariales Sporochnales Desmarestiales Laminariales Sphacelariales Dictyotales Fucales
Rhodophyceae (red algae)	7	Bangiales Nemalionales Gelidiales Cryptonemiales Gigartinales Rhodymeniales Ceramiales
Myxophyceae (Cyanophyceae, blue-green algae)	5	Chroococcales Chamaesiphonales

(continued)

Table 7.1 (continued)

Class	Number of orders	Details of orders
		Pleurocapsales Nostocales Stigonemales

4. Chrysophyta
5. Phaeophyta
6. Cyanophyta
7. Rhodophyta

These divisions were further divided into classes.

Prescott (1969) stressed on the presence or absence of true nucleus along with other traits like assimilatory product, pigments, and cell wall components as the classifying criteria. He classified algae into nine phyla that were further divided into fourteen classes and seventy-four orders. The most recent classification of Lee (2008) divided algae into four groups distributed between Prokaryota and Eukaryota.

Group 1: Algae with prokaryotic cell structure: Cyanophyta.

Group 2: Members have two-membraned chloroplast envelope: Chlorophyta (green algae), Rhodophyta (red algae), and Glaucophyta.

Group 3: Eukaryotic algae having chloroplasts surrounded by a single membrane of chloroplast endoplasmic reticulum: Dinophyta (dinoflagellates), Euglenophyta (euglenoids), and Apicomplexa.

Group 4: Eukaryotic algae having chloroplast enveloped by two membranes of chloroplast endoplasmic reticulum: Cryptophyta and Heterokontophyta.

7.3 Use of Algae in Nanoparticle Synthesis

Although algae have been utilized for green production of nanoparticles in the last few years, their usage has been less as compared to other living forms like fungi, bacteria, and higher plants. It is believed that different functional groups present in components of the algal cell walls act as efficient reducing agents for fabrication of different types of nanomaterials. Most of the major algal groups have been reported to efficiently synthesize nanoparticles. Biological synthesis of nanoparticles of varying sizes is dependent on the following factors:

1. Type of algae used
2. Concentration of algal extract/biomass
3. pH of the medium
4. Temperature
5. Incubation time

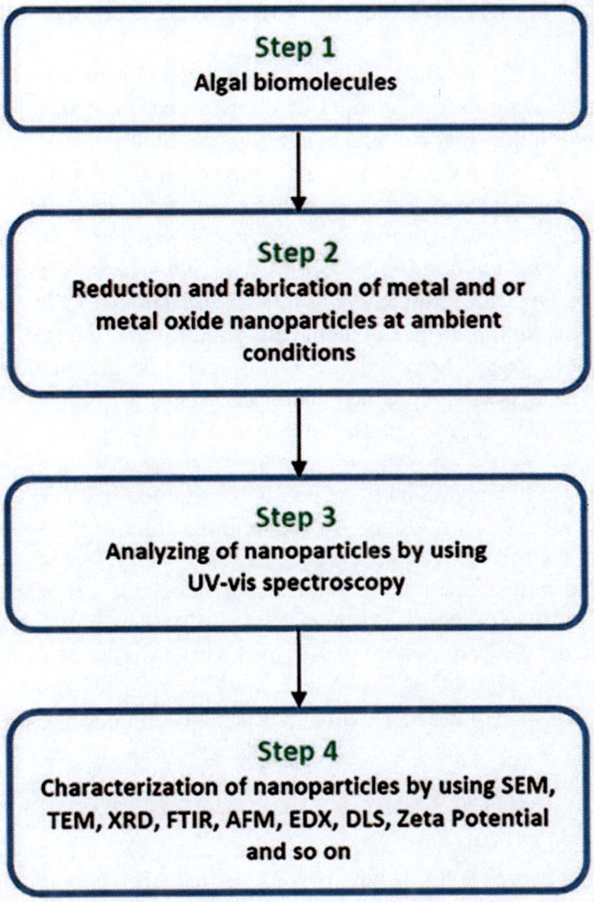

Fig. 7.2 Green synthesis and characterization of nanoparticles using algae (Siddiqi and Husen 2016)

6. Presence and absence of light
7. Type of metal salt used

The abovementioned can be altered to prevent agglomeration and get the nanoparticle of desired dimensions. For example, the rate of reaction increases on increase in temperature during formation of silver nanoparticles as a result of rapid utilization of the reactants that leads to synthesis of small-sized nanoparticles. The different steps in the production of nanoparticles using algae have been provided in Fig. 7.2.

7.4 Mechanism of Nanoparticle Synthesis by Algae

Biological production using algae can be either intracellular or extracellular depending on the location of the products formed. Intracellular production occurs within the algal cell and is a dose-dependent process wherein NADPH or NADPH-dependent reductase released during algal metabolic processes act as reducing agents. Green synthesis of gold nanoparticles has been accomplished intracellularly using *Ulva intestinalis, Rhizoclonium fontinale, Klebsormidium flaccidum, Plectonema boryanum, Spirogyra submaxima, Desmodesmus*, and *Tetraselmis kochinensis*. Extracellular synthesis of nanoparticles takes place when metal ions attach to the surface of algal cells and different metabolites act as reducing agents at the surface. The metabolic agents might be like nucleic acids, proteins, oil, lipid, carbohydrates, pigments, and/or enzymes.

There are three phases in the bio-reduction process using algal cells:

Phase I: Activation phase: involves metal ion reduction and nucleation as a result of algal cell secretions. A change in the color of the solution is characteristic of this phase.

Phase II: Growth phase: nucleated metal elements coalesce and form thermodynamically stable nanoparticles having different sizes and shapes.

Phase III: Termination phase: nanoparticles assume their final shape in this phase.

In spite of the fact that pretreatment techniques like washing and thorough mixing are a prerequisite, extracellular synthesis is much more convenient since nanoparticles can be easily obtained in a purified state. Various physiochemical conditions like pH, temperature, metal salt used, and initial concentration of the metal as well as the substrate influence the size, morphology, and agglomeration of nanoparticles. A high pH facilitates the capping and stabilization of the nanomaterials as a result of interaction between the amine group of proteins and their residual amino acids. At the same time, it reduces the tendency of nanoparticles to clump through enhancement of the reducing power of functional groups.

7.5 Blue-Green Algae or Cyanobacteria in Nanoparticle Production

Blue-green algae or Cyanophyceae are photosynthetic prokaryotes lacking membrane bound organelles. Blue-green algae are also referred to as Cyanobacteria (blue-green bacteria) which denote that this assemblage shows more affinity to the prokaryotic bacteria as compared to eukaryotic algae. This group constitutes the most diverse, widely distributed and ecologically most successful group that is found in both freshwater and marine habitats, as planktons (free floating), mats, and periphyton (attached to surfaces). These are found in all types of ecological niches over a wide range of temperatures from the frigid zone of Antarctica to the hot springs in Himalayas, Alps, New Zealand, and North America. The distinguishing feature of the group is the lack of organized chloroplasts, presence of phycobiliprotein pigments, cyanophycean starch as the storage product, and peptidoglycan matrices or walls. Cyanophyceae is considered as one of the most primitive forms on earth having a rich history of about 3.5 billion years and are presumed to be the ancestors of all plastids. Cyanobacteria were initially classified as algae but were later regrouped using morphological and reproductive traits. However, their systematic position is still a matter of debate. Most of the recent classifications of algae put forward by taxonomists during the last 50 years have placed blue-green algae into algae. These are the classifications proposed by:

1. F.E. Round: 1973
2. V.J. Chapman and D.J. Chapman: 1973
3. H.C. Bold and M. J. Wynne: 1978
4. S.P. Parker: 1982
5. R.E. Lee: 2008

Apart from having edible use and being an important component in the treatment of wastewaters, members of this group are of immense agricultural significance and play a major role in nitrogen fixation and soil enumeration (Fig. 7.3).

Members of Cyanophyceae can be easily cultured in an economical manner, which makes them potential contender for large-scale production of nanomaterials. The common cyanobacterial strains used in the green synthesis of nanostructures ranges from unicellular and colonial ones. Some studies have pointed toward phycoerythrin, the primary phycobiliprotein (pigment molecules; molecular weight, 7–15 MDa; light absorption range, 450–670 nm), as a major player in nanoparticle synthesis. Phycoerythrin, the pigment obtained from *Nostoc carneum* (family: Nostocaceae), has been utilized for reduction of silver nitrate ($AgNO_3$) into silver nanoparticles. Cyanobacteria have been thoroughly harnessed for the production of gold nanoparticles. Table 7.2 provides information on the use of different genera of blue-green algae in metal nanoparticle synthesis and the size of the nanoparticles produced.

The initial experiments on metal nanoparticle synthesis using cyanobacteria were carried out on the filamentous cyanobacterium *Plectonema* by a research group headed by M.F. Lengke in the year 2006 when he demonstrated that gold nanoparticles synthesized by *P. boryanum* using gold salts had variation in particle morphology. *P. boryanum* was exposed to the aqueous gold(I) thiosulfate [$Au(S_2O_3)_2^{3-}$] at 25–100 °C for a month. Another experimental was set up with incubation of algal cells in gold(III) chloride [$AuCl_4^-$] solution at 200 °C for 1 day. The interaction of cyanobacteria with aqueous $Au(S_2O_3)_2^{3-}$ led to the formation of cubic gold nanoparticles having size of <10–25 nm at membrane vesicles, within the algal cells as well as encrusted on the cyanobacteria. Interaction of the cyanobacteria with $AuCl_4^-$ gave rise to gold nanoparticles having dimensions <10 nm within the algal cells (Fig. 7.4). In another study, the same group reported the formation of palladium nanoparticles by mixing palladium (II) chloride solution ($PdCl_2$) with *P. boryanum* strain UTEX 485 and incubation at 25, 60, and 100 °C.

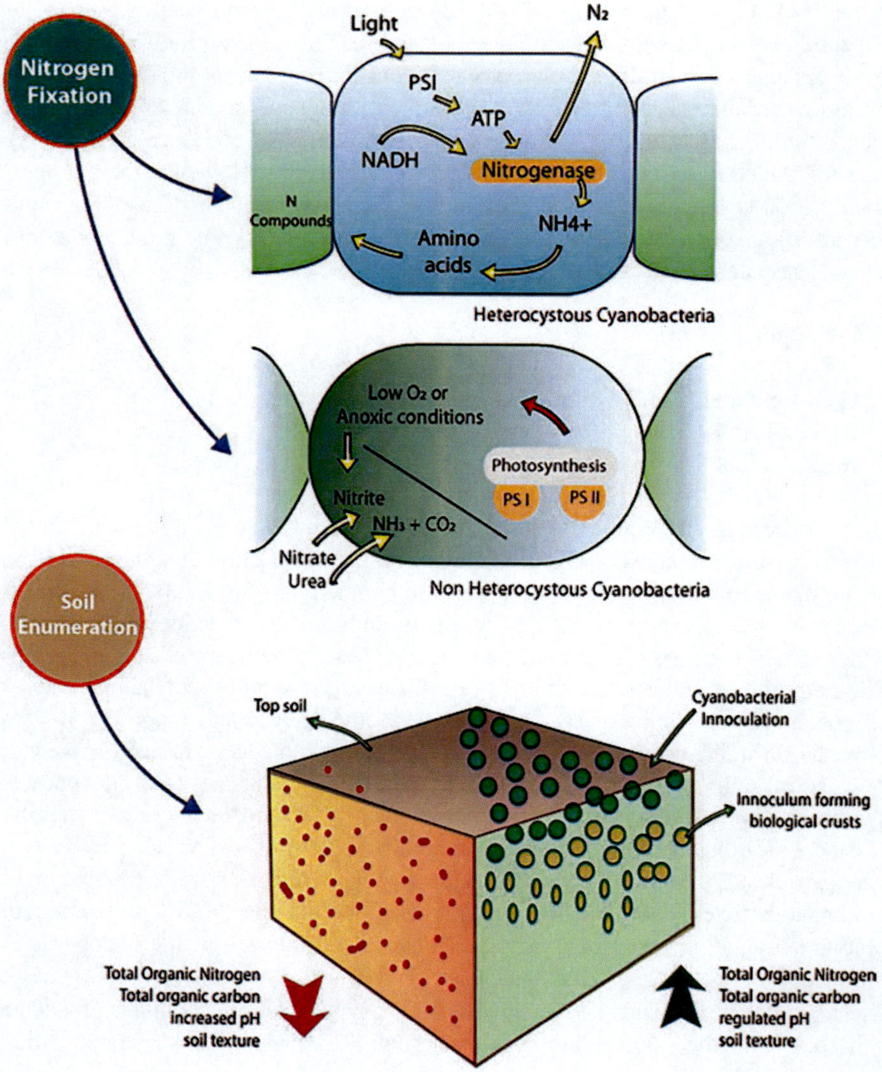

Fig. 7.3 Applications of Cyanophyceae in agriculture (Reprinted with permission from Garlapati et al., 2019)

The experiments led to synthesis of spherical and elongated palladium nanoparticles having size equal or less than 30 nm which were precipitated outside the cell. The exposure of aqueous $PdCl_2$ to the cyanobacteria caused the release of organic substances from the cyanobacteria and subsequent precipitation of dispersed palladium metal on the bacterial cells.

Brayner et al. (2007) demonstrated the formation of gold, palladium, silver, and platinum nanoparticles using *Anabaena, Calothrix*, and *Leptolyngbya* as recyclable

Table 7.2 Blue-green algae-mediated green synthesis of nanoparticles

Algae	Metal	Size (nm)	Reference
Anabaena doliolum	Silver	10–50	Singh et al. (2014)
Anabaena cylindrica	Gold	5.9–9.5	Rösken et al. (2016)
Anabaena variabilis	Silver	26.4	Ismail et al. (2020)
Calothrix sp.	Gold	30–120	Kumar et al. (2016)
Chroococcus minutus	Silver	nr	Sahoo et al. (2020)
Cyanothece sp.	Gold	80–129	Younis et al. (2019)
Cylindrospermum stagnale	Silver	38–40	Husain et al. (2015)
Desertifilum	Silver	4.5–26	Hamida et al. (2020)
Gloeocapsa sp.	Silver	nr	Al-Katib et al. (2015)
Limnothrix sp.	Silver	31.86	Patel et al. (2014)
Lyngbya majuscula	Gold	34.9–45.5	Bakir et al. (2018)
Leptolyngbya	Gold	100–200	Zada et al. (2018)
Microchaete	Silver	60–80	Husain et al. (2019)
Microcoleus chthonoplastes	Gold	<100	Parial et al. (2012)
Nostoc sp.	Silver	30–50	Bin-Meferij and Hamida (2019)
Nostoc carneum	Silver	7.1–26.68	El-Naggar et al. (2018)
Nostoc ellipsosporum	Gold	20–40	Parial et al. (2016)
Oscillatoria limnetica	Silver	3.30–17.97	Hamouda et al. (2019a, b)
Oscillatoria willei	Silver	100–200	Mubarak et al. (2011)
Plectonema boryanum	Gold Palladium	< 10 1–20	Lengke et al. (2006a) Lengke et al. (2007)
Phormidium tenue	Gold	14.84	Parial et al. (2012)
Phormidium valderianum	Gold	<100	Parial et al. (2012)
Spirulina platensis	Silver	7–16	Govindaraju et al. (2008)
	Gold	20–30	Kalabegishvili et al. (2013)
	Selenium	–	Zinicovscaia et al. (2017)
Spirulina subsalsa	Gold	<20 nm	Chakraborty et al. (2009)
Synechocystis sp.	Gold	13	Focsan et al. (2011)
	Silver	14.64	Patel et al. (2014)

nr not reported.

bioreactors. The three genera were initially cultivated for 6–30 days, and thereafter different concentrations (10^{-3}, 10^{-4} and 10^{-6} M) of metal salts were added in growth media devoid of nitrogen. The nitrogenase enzyme played an active role in the reduction of metallic salts to their respective nanoparticles. It was speculated that the enzyme nitrogenase was involved in nanoparticle production since most of the nanoparticles were formed inside specialized cells called heterocysts. This was perhaps the first report that elaborated the screening of cyanobacteria for nanoparticle production over a wide range of metal ion concentration keeping in mind the structural and functional integrity of the cells.

Phycoerythrin-induced production of cadmium sulfide (CdS) nanoparticles was carried out in *Phormidium tenue* NTDM05 by MubarakAli et al. (2012). The phycoerythrin extract was thoroughly mixed with aqueous solutions of 0.25 mM

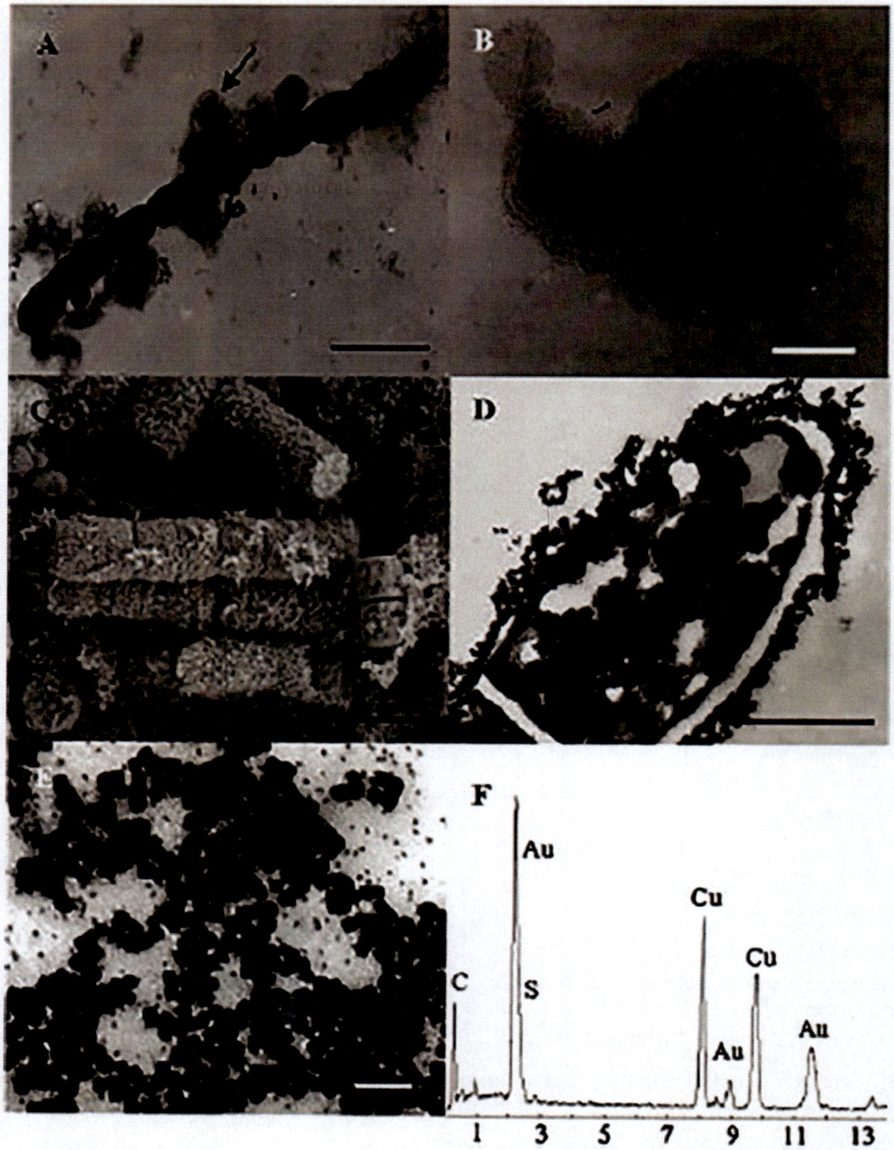

Fig. 7.4 TEM micrographs of whole mounts of cyanobacterial cells cultured in the presence of Au $(S_2O_3)_2^{3-}$ at incubation temperatures of 25 and 100 °C and 30 days: (**a**) membrane vesicle on the exterior cell wall of cyanobacteria at 25 °C; (**b**) precipitation of thin nanoscale gold particles on the membrane vesicle surface at 25 °C; (**c**) SEM micrograph of a cyanobacteria cell encrusted with gold and a gold sulfide nanoparticle at 100 °C; (**d**) TEM micrographs of a thin section of cyanobacteria cells with nanoparticles of gold and gold sulfide deposited on the sheaths and inside the cell at 100 °C; (**e**) cubic nanoparticles of gold precipitated in solution at 100 °C with smaller particles of gold sulfide (<10 nm) in the background; and (**f**) TEM-EDS for area e. Scale bars in a–e are 5, 0.5, 2, and 1 μm and 100 nm, respectively (Reprinted with permission from Lengke et al., 2006a)

cadmium chloride ($CdCl_2$) and 1 mM sodium sulfide (Na_2S). The initial step in this reaction was the formation of a complex comprising C-PE-Cd^{2+} which reacted with Na_2S to produce CdS nanoparticles. TEM results showed that the nanoparticles were spherical in shape and measured around 5 nm. The pigment played a dual role, both in synthesis as well as the stabilization of the nanoentities. The authors were of the view that this simple procedure was ideal for mass production of stable cadmium sulfide nanoparticles. The same research group also reported green synthesis of gold nanoparticles using the cyanobacterium *Phormidium* sp. The cyanobacterium was mixed with chloroauric acid ($HAuCl_4$), incubated in dark for about 60 minutes. Nanotriangles of about 25 nm were produced wherein the cellular proteins of the algae played a major role. It was observed that nanotriangles were increasingly formed due to high reduction rate at low pH.

Sudha et al. (2013) isolated nine cyanobacteria from mangrove forests and screened them for nanoparticle production and antibacterial activity. It was observed that *Microcoleus sp.* reduced silver ions to spherical silver nanoparticles. The nanoentities had size of 40–80 nm, an average size of 55 nm, and showed little aggregation. The silver nanoparticles also exhibited strong antimicrobial activity against eight pathogenic bacteria.

Adebayo-Tayo et al. (2019) carried out green synthesis of silver nanoparticles by mixing 1 g/mL methanol extract of *Oscillatoria* sp. with 0.1 mM of $AgNO_3$ solution. The silver nanoparticles were characterized using different techniques: UV-Vis spectrophotometry, FTIR, SEM, thermogravimetry, energy dispersive X-ray, and dynamic light scattering (DLS). The SEM images demonstrated that the nanoparticles were spherical in shape with size dimensions of 10 nm and partially thermostable and had a polydispersity index of 0.580. These biologically synthesized nanoparticles exhibited significant antibacterial and antibiofilm activity against a range of pathogenic microorganisms.

The green synthesis of metal oxide nanoparticles has also been explored using members of class Cyanophyceae. Rahman et al. (2009) synthesized copper oxide nanoparticles extracellularly under aerobic conditions by incubating *Phormidium cyanobacterium* at 28 °C with aqueous $CuSO_4$ salt solutions. TEM images showed quasi-spherical nanoparticles with size range of 10–40 nm. Fourier transform infrared (FTIR) spectroscopy performed after 36 h of bacterial reaction showed two bands at 1650 and 1550 cm^{-1} which indicated the presence of amide I and amide II bands of protein capped nanoparticles. Ali et al. (2015) carried out extracellular synthesis process of zinc oxide nanoparticles from zinc nitrate solution using cell-free aqueous extract of *Spirulina platensis*, an edible microalga. The nanoparticles were scattered and highly stable and did not agglomerate. TEM analysis showed the nanoparticles to be spherical shaped with a size range of 40–50 nm. Zinc oxide nanoparticles of 50–80 nm were synthesized by Ebadi et al. (2019) using the cell extract of *Nostoc* sp. SEM micrograph of the synthesized entities showed the presence of star-shaped nanoparticles having average diameter of 60 nm. These particles were also tested for their antimicrobial activity against *E. coli*, *P. aeruginosa*, and *S. aureus*. It was noticed that higher concentrations of zinc oxide nanoparticles significantly reduced cell viability of *P. aeruginosa*, *E. coli*, and *S. aureus*. Recently, He et al. (2019) utilized the freshwater cyanobacteria *Microcystis* to synthesize graphene-wrapped ZnO nanospheres. *Microcystis*, known to give rise to algal blooms, was used for green synthesis of ZnO nanoparticles having diverse industrial applications.

7.6 Green Algae in Nanoparticle Production

Chlorophyta or Prasinophyta is a large taxonomically diverse group of photosynthetic eukaryotic life forms that comprise about 9000–12,000 species and are believed to be descendent of the original plastid-harboring eukaryote. Members of chlorophyta mostly occur in fresh water (ponds, lakes, and streams), but many are also found in marine and terrestrial environments. The group is characterized by presence of pigments like chlorophyll a and b, carotenoids and xanthophylls, cellulosic cell wall, starch as the storage photosynthate, isokont flagella, and pyrenoids. The emergence and proliferation of chlorophyta on land about 470 million years ago is considered as a remarkable moment in evolutionary history since they played a major role in the colonization and subsequent evolution of modern land plants. This led to significant changes in the atmosphere and terrestrial substrates, which facilitated the evolutionary process.

Green algae have been categorized into micro and macro green algae. Of these the micro green algae primarily occur in freshwater habitats, while multicellular macro green algae are found primarily in marine environments. Of the two groups, micro algae are a luxuriant source of varied secondary metabolites and functional groups and therefore have been effectively utilized for biosynthesis of metal as well as semiconductor nanoparticles (Table 7.3). The nanoparticles synthesized from different species exhibit diverse physicochemical properties. Several macro algae like *Ulva fasciata, Chaetomorpha linum, Prasiola crispa*, and *Rhizoclonium fontinale* have also been successfully used to produce metal nanoparticles primarily of silver and gold.

7.7 Diatoms in Nanoparticle Production

Diatoms are unicellular, photosynthetic microalgae of class Bacillariophyceae, which ubiquitously occur in both marine and freshwater environments. These photosynthetic microalgae exist worldwide with more than 200 living diatom genera with about 100,000 species. The unique feature of diatoms is the hydrated amorphous silica exoskeletons of diverse size and shape composed of micro to nanoscale biosilica that is specific for each species. The cell wall of diatoms, known as frustules, are mesoporous materials and the most impressive example of 3-D structures occurring in nature. On the basis of frustule symmetry, diatoms are divided into two groups:

Table 7.3 List of nanoparticles synthesized by micro green algae (Modified from Chaudhary et al. 2020)

Algae	Type of NP	Shape and size (nm)	Reaction conditions	Reference
Acutodesmus dimorphus	Ag	Spherical, 2–20 nm	24 h incubation, centrifugation at 16,000 rpm for 20 min at 4 °C	Chokshi et al. (2016)
Botryococcus braunii	Ag	Spherical, 16 nm	3 h continuous stirring at room temperature, dried AgNPs at 55 °C for 5 h	Arya et al. (2019)
Caulerpa racemosa	Ag	FCC, 5–25 nm	24 h incubation at room temperature	Kathiraven et al. (2015)
Caulerpa serrulata	Ag	Spherical and FCC, 10 nm	Different time and pH ranges were studied	Aboelfetoh et al. (2017)
Chaetomorpha linum	Ag	Coalescence, 3–44 nm	30 min of incubation	Kannan et al. (2013a)
Chlamydomonas reinhardtii	Ag	Rectangular and rounded, 1–15 nm	5–10 h of incubation	Barwal et al. (2011)
Chlorella pyrenoidusa	Au	Icosahedral and spherical, 25–30 nm	Incubation at 100 °C at 100 rpm, pH 8	Oza et al. (2012)
Chlorella pyrenoidosa	Ag	FCC, 5–20 nm	24 h incubation at 28 °C, centrifugation at 10,000 rpm for 15 min	Aziz et al. (2015)
Chlorella vulgaris	Ag	Triangular, 28 nm	12 h incubation at room temperature	Xie et al. (2007a, b)
Chlorella vulgaris	Au	Spherical self-assembled cores, 2–10 nm	Incubation at 24 °C	Annamalai and Nallamuthu (2015)
Chlorococcum humicola	Ag	Spherical, 16 nm	72 h incubation at 28 °C, centrifugation at 12,000 rpm	Jena et al. (2013)
Codium capitatum	Ag	Coalescence, 3–44 nm	48 h room temperature incubation	Kannan et al. (2013b)
Coelastrella sp.	Au	Spherical, 30 nm	1 h incubation in dark, centrifugation at 6000 rpm for 7 min	MubarakAli et al. (2013)
Cosmarium impressulum	Au	Spherical, 5.7 nm	Incubation of 16 h light and 8 h dark at 20 °C	Dahoumane et al. (2012)
Enteromorpha flexuosa	Ag	Spherical, 2–32 nm	1 h incubation	Yousefzadi et al. (2014)
Klebsormidium flaccidum	Au	10–20 nm	Total 100 h incubation	Sicard et al. (2010)
Pithophora oedogonia	Ag	Cubical and hexagonal, 25–44 nm	Incubation at room temperature, centrifugation at 15,000 rpm for 15 min	Sinha et al. (2015)
Prasiola crispa	Au	FCC, facile, and spherical, 25 nm	12 h incubation, centrifugation at 10,000 rpm	Sharma et al. (2014a, b)

(continued)

Table 7.3 (continued)

Algae	Type of NP	Shape and size (nm)	Reaction conditions	Reference
Scenedesmus sp.	Ag	Spherical, 30–150 nm	72 h incubation at 28 °C, centrifugation at 12,000 rpm	Jena et al. (2014)
Spirogyra submaxima	Au	Triangular and spherical	–	Roychoudhury and Pal (2014)
Tetraselmis kochinensis	Au	Triangular, FCC, and spherical, 5–35 nm	48 h incubation at 28–29 °C on 200 rpm shaker	Senapati et al. (2012)
Tetraselmis suecica	Au	Spherical, 120 nm	5 min incubation at 90 °C	Shakibaie et al. (2010)
Ulva lactuca	Ag	Spherical, 20–50 nm	–	Bhimba and Kumari (2014)

1. Centrics: radially symmetrical arranged with respect to a point, the annulus, or a central areola.
2. Penates: bipolar symmetrical with the longitudinal axis running parallel to the plane of symmetry.

Diatoms are a storehouse of several photosynthetic pigments like chlorophyll a, chlorophyll c, and fucoxanthin which play major role in photo assimilation and contribute about 20% of the worldwide carbon fixation. Diatoms, both living and dead, have been successfully utilized for production of diverse nanomaterials either through metabolic insertion or through in situ coating/fabrication. However, recently they have been researched for biosynthesis of nanomaterials (Table 7.4).

Gutu et al. (2009) coated the intact surface of *Pinnularia* sp. biosilica frustules with a polycrystalline cadmium sulfide (CdS) nanofilm using the chemical bath deposition (CBD) technique. The cadmium sulfide film comprised of spherical nanoparticles having size of 75 nm which induced photoluminescence in the diatom nanostructure that were determined by the extent of covering of cadmium sulfide over the frustules surface. It was noticed that the size of cadmium sulfide nanoparticles was directly proportional to the concentration of cadmium chloride in the medium. It was also observed that the diatom frustules possessed an intense yellow emission band around 2.11 eV for cadmium sulfide nanoparticles with a density of <25% on frustules surface.

Jena et al. (2015) carried out green synthesis of silver nanoparticles by a light-dependent reaction using *Amphora* which is one of the largest genera of diatoms found in both marine and fresh water. The aqueous cell extract of the diatom was treated with 2×10^{-3} M $AgNO_3$ solution and incubated at 35–40 °C. TEM images showed that the silver nanoparticles formed were spherical, polydispersed, well scattered, having size ranging from 5 to 70 nm, and having an average of about 20–25 nm (Fig. 7.5). The polycrystalline nature of the nanoparticles was confirmed by XRD and SAED analysis. EDAX spectra showed strong characteristic signal of

Table 7.4 Use of diatoms in the biosynthesis of nanoparticles

Diatom	Type of nanoparticle	Size (nm)	Reference
Amphora sp.	Silver	5–70	Jena et al. (2015)
Chaetoceros sp.	Silver	149.03	Mishra et al. (2020)
Eolimna minima	Silver	5–80	Feurtet-Mazel et al. (2015)
Diadesmis gallica	Gold	22	Schröfel et al. (2011)
Navicula atomus	Gold	9	Schröfel et al. (2011)
	Silver	70–80	Chetia et al. (2017)
Navicula cincta	Silver	32	Boobalakrishnan et al. (2020)
Nitzschia frustulum	Silicon-germanium oxide	nr	Qin et al. (2008)
Phaeodactylum tricornutum	Cadmium sulfide	nr	Scarano and Morelli (2003)
Pinnularia sp.	Cadmium sulfide	75	Gutu et al. (2009)
Skeletonema sp.	Silver	186.73	Mishra et al. (2020)
Stephanopyxis turris	Gold	10–30	Pytlik et al. (2017)
Thalassiosira sp.	Silver	239.46	Mishra et al. (2020)

nr not reported.

silver, while the XRD spectra showed four intense diffraction peaks at 2θ values of 38.48°, 44°, 64.74°, and 77.4° which corresponded to (1 1 1), (2 0 0), (2 2 0), and (3 1 1) plane of (fcc) crystalline silver nanoparticles. The experiment demonstrated the potential of aqueous extract of *Amphora* in reducing silver ions and converting them into well-dispersed silver nanoparticles.

Green synthesis of gold nanoparticles by *Stephanopyxis turris* (family: Stephanopyxidaceae) was accomplished using 5, 10, 50, 100, 200, and 500 μM $HAuCl_4$ solution by Pytlik et al. (2017). At 200 μM concentration, the gold nanoparticles obtained were classified into two categories, viz., small and large. Most of the nanoparticles were small with diameters ranging from 10 to 20 nm and were present inside the pores distributed over the whole cell. Large particles having a diameter of about 50 nm were lesser in number and agglomerated and were arranged close to the biosilica pore structure. A strong gold signal (orange) exactly at the position of the bright particle confirmed the presence of gold nanoparticles. The TEM analysis revealed the presence of spherical and triangular shaped gold nanoparticles. The diameter of most of the nanoparticles varied between 10 and 30 nm.

A cost-effective, non-toxic, eco-friendly, photo-induced approach for bioconversion of metallic silver into silver nanoparticles was developed by Chetia et al. (2017) utilizing the diatom *Navicula*, a genus having boat-shaped cells (Fig. 7.6). The UV-Vis spectra showed a broad absorption peak in the visible range due to excitation of free electrons that initiates surface plasmon resonance. The size of nanoparticles assessed through SEM ranged from 70 to 80 nm. The EDX spectra showed strong characteristic signal of silver around 3 keV for the sample, while the

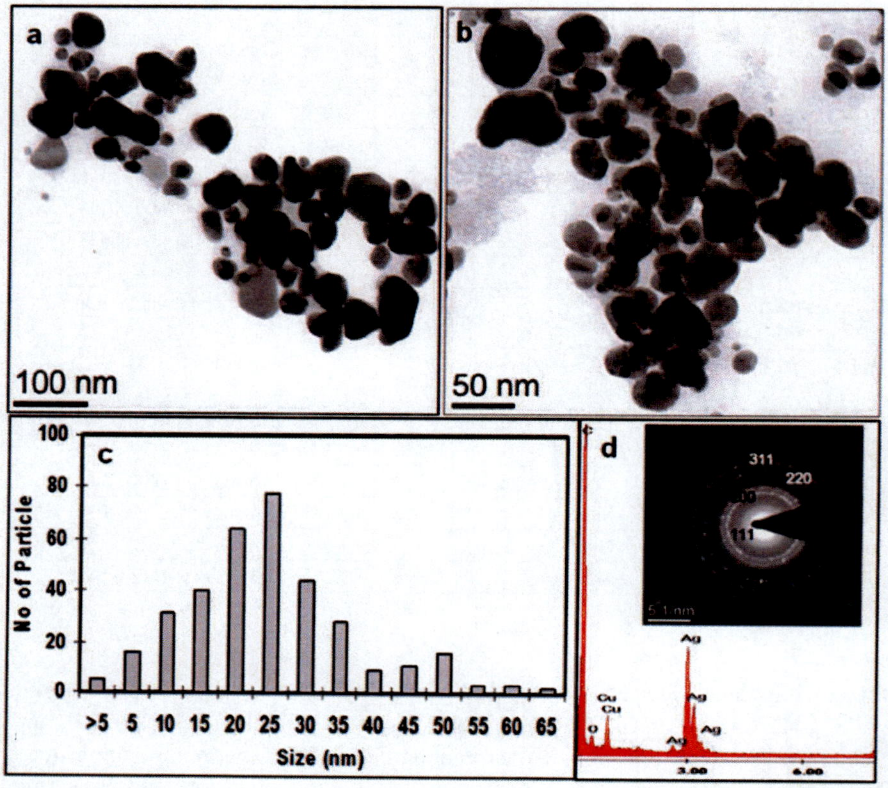

Fig. 7.5 (**a, b**) TEM image, (**c**) particle size distribution, (**d**) EDAX spectrum and SAED pattern of biosynthesized SNPs (Reprinted with permission from Jena et al., 2015)

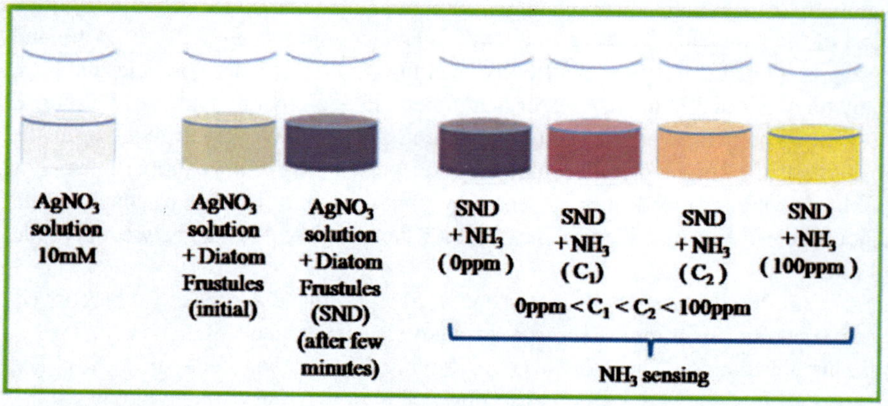

Fig. 7.6 Schematic illustration for synthesis of Ag-NPs using diatom frustules and visual sensing of ammonia (Reprinted with permission from Chetia et al., 2017)

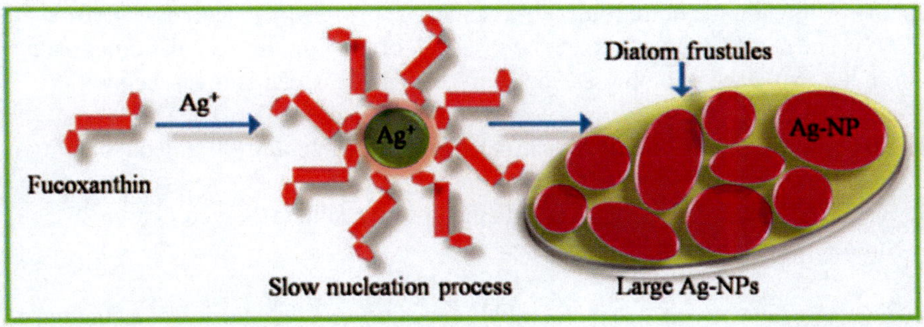

Fig. 7.7 Schematic showing the formation of Ag-NPs through slow nucleation process using fucoxanthin-contained diatom frustules (Reprinted with permission from Chetia et al., 2017)

XRD spectra confirmed the crystallinity of the silver nanoparticles. The average crystallite size of the silver nanoparticles attached on diatom cells by using Debye-Scherrer formula was calculated to be 30.35 nm. Figure 7.7 shows the formation of silver nanoparticles through the slow nucleation process utilizing fucoxanthin which is responsible for golden brown color of diatoms. In the solution, the metallic salt $AgNO_3$ interacts with the cellular organic compounds especially fucoxanthin which contains various oxygenic functional groups. The hydroxyl group present in the fucoxanthin structure makes it an ideal reducing agent along with protein and other biomolecules like carbohydrates, proteins, and silaffin which reduce the metal ion to stable nanoparticle through oxidation of hydroxyl to carbonyl groups. Apart from fucoxanthin, the hydroxyl group of silanol (Si-OH) in diatom frustules is also responsible for nanoparticle formation.

Apart from nanoparticles, diatoms have also been used to produce diverse nanostructures having industrial importance. The first study for fabrication of a photoluminescent nanocomb structure using cell culture system was carried out in the year 2008. A two-stage process was developed by Qin et al. (2008) to metabolically insert germanium into the frustule biosilica of the benthic diatom species *Nitzschia frustulum*. The process induced the formation of structures strongly resembling a double-sided nanocomb entity. The nanocomb structure had the following dimensions:

1. Overall length—8 μm
2. Rib width—200 nm
3. Rib length—500 nm
4. Slit width—100 nm

The frustules exhibited blue or blue-green photoluminescence at peak wavelengths of 450–480 nm resulting from contributions of nanostructured biosilica in the parent structure as well as the nanocomb daughter valves.

Similarly, the pioneer study involving the use of the diatom *Pinnularia* for fabricating semiconductor titanium dioxide (TiO_2) nanostructures in a controlled manner following the bottom-up self-assembly approach was done by Jeffryes et al. (2008). The study involved metabolic insertion of titanium into the diatom biosilica through a two-stage photobioreactor cultivation process that is given below:

1. Stage I—Culturing of *Pinnularia* cells in silicon medium till silicon starvation was achieved
2. Stage II—Addition of titanium and silicon to the silicon-starved diatom suspension for about 10 h

The experiment led to preferential deposition of titanium as a lining of nanodimensions at the base of frustule pore. Thermal annealing at intensely high temperatures (720 °C) transformed biogenic titanate to anatase titanium dioxide (TiO_2) having dimensions of 32 nm.

7.8 Brown Algae in Nanoparticle Production

Phaeophyceae or brown algae (or division Phaeophyta) is a class of algae comprising about 1500–2000 species ranging in size from small filamentous forms to large, complex seaweeds. Of the currently known species, less than 1% are known to inhabit fresh-water habitats. Members of this class have the following distinguishing characters:

- Chloroplasts have four surrounding membranes.
- Thylakoids in stacks of three.
- Presence of chlorophyll a and chlorophyll c. However, the presence of fucoxanthin gives a brownish appearance to the thallus.
- Reserve food material is laminarin, a soluble polysaccharide, and mannitol, a sugar alcohol.
- Presence of small refractile vesicles known as *fucosan vesicles having protective function*.
- Gametes generally have a pair of laterally inserted unequal flagella: a long anterior pantonematic and a small posterior acronematic.

The use of brown algae in biosynthesis of nanoparticles is evident in Table 7.5. It is evident from the table that only a few species of brown algae have been harnessed in the green synthesis of nanoparticles. *Sargassum* tops the list and has been utilized for the production of gold, silver, zirconia, palladium, and several metal oxide (iron oxide, zinc oxide, and magnesium oxide) nanoparticles. Apart from using the aqueous extracts of brown algae, several studies have been carried out for biosynthesis of nanoparticles using different components of the algal cells.

Chen et al. (2008) followed a simple method for synthesis of selenium nanoparticles using sodium selenite, polysaccharides of *Undaria pinnatifida*, and ascorbic acid. This edible seaweed is native to cold, temperate coasts of the northwest Pacific Ocean. TEM and HRTEM studies validated the presence of uniform

Table 7.5 Brown algae-mediated green synthesis of nanoparticles

Algae	Type of nanoparticle	Size (nm)	Reference
Bifurcaria bifurcata	Copper oxide	5–45	Abboud et al. (2014)
Cladosiphon okamuranus	Gold	8.54–10.74	Soisuwan et al. (2010)
Cystophora moniliformis	Silver	50–100	Prasad et al. (2013)
Cystoseira baccata	Gold	8.4	González-Ballesteros et al. (2017)
Cystoseira trinodis	Cupric oxide	6–7.8	Gu et al. (2018)
Desmarestia antarctica	Gold Silver	12.6 13.7	González-Ballesteros et al. (2021) González-Ballesteros et al. (2021)
Desmarestia menziesii	Gold Silver	11.5 17.8	González-Ballesteros et al. (2018) González-Ballesteros et al. (2018)
Dictyopteris divaricata	Gold	62.6	Young Park et al. (2019)
Ecklonia cava	Gold	30	Venkatesan et al. (2014)
Ecklonia stolonifera	Gold	27.9	Jun et al. (2020)
Egregia sp.	Gold	2–20	Colin et al. (2018)
Fucus vesiculosus	Gold	Varied	Mata et al. (2009)
Laminaria japonica	Gold	15–20	Ghodake and Lee (2011)
Padina boryana	Palladium	8.7	Sonbol et al. (2021)
Padina gymnospora	Gold Platinum	53–67 5–50	Singh et al. (2013) Ramkumar et al. (2017)
Padina pavonica	Iron oxide Silver	10–19.5 20–70	El-Kassas et al. (2016) Sudha and Balasundaram (2018)
Padina tetrastromatica	Gold	8–10	Rajeshkumar et al. (2017)
Sargassum acinarium	Iron oxide	21.6–27.4	El-Kassas et al. (2016)
Sargassum bovinum	Palladium	5–10	Momeni et al. (2015)
Sargassum glaucescens	Gold	3.65	Ajdari et al. (2016)
Sargassum incisifolium	Gold Silver	12.38 20	Mmola et al. (2016) Mmola et al. (2016)
Sargassum muticum	Gold Iron oxide Silver Zinc oxide	5.42 18 43–79 30–57	Namvar et al. (2015) Mahdavi et al. (2013) Madhiyazhagan et al. (2015) Azizi et al. (2014)
Sargassum myriocystum	Gold Zinc oxide	15 76–186	Stalin Dhas et al. (2012) Nagarajan and Kuppusamy (2013)
Sargassum plagiophyllum	Silver chloride	18–42	Stalin Dhas et al. (2014)
Sargassum swartzii	Gold	35	Dhas et al. (2014)
Sargassum tenerrimum	Silver	20	Kumar et al. (2012)
Sargassum wightii	Gold Magnesium oxide	8–12 68.06	Singaravelu et al. (2007) Pugazhendhi et al. (2019)

(continued)

Table 7.5 (continued)

Algae	Type of nanoparticle	Size (nm)	Reference
	Silver Zinc oxide Zirconia	8–27 40–50 4.8	Govindaraju et al. (2009) Ishwarya et al. (2018) Kumaresan et al. (2018)
Stoechospermum marginatum	Gold	18.7–93.7	Rajathi et al. (2012)
Turbinaria conoides	Gold Zinc oxide	6–10 Nr	Rajeshkumar et al. (2013) Raajshree and Brindha (2018)
Turbinaria ornata	Silver	20–32	Deepak et al. (2017)

spherical nanoparticles having particle size of 44–92 nm and mean diameter of 59 nm.

Fucoidan, an anionic, sulfated polysaccharide, was extracted from the seaweed *Padina tetrastromatica* by Rajeshkumar (2017) and used for biological production of silver nanoparticles. The fucoidan extracted from *P. tetrastromatica* was incubated with 1 mM $AgNO_3$ solution, and the characterization done using ultraviolet-visible (UV-Vis) spectrophotometer wavelength ranging from 350 to 650 nm, SEM, TEM, SAED, and XRD analysis. The UV-Vis spectroscopy revealed the presence of strong surface plasmon resonance band at 440 nm corresponding to the characteristic peaks of silver nanoparticles. TEM images showed the presence of highly variable, spherical to triangular-shaped nanoentities. The congregations were aggregates of nanoparticles ranging from 4 to 25 nm with the average diameter of 17 nm. The nanoparticles were not in contact within the aggregates, which indicated their stabilization by fucoidan as the capping agent. The SEM images also showed some undefined and rectangle-shaped silver nanoparticles. Similarly, fucoidan was isolated from *Spatoglossum asperum*, a marine brown algae of family Dictyotaceae, and different concentrations (1–10 mg/ml) of the same were treated with 1 mM $AgNO_3$ solution followed by incubation at room temperature. The SEM images showed silver nanoparticles that were spherical to oval in shape having size ranging from 32 to 51 nm. The TEM images showed the presence of well-separated spherical nanoparticles ranging from 20 to 46 nm.

A simple and eco-friendly method was developed by Yugay et al. (2020) for the preparation of silver nanoparticles using different polysaccharides, viz., laminaran (ScL) and fucoidan (ScF) isolated from *Saccharina cichorioides*, and fucoidan (FeF) and alginate (FeA) obtained from *Fucus evanescens*, both representatives of brown algae. The UV-Vis spectra of the synthesized Ag-NPs-FeA (silver nanoparticles synthesized using alginate from *F. evanescens*), Ag-NPs-FeF (silver nanoparticles synthesized using fucoidan from *F. evanescens*), Ag-NPs-ScF (silver nanoparticles synthesized using fucoidan from *Saccharina cichorioides*), and Ag-NPs-ScL (silver nanoparticles synthesized using laminaran from *Saccharina cichorioides*) demonstrated the maximum peaks at 408, 402, 412, and 414 nm, respectively, corresponding to characteristic excitation of silver nanoparticles surface plasmons in the range of 400–490 nm. The TEM images of Ag-NPs-FeA, Ag-NPs-FeF, Ag-NPs-ScF, and Ag-NPs-ScL confirmed the prevalence of spherical nanoparticles having dimensions of 57 ± 9, 64 ± 6, 53 ± 4, and 45 ± 2 nm, respectively.

7.9 Red Algae in Nanoparticle Production

Rhodophyta or red algae are an anciently derived lineage of eukaryotic algae constituting an algal division that comprise of the following characters:

- Eukaryotic cells
- Absence of flagellated structures during the life cycle
- Floridean starch
- Presence of phycobiliprotein pigments (phycocyanin, allophycocyanin, and phycoerythrin)
- Unstacked thylakoids
- Chloroplasts lacking an external endoplasmic reticulum

Members of this group are primarily marine in distribution, with about 3% of the total 6500 species occurring in freshwater habitats mainly as macroscopic and benthic forms.

Table 7.6 provides information on the use of red algae in nanoparticle production, primarily silver gold and zinc nanoparticles. However, the secondary metabolites and the by-products of red algae also serve as important material for biosynthesis of nanoparticles. One of the earliest reports using agar for nanoparticle production was put forward by Shukla et al. (2012). The agar polysaccharide obtained from the red alga *Gracilaria dura* was used for biological synthesis of spherical silver nanoparticles (size: 6.0 ± 2 nm) using 5 mM $AgNO_3$ solution. Later these nanoparticles were entrapped in the agar for synthesis of agar/silver nanocomposite film by pouring the heated solution of agar/silver nanocomposite solution and subsequent drying. This nanocomposite film showed significant bactericidal effect against *Bacillus pumilus*, a Gram-positive, rod-shaped, endospore forming bacteria. Similarly, pigments of red algae like r-phycoerythrin have been utilized as template and reducing agent to generate fluorescent silver nanoparticles by mixing with 24 mM $AgNO_3$ solution (pH 5.0, 30 °C, 1000 rpm). The nanoparticles formed were spherical, well dispersed, and uniform with an average size of about 7.2 ± 3.5 nm in diameter. Bekasova et al. (2018) produced a bioconjugate of R-phycoerythrin with silver nanoparticles of 6.5 nm diameter by mixing aqueous solutions of $AgNO_3$ and R-phycoerythrin. The bioconjugate self-assembled into nanorods of 25 ± 5 nm long till 90 °C. Further heating to 100 °C led to the formation of 0.5 ± 0.2 μm-long filaments and spherical aggregates having a diameter of 60 ± 10 nm.

Table 7.6 Red algae-mediated green synthesis of nanoparticles

Algae	Type of nanoparticle	Size (nm)	Reference
Acanthophora spicifera	Gold Silver	<20 48	Babu et al. (2020) Kumar et al. (2012)
Amphiroa fragilissima	Silver	nr	Sajidha Parveen and Lakshmi (2016)
Amphiroa rigida	Silver	25	Gopu et al. (2021)
Centroceras clavulatum	Silver	35–65	Murugan et al. (2016)
Chondrus crispus	Gold	30–50	Castro et al. (2013)
Corallina elongata	Silver	12–20	Hamouda et al. (2019a, b)
Corallina officinalis	Gold	14.6	El-Kassas and El-Sheekh (2014)
Galaxaura elongata	Gold	3.85–77.13	Abdel-Raouf et al. (2017)
Gracilaria crassa	Silver	122.7	Lavakumar et al. (2015)
Gracilaria corticata	Gold	45–57	Naveen and Prakash (2013)
Gracilaria birdiae	Silver	20.3	de Aragao et al. (2019)
Gracilaria dura	Silver	6	Shukla et al. (2012)
Gracilaria edulis	Silver Zinc oxide	55–99 20–50	Priyadharshini et al. (2014) Mohamed Asik et al. (2019)
Gelidiella acerosa	Silver	20–50	Satish et al. (2017)
Gelidium amansii	Gold Silver	4–7 27–54	Murphin Kumar et al. (2017) Pugazhendhi et al. (2018)
Hypnea musciformis	Silver	40–65	Roni et al. (2015)
Hypnea valencia	Zinc oxide	76–186	Nagarajan and Arumugam Kuppusamy (2013)
Iridaea cordata	Gold Silver	12.3 17.5	González-Ballesteros et al. (2021) González-Ballesteros et al. (2021)
Kappaphycus alvarezii	Gold Silver	10–40 73	Rajasulochana et al. (2012) Ganesan et al. (2013)
Lemanea fluviatilis	Gold	5–15	Sharma et al. (2014a, b)
Porphyra vietnamensis	Silver	13	Venkatpurwar and Pokharkar (2011)
Portieria hornemannii	Silver	60–70	Fatima et al. (2020)
Pterocladiella capillacea	Silver	11.4	El Kassas and Attia (2014)
Pyropia yezoensis	Silver	20–22	Ulagesan et al. (2021)
Kappaphycus alvarezii	Gold	10–40	Rajasulochana et al. (2011)

Red algae are a rich source of a number of polysaccharides like carrageenans, agarose, and porphyran which can be harnessed for green synthesis of polysaccharide based nanomaterials that find applications in biomedical field due to their biocompatibility, biodegradability, low cost, and nontoxic nature. Carrageenan is a family of sulfated galactans normally isolated from marine red algae that is composed of d-galactose residues linked alternately by (1 → 3)-linked *β*-d-galactopyranose (unit G) and (1 → 4)-linked *α*-d-galactopyranose (unit D). Chen et al. (2018) successfully accomplished biosynthesis of gold nanoparticles using carrageenan oligosaccharide as a reducing and capping agent. The carrageenan oligosaccharide was added to 6×10^{-4} mol/L of gold (III) chloride trihydrate ($HAuCl_4.3H_2O$) and was vigorously shaken at 50 °C for 3 h leading to a change in color from light yellow to ruby red indicating the synthesis of carrageenan oligosaccharide-gold nanoparticles (CAO-AuNPs) due to the reduction of Au^{3+} into Au^0 ions. The nanoparticles had average size of 35 ± 8 nm and a polydispersity index of 0.31 ± 0.02 which indicated homogeneous size distribution of the newly synthesized nanoparticles. The CAO-AuNPs showed significant cytotoxic activity toward human colon cancer and human breast cancer cell lines, which exhibited their promise as a potential nanomaterial for drug delivery.

El-Rafie et al. (2013) utilized water-soluble polysaccharides obtained from four algal species comprising two red algae (*Pterocladia capillacae* and *Jania rubins*), one brown alga (*Ulva fasciata*), and one green alga (*Colpmenia sinusa*) as reducing and stabilizing agents. The crude hot water soluble polysaccharides (CHWPS) extract of each alga was dissolved in distilled water at 70 °C, and 0.1 mM $AgNO_3$ solution was added with constant stirring. The reduction of silver nanoparticles was monitored through color change as well as UV-visible spectral analysis, FTIR, and TEM analysis. TEM images revealed that the nanoparticles were polydispersed and predominantly spherical and had the maximum diameter size of 7 and 12 nm for those prepared from the polysaccharides of *P. capillacae* and *J. rubins*, respectively.

7.10 Future Prospects

Thus, algae provides numerous advantages as "biological nanofactories" like being environmentally "clean," cost-effective, safe, and requiring less energy in comparison to the physical and chemical synthesis methods. Algae can act as natural nanofactories for metals, metal oxides, nitrides, carbides, and sulfide-based nanoparticles. However, in spite of the immense biomolecular diversity in several algal groups, phycosynthesis of nanoparticles is still in the nascent stages. Marine microalgae are more advantageous in production of metal nanoparticles due to their prior exposure to high salt concentrations present in seawater. However, in spite of the presence of long chain surfactant molecules, biosynthesis of monodispersed nanoparticles using marine algae is still a daunting task.

The current scenario points toward increased exploration of algae for green production of nanoparticles, but many intriguing issues need to be addressed for their efficient utilization:

- The mechanism of nanoparticle synthesis needs thorough elucidation especially the role of different proteins and enzymes involved.
- Development of simple cost-efficient techniques keeping in mind the commercial aspect.

- Standardization of protocols for controlling the size, shape, and monodispersity of the nanostructures which are dependent on the type of algae used, optimum growth stage, the cultivation medium, and the reaction medium.

References

Abboud Y, Saffaj T, Chagraoui A, El Bouari A, Brouzi K, Tanane O, Ihssane B (2014) Biosynthesis, characterization and antimicrobial activity of copper oxide nanoparticles (CONPs) produced using brown alga extract (*Bifurcaria bifurcata*). Appl Nanosci 4:571–576

Abdel-Rouf N, Al-Enazi NM, Ibraheem IBM (2017) Green biosynthesis of gold nanoparticles using *Galaxaura elongate* and characterization of their antibacterial activity. Arab J Chem 10(2): S3029–S3039

Aboelfetoh EF, El-Shenody RA, Ghobara MM (2017) Eco-friendly synthesis of silver nanoparticles using green algae (*Caulerpa serrulata*): reaction optimization, catalytic and antibacterial activities. Environ Monit Assess 189:349

Adebayo-Tayo B, Salaam A, Ajibade A (2019) Green synthesis of silver nanoparticle using *Oscillatoria* sp. extract, its antibacterial, antibiofilm potential and cytotoxicity activity. Heliyon 5(10):e02502

Ajdari Z, Rahman H, Shameli K, Abdullah R, Abd Ghani M, Yeap S, Abbasiliasi S, Ajdari D, Ariff A (2016) Novel gold nanoparticles reduced by *Sargassum glaucescens*: preparation, characterization and anticancer activity. Molecules 21(3):123

Ali MI, Sharma G, Kumar M, Jasuja ND, Rajgovind (2015) Biological approach of zinc oxide nanoparticles synthesis by cell free extract of *Spirulina platensis*. Int J Cur Eng Technol 5(4):2531–2534

Al-Katib M, Al-Shahri Y, Al-Niemi A (2015) Biosynthesis of silver nanoparticles by cyanobacterium *Gloeocapsa* sp. Int J Enhanced Res Sci Technol Eng 4:60–73

Annamalai J, Nallamuthu T (2015) Characterization of biosynthesized gold nanoparticles from aqueous extract of *Chlorella vulgaris* and their anti-pathogenic properties. Appl Nanosci 5:603–607

Arya A, Mishra V, Chundawat TS (2019) Green synthesis of silver nanoparticles from green algae (*Botryococcus braunii*) and its catalytic behavior for the synthesis of benzimidazoles. Chem Data Collect 20:100190

Aziz N, Faraz M, Pandey R, Shakir M, Fatma T, Varma A, Barman I, Prasad R (2015) Facile algae-derived route to biogenic silver nanoparticles: synthesis, antibacterial, and photocatalytic properties. Langmuir 31:11605–11612

Azizi S, Ahmad MB, Namvar F, Mohamad R (2014) Green biosynthesis and characterization of zinc oxide nanoparticles using brown marine macroalga *Sargassum muticum* aqueous extract. Mater Lett 116:275–277

Babu B, Palanisamy S, Vinosha M, Anjali R, Kumar P, Pandi B, Tabarsa M, You S, Prabhu NM (2020) Bioengineered gold nanoparticles from marine seaweed *Acanthophora spicifera* for pharmaceutical uses: antioxidant, antibacterial, and anticancer activities. Bioprocess Biosyst Eng 43(12):2231–2242

Bakir EM, Younis NS, Mohamed ME, El Semary NA (2018) Cyanobacteria as nanogold factories: chemical and anti–myocardial infarction properties of gold nanoparticles synthesized by *Lyngbya majuscula*. Mar Drugs 16:217

Barsanti L, Gualtieri P (2014) Algae: anatomy, biochemistry and biotechnology. CRC Press, New York

Barwal I, Ranjan P, Kateriya S, Yadav SC (2011) Cellular oxido-reductive proteins of *Chlamydomonas reinhardtii* control the biosynthesis of silver nanoparticles. J Nanobiotechnol 9:56

Bekasova OD, Stein-Margolina VA, Kurganov BI (2018) Ag^0 nanoparticles synthesized in R-phycoerythrin: change in bioconjugate properties upon ripening of nanoparticles. Curr Pharm Biotechnol 19(5):422–427

Bhimba B, Kumari P (2014) Phytosynthesis of silver nanoparticles from the extracts of seaweed *Ulva lactuca* and its antimicrobial activity. Int J Pharm BioSci 5:666–677

Bin-Meferij MM, Hamida RS (2019) Biofabrication and antitumor activity of silver nanoparticles utilizing novel *Nostoc sp*. Int J Nanomedicine 14:9019–9029

Bold HC, Wynne MJ (1978) Introduction to algae: structure and reproduction. Prentice Hall of India, New Delhi

Boobalakrishnan S, Mani P, Pottail L, Sundarajan SK (2020) Synthesis of nano-silver from *Navicula cincta* and the evaluation of its antimicrobial activity. Micro Nano Lett 15:119–124

Brayner R, Barberousse H, Hernadi M, Djedjat C, Yepremian C, Coradin T, Livage J, Fievet F, Coute A (2007) Cyanobacteria as bioreactors for the synthesis of au, Ag, Pd, and Pt nanoparticles via an enzyme-mediated route. J Nanosci Nanotechnol 7:2696–2708

Castenholz RW, Phylum BX (2001) Cyanobacteria. In: Oxygenic photosynthetic bacteria. Springer, New York, pp 473–599

Castro L, Blázquez ML, Muñoz JA, González F, Ballester A (2013) Biological synthesis of metallic nanoparticles using algae. IET Nanobiotechnol 7:109–116

Chakraborty N, Banerjee A, Lahiri S, Panda A, Ghosh AN, Pal R (2009) Biorecovery of gold using cyanobacteria and an eukaryotic alga with special reference to nanogold formation – a novel phenomenon. J Appl Phycol 21:145–152

Chapman DJ, Chapman VJ (1973) The algae, 2nd edn. Macmillan, London

Chaudhary R, Nawaz K, Khan AK, Hano C, Abbasi BH, Anjum S (2020) An overview of the algae-mediated biosynthesis of nanoparticles and their biomedical applications. Biomol Ther 10(11):1498

Chen T, Wong Y-S, Zheng W, Bai Y, Huang L (2008) Selenium nanoparticles fabricated in *Undaria pinnatifida* polysaccharide solutions induce mitochondria-mediated apoptosis in A375 human melanoma cells. Colloids Surf B Biointerfaces 67(1):26–31

Chen X, Zhao X, Gao Y, Yin J, Bai M, Wang F (2018) Green synthesis of gold nanoparticles using carrageenan oligosaccharide and their in vitro antitumor activity. Mar Drugs 16(8):277

Chetia L, Kalita D, Ahmed GA (2017) Synthesis of Ag nanoparticles using diatom cells for ammonia sensing. Sensing Bio-sensing Res 16:55–61

Chokshi K, Pancha I, Ghosh T, Paliwal C, Maurya R, Ghosh A, Mishra S (2016) Green synthesis, characterization and antioxidant potential of silver nanoparticles biosynthesized from de-oiled biomass of thermotolerant oleaginous microalgae *Acutodesmus dimorphus*. RSC Adv 6:72269–72274

Colin JA, Pech-Pech IE, Oviedo M, Águila SA, Romo-Herrera JM, Contreras OE (2018) Gold nanoparticles synthesis assisted by marine algae extract: biomolecules shells from a green chemistry approach. Chem Phys Lett 708:210–215

Dahoumane SA, Djediat C, Yéprémian C, Couté A, Fiévet F, Coradin T, Brayner R (2012) Species selection for the design of gold nanobioreactor by photosynthetic organisms. J Nanopart Res 14:883

de Aragao AP, de Oliveira TM, Quelemes PV, Perfeito MLG, Araujo MC, Santiago JDAS, Cardoso VS, Quaresma P, de Almeida JRDS, da Silva DA (2019) Green synthesis of silver nanoparticles using the seaweed *Gracilaria birdiae* and their antibacterial activity. Arab J Chem 12:4182–4188

De Clerck O, Bogaert K, Leliaert F (2012) Diversity and evolution of algae: primary endosymbiosis. Adv Bot Res 64:55–86

Deepak P, Sowmiya R, Ramkumar R, Balasubramani G, Aiswarya D, Perumal P (2017) Structural characterization and evaluation of mosquito-larvicidal property of silver nanoparticles synthesized from the seaweed, *Turbinaria ornata* (turner). Artif Cells Nanomed Biotechnol 45(5):990–998

Dhas TS, Kumar VG, Karthick V, Govindaraju K, Shankara NT (2014) Biosynthesis of gold nanoparticles using *Sargassum swartzii* and its cytotoxicity effect on HeLa cells. Spectrochim Acta A Mol Biomol Spectrosc 133:102–106

Domozych DS, Ciancia M, Fangel JU, Mikkelsen MD, Ulvskov P, Willats WGT (2012) The cell walls of green algae: a journey through evolution and diversity. Front Plant Sci 3:82

Ebadi M, Zolfaghari MR, Aghaei SS, Zargar M, Shafiei M, Zahiri HS, Noghabi KA (2019) A bio-inspired strategy for the synthesis of zinc oxide nanoparticles (ZnO NPs) using the cell extract of cyanobacterium *Nostoc* sp. EA03: from biological function to toxicity evaluation. RSC Adv 9:23508–23525

El Kassas HY, Attia AA (2014) Bactericidal application and cytotoxic activity of biosynthesized silver nanoparticles with an extract of the red seaweed *Pterocladiella capillacea* on the HepG2 cell line. Asian Pac J Cancer Prev 15(3):1299–1306

El-Kassas HY, Aly-Eldeen MA, Gharib SM (2016) Green synthesis of iron oxide (Fe_3O_4) nanoparticles using two selected brown seaweeds: characterization and application for lead bioremediation. Acta Oceanol Sin 35:89–98

El-Kassas HY, El-Sheekh MM (2014) Cytotoxic activity of biosynthesized gold nanoparticles with an extract of the red seaweed *Corallina officinalis* on the MCF-7 human breast cancer cell line. Asian Pac J Cancer Prev 15(10):4311–4317

El-Naggar NE, Hussein MH, El-Sawah AA (2018) Phycobiliprotein-mediated synthesis of biogenic silver nanoparticles, characterization, in vitro and in vivo assessment of anticancer activities. Sci Rep 8(1):8925

El-Rafie H, El-Rafie M, Zahran M (2013) Green synthesis of silver nanoparticles using polysaccharides extracted from marine macro algae. Carbohydr Polym 96:403–410

Fatima R, Priya M, Indurthi L, Radhakrishnan V, Sudhakaran R (2020) Biosynthesis of silver nanoparticles using red algae *Portieria hornemannii* and its antibacterial activity against fish pathogens. Microb Pathog 138:103780

Feurtet-Mazel A, Mornet S, Charron L, Mesmer-Dudons N, Maury-Brachet R, Baudrimont M (2015) Biosynthesis of gold nanoparticles by the living freshwater diatom *Eolimna minima*, a species developed in river biofilms. Environ Sci Pollut Res 23(5):4334–4339

Focsan M, Ardelean II, Craciun C, Astilean S (2011) Interplay between gold nanoparticle biosynthesis and metabolic activity of cyanobacterium *Synechocystis* sp. PCC 6803. Nanotechnology 22(48):485101

Fritsch FE (1945) The structure and reproduction of the algae I & II. Cambridge University Press, Cambridge

Ganesan V, Aruna DJ, Astalakshmi A, Nima P, Thangaraja A (2013) Eco-friendly synthesis of silver nanoparticles using a sea weed, *Kappaphycus alvarezii* (Doty) Doty ex PC Silva. Int J Eng Adv Technol 2:559–563

Garg D, Sarkar A, Chand P, Bansal P, Gola D, Sharma S, Khantwal S, Surabhi MR, Chauhan N, Bharti RK (2020) Synthesis of silver nanoparticles utilizing various biological systems: mechanisms and applications-a review. Prog Biomater 9(3):81–95

Garlapati D, Chandrasekaran M, Devanesan A, Mathimani T, Pugazhendhi A (2019) Role of cyanobacteria in agricultural and industrial sectors: an outlook on economically important byproducts. Appl Microbiol Biotechnol 103:4709–4721

Ghodake G, Lee DS (2011) Biological synthesis of gold nanoparticles using the aqueous extract of the brown algae *Laminaria japonica*. J Nanoelectron Optoelectron 6:1–4

González-Ballesteros N, Prado-López S, Rodríguez-González JB, Lastra M, Rodríguez-Argüelles MC (2017) Green synthesis of gold nanoparticles using brown algae *Cystoseira baccata*: its activity in colon cancer cells. Colloids Surf B Biointerfaces 153:190–198

González-Ballesteros N, Rodríguez-Argüelles MC, Lastra-Valdor M (2021) Evaluation of the antioxidant capacities of antarctic macroalgae and their use for nanoparticles production. Molecules 26(4):1182

González-Ballesteros N, Rodríguez-González JB, Lastra-Valdor M, Rodríguez-Argüelles MC (2018) New application of two Antarctic macroalgae *Palmaria Decipiens* and *Desmarestia menziesii* in the synthesis of gold and silver nanoparticles. Pol Sci 15:49–54

Gopu M, Kumar P, Selvankumar T, Senthilkumar B, Sudhakar C, Govarthanan M, Selva Kumar R, Selvam K (2021) Green biomimetic silver nanoparticles utilizing the red algae *Amphiroa rigida* and its potent antibacterial, cytotoxicity and larvicidal efficiency. Bioprocess Biosyst Eng 44(2):217–223

Govindaraju K, Kiruthiga V, Kumar VG, Singaravelu G (2009) Extracellular synthesis of silver nanoparticles by a marine alga, *Sargassum wightii* Grevilli and their antibacterial effects. J Nanosci Nanotechnol 9(9):5497–5501

Govindraju K, Basha K, Ganesh S, Kumar V, Singaravelu G (2008) Silver, gold and bimetallic nanoparticles production using single cell protein (*Spirulina platensis*) Geitler. J Mater Sci 43: 5115–5122

Gu H, Chen X, Chen F, Zhou X, Parsaee Z (2018) Ultrasound-assisted biosynthesis of CuO-NPs using brown alga *Cystoseira trinodis*: characterization, photocatalytic AOP, DPPH scavenging and antibacterial investigations. Ultrason Sonochem 41:109–119

Gutu T, Debra KG, Jeffryes C, Gregory LR, Jiao J (2009) Electron microscopy and optical characterization of cadmium sulphide nanocrystals deposited on the patterned surface of diatom biosilica. J Nanomater 2009:860536

Hamida RS, Abdelmeguid NE, Ali MA, Bin-Meferij MM, Khalil MI (2020) Synthesis of silver nanoparticles using a novel cyanobacteria *Desertifilum* sp. extract: their antibacterial and cytotoxicity effects. Int J Nanomedicine 15:49–63

Hamouda RA, Abd El-Mongy M, Eid KF (2019a) Comparative study between two red algae for biosynthesis silver nanoparticles capping by SDS: insights of characterization and antibacterial activity. Microb Pathog 129:224–232

Hamouda RA, Hussein MH, Abo-Elmagd RA, Bawazir SS (2019b) Synthesis and biological characterization of silver nanoparticles derived from the cyanobacterium *Oscillatoria limnetica*. Sci Rep 9(1):13071

Harman C, Bekkby T, Calabrese S, Trannum H, Oug E, Hagen AG, Green N, Kaste O, Frigstad H (2019) The environmental status of Norwegian coastal waters. In: Sheppard C (ed) World seas: an environmental evaluation. Academic Press, Amsterdam, pp 69–84

He J, Jiang L, Chen Y, Luo Z, Yan Z, Wang J (2019) Facile direct synthesis of graphene-wrapped ZnO nanospheres from cyanobacterial cells. Chem Commun (Camb) 55(76):11410–11413

Husain S, Afreen S, Hemlata D, Yasin B, Fatma AT (2019) Cyanobacteria as a bioreactor for synthesis of silver nanoparticles-an effect of different reaction conditions on the size of nanoparticles and their dye decolorization ability. J Microbiol Methods 162:77–82

Husain S, Sardar M, Fatma T (2015) Screening of cyanobacterial extracts for synthesis of silver nanoparticles. World J Microbiol Biotechnol 31:1279–1283

Ishwarya R, Vaseeharan B, Subbaiah S, Nazar AK, Govindarajan M, Alharbi NS, Kadaikunnan S, Khaled JM, Al-Anbr MN (2018) Sargassum wightii-synthesized ZnO nanoparticles – from antibacterial and insecticidal activity to immunostimulatory effects on the green tiger shrimp *Penaeus semisulcatus*. J Photochem Photobiol B Biol 183:318–330

Ismail GA, Allam NG, El-Gemizy WM, Salem MA (2020) The role of silver nanoparticles biosynthesized by *Anabaena variabilis* and *Spirulina platensis* cyanobacteria for malachite green removal from wastewater. Environ Technol 23:1–15

Jacob JM, Ravindran R, Narayanan M, Samuel SM, Pugazhendhi A, Kumar G (2021) Microalgae: a prospective low cost green alternative for nanoparticle synthesis. Curr Opin Environ Sci Health 20:100163

Jeffryes C, Gutu T, Jiao J, Rorrer GL (2008) Metabolic insertion of nanostructured TiO2 into the patterned biosilica of the diatom *Pinnularia* sp. by a two-stage bioreactor cultivation process. ACS Nano 2(10):2103–2112

Jena J, Pradhan N, Dash BP, Panda PK, Mishra BK (2015) Pigment mediated biogenic synthesis of silver nanoparticles using diatom *amphora* sp. and its antimicrobial activity. J Saudi Chem Soc 19:661–666

Jena J, Pradhan N, Dash BP, Sukla LB, Panda PK (2013) Biosynthesis and characterization of silver nanoparticles using microalga *Chlorococcum humicola* and its antibacterial activity. Int J Nanomater Biostruct 3:1–8

Jena J, Pradhan N, Nayak RR, Dash BP, Sukla LB, Panda PK, Mishra BK (2014) Microalga *Scenedesmus* sp.: a potential low-cost green machine for silver nanoparticle synthesis. J Microbiol Biotechnol 24:522–533

Jun ES, Kim YJ, Kim HH, Park SY (2020) Gold nanoparticles using *Ecklonia stolonifera* protect human dermal fibroblasts from UVA-induced senescence through inhibiting MMP-1 and MMP-3. Mar Drugs 18(9):433

Kalabegishvili T, Kirkesali EI, Rcheulishvili AN, Ginturi E, Murusidze I, Kuchava N, Bagdavadze N, Tsertsvadze G, Gabunia V, Frontasyeva M (2013) Synthesis of gold nanoparticles by blue-green algae *Spirulina platensis*. J Appl Microbiol Biotechnol 4:31

Kannan R, Stirk W, Van Staden J (2013b) Synthesis of silver nanoparticles using the seaweed *Codium capitatum* PC Silva (Chlorophyceae). S Afr J Bot 86:1–4

Kannan RRR, Arumugam R, Ramya D, Manivannan K, Anantharaman P (2013a) Green synthesis of silver nanoparticles using marine macroalga *Chaetomorpha linum*. Appl Nanosci 3:229–233

Kathiraven T, Sundaramanickam A, Shanmugam N, Balasubramanian T (2015) Green synthesis of silver nanoparticles using marine algae *Caulerpa racemosa* and their antibacterial activity against some human pathogens. Appl Nanosci 5:499–504

Khanna P, Kaur A, Goyal D (2019) Algae-based metallic nanoparticles: synthesis, characterization and applications. J Microbiol Methods 163:105656

Khosravi-Darani K, Gomes da Cruz A, Shamloo E, Abdimoghaddam Z, Mozafari MR (2019) Green synthesis of metallic nanoparticles using algae and microalgae. Lett Appl NanoBioSci 8: 666–670

Kumar B, Smita K, Sanchez E, Guerra S, Cumbal L (2016) Ecofriendly ultrasound-assisted rapid synthesis of gold nanoparticles using *Calothrix* algae. Adv Nat Sci Nanosci Nanotechnol 7:2

Kumar P, Senthamil Selvi S, Lakshmi Praba A, Prem Kumar K, Ganeshkumar RS, Govindaraju M (2012b) Synthesis of silver nanoparticles from *Sargassum tenerrimum* and screening phytochemicals for its anti-bacterial activity. Nano Biomed Eng 4:12–16

Kumar P, Senthamilselvi S, Lakshmipraba A, Premkumar K, Muthukumaran R, Visvanathan P, Ganeshkumar R, Govindaraju M (2012) Efficacy of bio-synthesized silver nanoparticles using *Acanthophora spicifera* to encumber biofilm formation. Dig J Nanomater Biostruct 7:511–522

Kumaresan M, Vijai Anand K, Govindaraju K, Tamilselvan S, Ganesh KV (2018) Seaweed *Sargassum wightii* mediated preparation of zirconia (ZrO_2) nanoparticles and their antibacterial activity against gram positive and gram negative bacteria. Microb Pathog 124:311–315

Lavakumar V, Masilamani K, Ravichandiran V, Venkateshan N, Saigopal DV, Ashok Kumar CK, Sowmya C (2015) Promising upshot of silver nanoparticles primed from *Gracilaria crassa* against bacterial pathogens. Chem Cent J 9:42

Lee RE (2008) Phycology, 4th edn. Cambridge University Press, London

Lengke MF, Fleet ME, Southam G (2006a) Morphology of gold nanoparticles synthesized by filamentous cyanobacteria from gold(I)-thiosulfate and gold(III)-chloride complexes. Langmuir 22:2780–2787

Lengke MF, Fleet ME, Southam G (2006b) Bioaccumulation of gold by filamentous cyanobacteria between 25 and 200°C. Geomicrobiol J 23:591–597

Lengke MF, Fleet ME, Southam G (2007) Synthesis of palladium nanoparticles by reaction of filamentous cyanobacterial biomass with a palladium(II) chloride complex. Langmuir 23(17):8982–8987

Madhiyazhagan P, Murugan K, Naresh Kumar A, Nataraj T, Dinesh D, Panneerselvam C, Subramaniam J, Mahesh Kumar P, Suresh U, Roni M, Nicoletti M, Alarfaj AA, Higuchi A, Munusamy MA, Benelli G (2015) Sargassum muticum-synthetized silver nanoparticles: an effective control tool against mosquito vectors and bacterial pathogens. Parasitol Res 114(11):4305–4317

Mahdavi M, Namvar F, Ahmad MB, Mohamad R (2013) Green biosynthesis and characterization of magnetic iron oxide (Fe_3O_4) nanoparticles using seaweed (*Sargassum muticum*) aqueous extract. Molecules 18:5954–5964

Manivasagan P, Oh J (2016) Marine polysaccharide-based nanomaterials as a novel source of nanobiotechnological applications. Int J Biol Macromol 82:315–327

Mata YN, Torres E, Blazquez ML, Ballester A, Gonzalez F, Munoz JA (2009) Gold(III) biosorption and bioreduction with the brown alga *Fucus vesiculosus*. J Hazard Mater 166:612–618

Mishra B, Saxena A, Tiwari A (2020) Biosynthesis of silver nanoparticles from marine diatoms *Chaetoceros* sp., *Skeletonema* sp., *Thalassiosira* sp., and their antibacterial study. Biotechnol Rep (Amst) 28:e00571

Mmola M, Roes-Hill ML, Durrell K, Bolton JJ, Sibuyi N, Meyer ME, Beukes DR, Antunes E (2016) Enhanced antimicrobial and anticancer activity of silver and gold nanoparticles synthesized using *Sargassum incisifolium* aqueous extracts. Molecules 21:1633

Mohamed Asik R, Gowdhami B, Mohamed Jaabir MS, Archunan G, Suganthy N (2019) Anticancer potential of zinc oxide nanoparticles against cervical carcinoma cells synthesized via biogenic route using aqueous extract of *Gracilaria edulis*. Korean J Couns Psychother 103: 109840

Mohseniazar M, Barin M, Zarredar H, Alizadeh S, Shanehbandi D (2011) Potential of microalgae and lactobacilli in biosynthesis of silver nanoparticles. Bioimpacts 1:149

Momeni S, Nabipour I (2015) A simple green synthesis of palladium nanoparticles with *Sargassum* alga and their electrocatalytic activities towards hydrogen peroxide. Appl Biochem Biotechnol 176(7):1937–1949

Moreira C, Vasconcelos V, Antunes A (2013) Phylogeny and biogeography of cyanobacteria and their produced toxins. Mar Drugs 11(11):4350–4369

Mubarak DA, Sasikala M, Gunasekaran M, Thajuddin N (2011) Biosynthesis and characterization of silver nanoparticles using marine cyanobacterium, *Oscillatoria willei* NTDM01. Dig J Nanomater Biostruct 6:385–390

MubarakAli D, Arunkumarb J, Nagc KH, SheikSyedIshackd KA, Baldeva E, Pandiaraje D, Thajuddina N (2013) Gold nanoparticles from pro and eukaryotic photosynthetic microorganisms- comparative studies on synthesis and its application on biolabelling. Colloids Surf B Biointerfaces 103:166–173

MubarakAli D, Gopinath V, Rameshbabu N, Thajuddin N (2012) Synthesis and characterization of CdS nanoparticles using C-phycoerythrin from the marine cyanobacteria. Mater Lett 74:8–11

Murphin Kumar PS, MubarakAli D, Saratale RG, Saratale GD, Pugazhendhi A, Gopalakrishnan K, Thajuddin N (2017) Synthesis of nano-cuboidal gold particles for effective antimicrobial property against clinical human pathogens. Microb Pathog 113:68–73

Murugan K, Aruna P, Panneerselvam C, Madhiyazhagan P, Paulpandi M, Subramaniam J, Rajaganesh R, Wei H, Alsalhi MS, Devanesan S, Nicoletti M, Syuhei B, Canale A, Benelli G (2016) Fighting arboviral diseases: low toxicity on mammalian cells, dengue growth inhibition (in vitro), and mosquitocidal activity of *Centroceras clavulatum*-synthesized silver nanoparticles. Parasitol Res 115(2):651–662

Nagarajan S, Kuppusamy KA (2013) Extracellular synthesis of zinc oxide nanoparticle using seaweeds of gulf of Mannar, India. J Nanobiotechnol 11:39

Namvar F, Azizi S, Ahmad M, Shameli K, Mohamad R, Mahdavi M, Tahir P (2015) Green synthesis and characterization of gold nanoparticles using the marine macroalgae *Sargassum muticum*. Res Chem Intermed 41:5723–5730

Naveen BE, Prakash S (2013) Biological synthesis of gold nanoparticles using algae *Gracilaria corticata* and its application as a potent antimicrobial and antioxidant agent. Asian J Pharm Clin Res 6(2):179–182

Oza G, Pandey S, Mewada A, Kalita G, Sharon M, Phata J, Ambernath W, Sharon M (2012) Facile biosynthesis of gold nanoparticles exploiting optimum pH and temperature of fresh water algae chlorella pyrenoidusa. Adv Appl Sci Res 3:1405–1412

Parial D, Gopal PK, Paul S, Pal R (2016) Gold (III) bioreduction by cyanobacteria with special reference to in vitro biosafety assay of gold nanoparticles. J Appl Phycol 2016(28):3395–3406

Parial D, Pal R (2015) Biosynthesis of monodisperse gold nanoparticles by green alga *Rhizoclonium* and associated biochemical changes. J Appl Phycol 27(2):975–984

Parial P, Patra HK, Dasgupta AKP, Pal P (2012) Screening of different algae for synthesis of gold nanoparticles. Eur J Phycol 47:22–29

Parker SP (1982) Synapsis and classification of living organisms. McGraw Hill, New York

Pascher A (1914) Uber flagellaten and algen. Ber Deutsch Bot Ges 32:136–160

Pascher A (1931) Systematische Ubersicht uber die mit Flagellaten in Zusamennhang stehenden Algentriehen und versuch einer Einreihung dieser Algenstame in die stamme des Pflanzen – reiches. Beibefte Botanischen Centralblatt 48:317–332

Patel V, Berthold D, Puranik P, Gantar M (2014) Screening of cyanobacteria and microalgae for their ability to synthesize silver nanoparticles with antibacterial activity. Biotechnol Rep (Amst) 5:112–119

Prasad TNVKV, Kambala VSR, Naidu R (2013) Phyconanotechnology: synthesis of silver nanoparticles using brown marine algae *Cystophora moniliformis* and their characterisation. J Appl Phycol 25:177–182

Priyadharshini RI, Prasannaraj G, Geetha N, Venkatachalam P (2014) Microwave-mediated extracellular synthesis of metallic silver and zinc oxide nanoparticles using macro-algae (*Gracilaria edulis*) extracts and its anticancer activity against human PC3 cell lines. Appl Biochem Biotechnol 174(8):2777–2790

Pugazhendhi A, Prabakar D, Jacob JM, Karuppusamy I, Saratale RG (2018) Synthesis and characterization of silver nanoparticles using *Gelidium amansii* and its antimicrobial property against various pathogenic bacteria. Microb Pathog 114:41–45

Pugazhendhi A, Prabhu R, Muruganantham K, Shanmuganathan R, Natarajan S (2019) Anticancer, antimicrobial and photocatalytic activities of green synthesized magnesium oxide nanoparticles (MgONPs) using aqueous extract of *Sargassum wightii*. J Photochem Photobiol B 190:86–97

Pytlik N, Butscher D, Machill S, Brunner E (2018) Diatoms—a "green" way to biosynthesize gold-silica nanocomposites? Zeitschrift für Physikalische Chemie 232(9–11):1353–1368

Pytlik N, Kaden J, Finger M, Naumann J, Wanke S, Machill S, Brunner E (2017) Biological synthesis of gold nanoparticles by the diatom *Stephanopyxis turris* and in vivo SERS analyses. Algal Res 28:9–15

Qin T, Gutu T, Jiao J, Chang CH, Rorrer GL (2008) Biological fabrication of photoluminescent nanocomb structures by metabolic incorporation of germanium into the biosilica of the diatom *Nitzschia frustulum*. ACS Nano 2(6):1296–1304

Raajshree RK, Brindha D (2018) In vivo anticancer activity of biosynthesized zinc oxide nanoparticle using *Turbinaria conoides* on a Dalton's lymphoma ascites mice model. J Environ Pathol Toxicol Oncol 37(2):103–115

Ragni R, Cicco SR, Vona D, Farinola GM (2017) Multiple routes to smart nanostructured materials from diatom microalgae: a chemical perspective. Adv Mater 30(19):1704289

Rahman A, Ismail A, Jumbianti D, Magdalena S, Sudrajat H (2009) Synthesis of copper oxide nano particles by using *Phormidium cyanobacterium*. Indian J Chem 9(3):355–360

Rahman A, Kumar S, Nawaz T (2020) Biosynthesis of nanomaterials using algae. In: Microalgae cultivation for biofuels production. Academic Press, San Diego, pp 265–279

Rajasulochana P, Dhamotharan R, Murugakoothan P, Subbiah M, Krishnamoorthy P (2011) Biosynthesis and characterization of gold nanoparticles using the alga *Kappaphycus alvarezii*. Int J Nanosci 9:511–516

Rajasulochana P, Krishnamoorthy P, Dhamotharan R (2012) Potential application of *Kappaphycus alvarezii* in agricultural and pharmaceutical industry. J Chem Pharm Res 4:33–37

Rajathi FAA, Parthiban C, Kumar VG, Anantharaman P (2012) Biosynthesis of antibacterial gold nanoparticles using brown alga, *Stoechospermum marginatum* (Ku ¨tzing). Spectrochim Acta Part A 99:166–173

Rajeshkumar S (2017) Phytochemical constituents of fucoidan (*Padina tetrastromatica*) and its assisted AgNPs for enhanced antibacterial activity. IET Nanobiotechnol 11(3):292–299

Rajeshkumar S, Kumar SV, Malaakodi C, Vanaja M, Paulkumar K, Annadurai G (2017) Optimized synthesis of gold nanoparticles using green chemical process and its in vitro anticancer activity against HepG2 and A549 cell lines, mechanics, materials science & engineering

Rajeshkumar S, Malarkodi C, Gnanajobitha G, Paulkumar K, Vanaja M, Kannan C, Annadurai G (2013) Seaweed-mediated synthesis of gold nanoparticles using *Turbinaria conoides* and its characterization. J Nanostruct Chem 3(1):44

Rajeshkumar S, Malarkodi C, Paulkumar K, Vanaja M, Gnanajobitha G, Annadurai G (2014) Algae mediated green fabrication of silver nanoparticles and examination of its antifungal activity against clinical pathogens. Int J Met 2014:692643

Ramkumar VS, Pugazhendhi A, Prakash S, Ahila NK, Vinoj G, Selvam S, Kumar G, Kannapiran E, Rajendran RB (2017) Synthesis of platinum nanoparticles using seaweed *Padina gymnospora* and their catalytic activity as PVP/PtNPs nanocomposite towards biological applications. Biomed Pharmacother 92:479–490

Roni M, Murugan K, Panneerselvam C, Subramaniam J, Nicoletti M, Madhiyazhagan P, Dinesh D, Suresh U, Khater HF, Wei H, Canale A, Alarfaj AA, Munusamy MA, Higuchi A, Benelli G (2015) Characterization and biotoxicity of *Hypnea musciformis* synthesized silver nanoparticles as potential eco-friendly control tool against *Aedes aegypti* and *Plutella xylostella*. Ecotoxicol Environ Saf 121:31–38

Rorrer GL, Chang CH, Liu SH, Jeffryes C, Jiao J, Hedberg JA (2005) Biosynthesis of silicon-germanium oxide nanocomposites by the marine diatom *Nitzschia frustulum*. J Nanosci Nanotechnol 5(1):41–49

Rösken LM, Cappel F, Körsten S, Fischer CB, Schönleber A, van Smaalen S, Geimer S, Beresko C, Ankerhold G, Wehner S (2016) Time-dependent growth of crystalline Au(0)-nanoparticles in cyanobacteria as self-reproducing bioreactors: 2 Anabaena Cylindrica Beilstein. J Nanotechnol 7:312–327

Round FE (1973) The biology of the algae. Edward Arnold Publishers, London

Roychoudhury P, Pal R (2014) Spirogyra submaxima—a green alga for nanogold production. J Algal Biomass Util 5:15–19

Sahoo CR, Maharana S, Mandhata CP, Bishoyi AK, Paidesetty SK, Padhy RN (2020) Biogenic silver nanoparticle synthesis with cyanobacterium *Chroococcus minutus* isolated from Baliharachandi Sea-mouth, Odisha, and *in vitro* antibacterial activity. Saudi J Biol Sci 27(6):1580–1586

Sajidha PK, Lakshmi D (2016) Biosynthesis of silver nanoparticles using red algae, *Amphiroa fragilissima* and its antibacterial potential against gram positive and gram negative bacteria. Int J Curr Sci 19:93–100

Satish L, Santhakumari S, Gowrishankar S, Pandian SK, Ravi AV, Ramesh M (2017) Rapid biosynthesized AgNPs from *Gelidiella acerosa* aqueous extract mitigates quorum sensing mediated biofilm formation of *vibrio* species—an in vitro and in vivo approach. Environ Sci Pollut Res 24(35):27254–27268

Scarano G, Morelli E (2003) Properties of phytochelatin-coated CdS nanocrystallites formed in a marine phytoplanktonic alga (*Phaeodactylum tricornutum*, Bohlin) in response to cd. Plant Sci 165:803–810

Schröfel A, Kratosova G, Bohunicka M, Dobrocka E, Vavra I (2011) Biosynthesis of gold nanoparticles using diatoms-silica-gold and EPS-gold bionanocomposite formation. J Nanopart Res 13(8):3207–3216

Senapati S, Syed A, Moeez S, Kumar A, Ahmad A (2012) Intracellular synthesis of gold nanoparticles using alga *Tetraselmis kochinensis*. Mater Lett 79:116–118

Shakibaie M, Forootanfar H, Mollazadeh-Moghaddam K, Bagherzadeh Z, Nafissi-Varcheh N, Shahverdi AR, Faramarzi MA (2010) Green synthesis of gold nanoparticles by the marine microalga *Tetraselmis suecica*. Biotechnol Appl Biochem 57:71–75

Sharma B, Purkayastha DD, Hazra S, Gogoi L, Bhattacharjee CR, Ghosh NN, Rout J (2014a) Biosynthesis of gold nanoparticles using a freshwater green alga, *Prasiola crispa*. Mater Lett 116:94–97

Sharma B, Purkayastha DD, Hazra S, Thajamanbi M, Bhattacharjee CR, Ghosh NN, Rout J (2014b) Biosynthesis of fluorescent gold nanoparticles using an edible freshwater red alga, *Lemanea fluviatilis* (L.) C.Ag. and antioxidant activity of biomatrix loaded nanoparticles. Bioprocess Biosyst Eng 37(12):2559–2565

Sheath R.G. 2003. Red algae. In: Wehr G.D. and Sheath R.G. (Eds.) Freshwater algae of North America. Elsevier, Amsterdam. pp. 197–224

Shukla MK, Singh RP, Reddy CRK, Jha B (2012) Synthesis and characterization of agar-based silver nanoparticles and nanocomposite film with antibacterial applications. Bioresour Technol 107:295–300

Sicard C, Brayner R, Margueritat J, Hémadi M, Couté A, Yéprémian C, Djediat C, Aubard J, Fiévet F, Livage J (2010) Nano-gold biosynthesis by silica-encapsulated micro-algae: a "living" bio-hybrid material. J Mater Chem 20:9342–9347

Siddiqi KS, Husen A (2016) Fabrication of metal and metal oxide nanoparticles by algae and their toxic effects. Nanoscale Res Lett 11:363

Singaravelu G, Arockiamary JS, Kumar VG, Govindaraju K (2007) A novel extracellular synthesis of monodisperse gold nanoparticles using marine alga, *Sargassum wightii* Greville. Colloids Surf B Biointerfaces 57:97–101

Singh G, Babele PK, Shahi SK, Sinha RP, Tyagi MB, Kumar A (2014) Green synthesis of silver nanoparticles using cell extracts of *Anabaena doliolum* and screening of its antibacterial and antitumor activity. J Microbiol Biotechnol 24(10):1354–1367

Singh M, Kalaivani R, Manikandan S, Saneetha N, Kumaraguru AK (2013) Facile green synthesis of variable metallic gold nanoparticles using *Padina gymnospora*, a brown marine macroalga. Appl Nanosci 3:145–151

Singh NK, Sonani RR, Rastogi RP, Madamwar D (2015) The phycobilisomes: an early requisite for efficient photosynthesis in cyanobacteria. EXCLI J 14:268–289

Sinha SN, Paul D, Halder N, Sengupta D, Patra SK (2015) Green synthesis of silver nanoparticles using fresh water green alga *Pithophora oedogonia* (Mont.) Wittrock and evaluation of their antibacterial activity. Appl Nano 5:703–709

Soisuwan S, Warisnoicharoen W, Lirdprapamongkol K, Svasti J (2010) Eco-friendly synthesis of fucoidan-stabilized gold nanoparticles. Am J Appl Sci 7(8):1038–1042

Sonbol H, Ameen F, AlYahya S, Almansob A, Alwakeel S (2021) Padina boryana mediated green synthesis of crystalline palladium nanoparticles as potential nanodrug against multidrug resistant bacteria and cancer cells. Sci Rep 11(1):5444

Stalin Dhas T, Ganesh Kumar V, Karthick V, Jini Angel K, Govindaraju K (2014) Facile synthesis of silver chloride nanoparticles using marine alga and its antibacterial efficacy. Spectrochim Acta A Mol Biomol Spectrosc 120:416–420

Stalin Dhas T, Ganesh Kumar V, Stanley Abraham L, Karthick V, Govindaraju K (2012) Sargassum myriocystum mediated biosynthesis of gold nanoparticles. Spectrochim Acta A Mol Biomol Spectrosc 99:97–101

Sudha G, Balasundaram A (2018) Synthesis and characterization of silver nanoparticles using *Padina pavonica* extract and evaluation of their antibacterial activity. J Nanosci Technol 4(4):424–426

Sudha SS, Karthic R, Rengaramanujam J (2013) Microalgae mediated synthesis of silver nanoparticles and their antibacterial activity against pathogenic bacteria. Indian J Exp Biol 52: 393–399

Terracciano M, De Stefano L, Rea I (2018) Diatoms green nanotechnology for biosilica-based drug delivery systems. Pharmaceutics 10(4):242

Tomitani A, Knoll AH, Cavanaugh CM, Ohno T (2006) The evolutionary diversification of cyanobacteria: molecular-phylogenetic and paleontological perspectives. Proc Natl Acad Sci USA 103(14):5442–5447

Ulagesan S, Nam TJ, Choi YH (2021) Biogenic preparation and characterization of *Pyropia yezoensis* silver nanoparticles (P.y AgNPs) and their antibacterial activity against Pseudomonas aeruginosa. Bioprocess Biosyst Eng 44(3):443–452

Venkatesan J, Manivasagan P, Kim SK, Kirthi AV, Marimuthu S, Rahuman AA (2014) Marine algae-mediated synthesis of gold nanoparticles using a novel *Ecklonia cava*. Bioprocess Biosyst Eng 37:1591–1597

Venkatpurwar V, Pokharkar V (2011) Green synthesis of silver nanoparticles using marine polysaccharide: study of in-vitro antibacterial activity. Mater Lett 65:999–1002

Wei D, Qian W (2008) Facile synthesis of ag and au nanoparticles utilizing chitosan as a mediator agent. Colloids Surf B Biointerfaces 62:136–142

Xie J, Lee JY, Wang DI, Ting YP (2007a) Identification of active biomolecules in the high-yield synthesis of single-crystalline gold nanoplates in algal solutions. Small 3:672–682

Xie J, Lee JY, Wang DI, Ting YP (2007b) Silver nanoplates: from biological to biomimetic synthesis. ACS Nano 1:429–439

Xu Y, Hou Y, Wang Y, Wang Y, Li T, Song C, Wei N, Wang Q (2019) Sensitive and selective detection of Cu^{2+} ions based on fluorescent Ag nanoparticles synthesized by R-phycoerythrin from marine algae *Porphyra yezoensis*. Ecotoxicol Environ Saf 168:356–362

Young Park S, Jin Kim Y, Park G, Kim HH (2019) Neuroprotective effect of *Dictyopteris divaricata* extract-capped gold nanoparticles against oxygen and glucose deprivation/reoxygenation. Colloids Surf B Biointerfaces 179:421–428

Younis NS, Bakir EM, Mohamed ME, El Semary NA (2019) Cyanobacteria as Nanogold factories II: chemical reactivity and anti-myocardial infraction properties of customized gold nanoparticles biosynthesized by *Cyanothece* sp. Mar Drugs 17(7):402

Yousefzadi M, Rahimi Z, Ghafori V (2014) The green synthesis, characterization and antimicrobial activities of silver nanoparticles synthesized from green alga *Enteromorpha flexuosa* (wulfen). J Agardh Mater Lett 137:1–4

Yugay YA, Usoltseva RV, Silant'ev VE, Egorova AE, Karabtsov AA, Kumeiko VV, Ermakova SP, Bulgakov VP, Shkryl YN (2020) Synthesis of bioactive silver nanoparticles using alginate, fucoidan and laminaran from brown algae as a reducing and stabilizing agent. Carbohydr Polym 245:116547

Zada S, Ahmad A, Khan S et al (2018) Biofabrication of gold nanoparticles by *Lyptolyngbya* JSC-1 extract as super reducing and stabilizing agents: synthesis, characterization and antibacterial activity. Microb Pathog 114:116–123

Zinicovscaia I, Chiriac T, Cepoi L et al (2017) Selenium uptake and assessment of the biochemical changes in *Arthrospira* (*spirulina*) *platensis* biomass during the synthesis of selenium nanoparticles. Can J Microbiol 63(1):27–34

Biological Synthesis of Nanoparticles: Bryophytes

8

Abstract

Bryophytes are leafy nonvascular embryophytes that compulsorily require water to complete their life cycle. This basal group among the terrestrial plants comprises of liverworts, hornworts, and mosses. Bryophytes rose before the divergence of extant tracheophytes and have a life cycle dominated by the gametophyte which is permanently associated with the sporophyte. Interest in bryophytes for biofabrication of nanoparticles has arisen since the last decade or so, but such studies are quite rare. A large number of genera have been untouched and provide scope for novel studies.

8.1 Introduction

Bryophytes constitute the second largest taxonomic group in the plant kingdom and incidentally the second most diverse land plant group after angiosperms. This basal plant group among the terrestrial plants include liverworts (approximately 6000–8000 species), hornworts (approximately 220 species), and mosses (approximately 10,000 species). Bryophytes are termed as "amphibians of the plant kingdom." Of these, the hornworts comprising of division Anthocerotophyta comprises of species having widespread geographical distribution primarily in the tropical regions. Bryophytes consist of seedless plants lacking a well-developed vascular system (xylem and phloem) which thrive in mostly damp habitats around the world. These haploid, rootless, early diverging green plants have an efficient internal conducting system and differ from other terrestrial flora by two plesiomorphic traits:

1. A long-lived haploid gametophyte which forms the dominant stage in the life cycle.

S. Srivastava, A. Bhargava, *Green Nanoparticles: The Future of Nanobiotechnology*,
https://doi.org/10.1007/978-981-16-7106-7_8

2. A short-lived, unbranched sporophyte that remains physically dependent on the parent plant.

Bryophytes rose before the divergence of extant tracheophytes and display life cycles dominated by a green gametophyte which is permanently associated with the sporophyte.

8.2 Evolutionary Trends in Bryophytes

Bryophytes are the oldest living remnants of the first true plants that colonized the Earth. These have one set of chromosomes and generally prefer moist shady habitats. The term "bryophyta" was introduced by Braun in 1864. Members of this group have unique features and adaptations in the gametophyte as well as the sporophyte which facilitated their successful colonization on land. Although bryophytes are considered to have been existed on Earth since antiquity, the paleontological evidence to support this statement is lacking. This is because the tissues of members of this group are very delicate and therefore have less chance of becoming fossilized. Apart from this, massive early extinctions, late radiations, limited morphological variation, and inconsistent research findings have maintained phylogenetic ambiguity within the group. Spores of first fossil land plants supposedly a liverwort have been traced to the middle Ordovician Period of about 475 million years ago. Anacrogynous (female sex organs arising from any cell below the apex of the stem) Jungermanniales are known to have existed in the middle Carboniferous Period that spanned from 359.2 to 299 million years ago. Mosses like prototypes

Fig. 8.1 Image of peat moss *Sphagnum flexuosum* (Source: Wikipedia)

existed in the upper Carboniferous, and traces of *Sphagnum* (peat moss) (Fig. 8.1) leaves have been found in the Cretaceous.

Cytological studies like determination of the chromosome number and karyotypic studies have also been used to infer phylogenetic relationship among the bryophytes. The presence of polyploidy in extant bryophytes, especially in mosses, has been of significant utility in deciphering the evolutionary mechanism of reproductive behavior in the group. Newton (1986) was of the view that the comparatively higher amount of heterochromatin sensu lato Y chromosome in bryophytes than the X chromosome indicated that monoecism has preceded dioecism in mosses as well as in liverworts.

However, the improvement and rapid strides in molecular techniques is expected to throw new light on the evolution and phylogenetic association within bryophytes in the coming years. A phylogenetic study by Nishiyama et al. (2004) using 51 genes from the chloroplast genome sequences of 20 species of plant kingdom indicated that extant bryophytes are monophyletic in nature and possibly sister to the living vascular plants. A recent study by Li et al. (2020) involving 742 mostly single-copy genes from 21 algae and plant genomes has shown that hornworts were a sister clade to liverworts and mosses (Fig. 8.2).

8.3 Bryophytes in Nanoparticle Production

Interest in bryophytes for biofabrication of nanoparticles has arisen since the last decade, or so but such studies are quite rare. Most of the available scientific literature at hand points toward availability of sporadic studies in this respect (Table 8.1).

The initial reports of the use of bryophytes in the nanofabrication came forth in 2011 when an Indian research group synthesized AgNPs for the first time using bryophytes. Kulkarni et al. (2011) synthesized AgNPs from the hornwort *Anthoceros*. The extract of thalli collected from Mahabaleshwar (Maharashtra, India) was tested against a range of concentration of silver nitrate solutions ranging from 0.5 to 5 mM. The characterization of silver nanoparticles was carried out using UV-Vis spectra analysis, SEM, and EDS. SEM images showed the presence of cuboidal and triangular nanoparticles having the size range of 20–50 nm. The AgNPs exhibited intense antibacterial activity against four bacteria, namely, *E. coli*, *K. pneumoniae*, *P. aeruginosa*, and *B. subtilis*. Using well diffusion assay, the width of zone of inhibition was measured as 11, 8, 12, and 9 mm for *E. coli*, *K. pneumoniae*, *P. aeruginosa*, and *B. subtilis*, respectively.

Srivastava et al. (2011) successfully carried out biosynthesis of AgNPs using the moss *Fissidens minutus* (family Fissidentaceae). Varying concentration of silver nitrate (0.5 mM, 1 mM, 2 mM, 3 mM, and 5 mM) was used for the experiment. The color change from light green to dark yellowish brown indicated the formation of nanoparticles. The optimum concentration of $AgNO_3$ for efficient nanoparticle synthesis was ~0.5 mM. After the confirmation of formation of the biologically synthesized nanoparticles, their antibacterial activity was tested against four bacterial species, viz., *E. coli*, *K. pneumoniae*, *P. aeruginosa*, and *B. subtilis*. Thus,

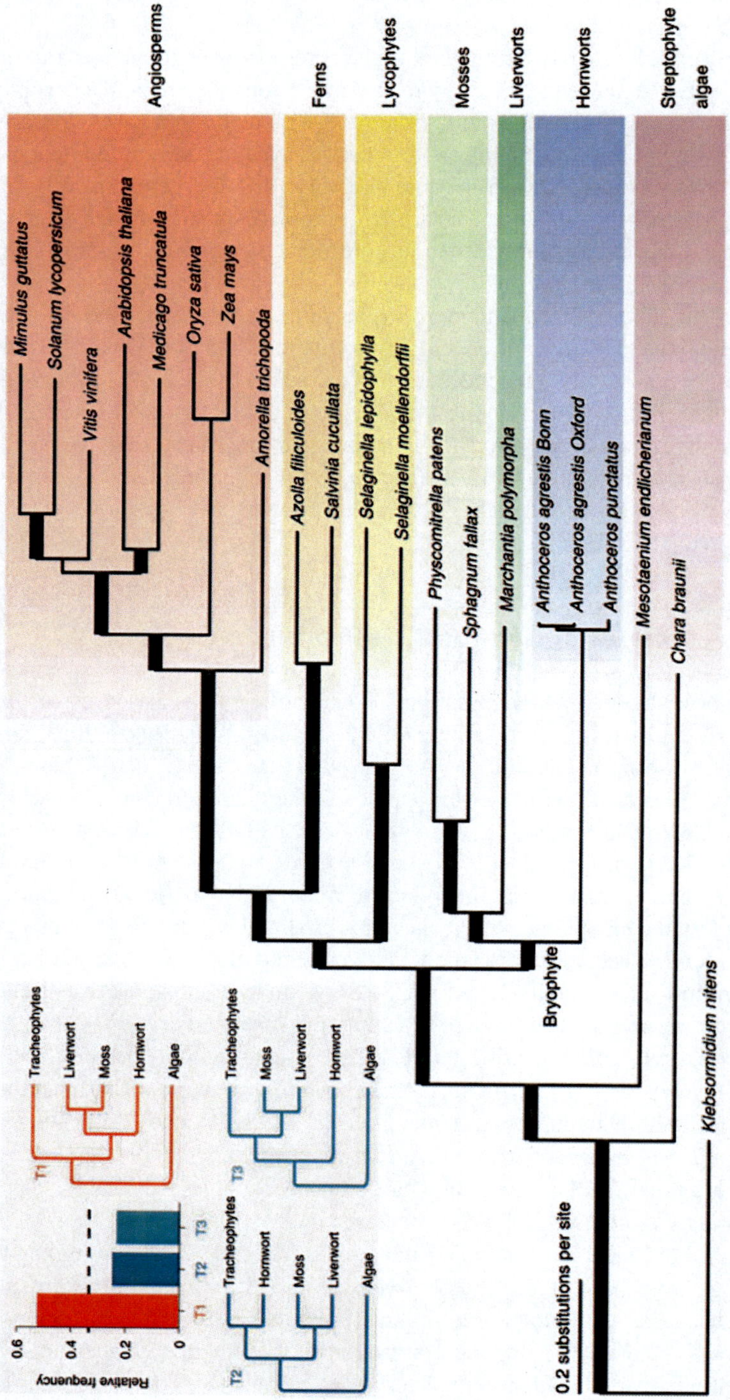

Fig. 8.2 Land plant phylogeny inferred from 742 mostly single-copy genes (Reprinted from Li et al. 2020)

Table 8.1 List of nanoparticles synthesized by bryophytes

Bryophyte	Type of nanoparticle	Reaction conditions	Size (nm)	References
Anthoceros	Silver	Room temperature	20–50	Kulkarni et al. (2011)
Bryum medianum	Silver	Light	85	Sathish et al. (2016)
Bryum salakense	Silver	Sunlight	35–40	Vimala and Sathish (2017)
Campylopus flexuosus	Silver	Room temperature	58	Vimala et al. (2017)
Fissidens minutus	Silver	Room temperature	Nr	Srivastava et al. (2011)
Marchantia	Silver	Room temperature	0.020–0.040	Fibriana et al. (2020)
Riccia sp.	Silver	Room temperature	20–50	Kulkarni et al. (2012)
Taxithelium nepalense	Gold	Room temperature	42–145	Acharya and Sarkar (2014)

nr not reported.

F. minutus swiftly reduced metal ions into nanoparticles that demonstrated strong inhibitory and bactericidal effects.

The gametophytic extract of the bryophyte *Taxithelium nepalense* (Schwagr.) (Family Sematophyllaceae) Broth. was utilized by Acharya and Sarkar (2014) for the production of gold nanoparticles using tetrachloroauric acid ($HAuCl_4.6H_2O$). The aqueous extract of the gametophyte was incubated with 3 mM of $HAuCl_4$ solution. The light yellow reaction mixture initially turned pinkish within an hour of incubation and later turned dark red pointing toward the formation of gold nanoparticles. TEM analysis revealed the presence of spherical, triangular, and hexagonal nanoparticles of 42–145 nm size and having mean diameter of around 70 nm. The EDX analysis exhibited strong signal from gold atoms and sharp optical absorption peak (2–3 keV) amply pointed toward the presence of gold nanostructures.

Vimala et al. (2017) synthesized and characterized AgNPs from *Campylopus flexuosus*, an acrocorpous moss, belonging to the family Dicranaceae (Fig. 8.3). An aqueous extract of the moss collected from Namakkal District of Tamil Nadu, India, was treated with 1 mM $AgNO_3$ solution. The initial color change pointed toward the possibility of nanoparticle synthesis. The characterization based on UV-Vis, FTIR, FESEM, zeta potential, particle size, and XRD analysis confirmed the formation of nanoparticles having mean size of 58 nm.

Fibriana et al. (2020) successfully carried out biofabrication of silver nanoparticles using of *Marchantia* (family: Marchantiaceae), a thalloid liverwort that produces umbrella-shaped structures. An absorbance peak at 422 nm was evident in the UV-Vis spectra, while electron microscopy depicted the presence of irregular-shaped AgNPs having size ranging between 0.020 and 0.040 nm. Such a small size of silver nanoparticles seems quite unrealistic and needs further detailed studies for confirmation. Isothiocyanate (–N=C=S) and amide (–C(=O)N=) were

Fig. 8.3 Image of the moss *Campylopus flexuosus* (Source: Wikimedia Commons)

the primary chemical/functional groups involved in the stabilization of the AgNPs. The AgNPs were tested for antimicrobial activity using agar well diffusion method and in the form of alcohol-free hand sanitizer gel, both providing encouraging results.

8.4 Future Prospects

Bryophytes seem to be the least explored group with reference to biological synthesis of nanoparticles. Also no studies have been carried out with respect to the mechanistic aspects of nanoparticle production. A large number of genera have been untouched and provide scope for novel studies. Thus, more studies need to be carried for exploring nanoparticles using members of this group and decipher the mechanism behind nanoparticle production.

References

Acharya K, Sarkar J (2014) Bryo-synthesis of gold nanoparticles. Int J Pharm Sci Rev Res 29:82–86

Asakawa Y, Ludwiczuk A (2018) Chemical constituents of bryophytes: structures and biological activity. J Nat Prod 81:641–660

Bennici A (2008) Origin and early evolution of land plants: problems and considerations. Commun Integr Biol 1(2):212–218

Fibriana F, Amalia AV, Muntamah S, Ulva L, Aryanti S (2020) Antimicrobial activities of green synthesized silver nanoparticles from *Marchantia* sp. extract: testing an alcohol-free hand sanitizer product formula. J Microbiol Biotechnol Food Sci 9:1034–1038

Graham LE (1984) Coleochaete and the origin of land plants. Am J Bot 71:603–608

Hao DC, Gu X-J, Xiao PG (2015) Chemotaxonomy: a phylogeny-based approach. In: Hao DC, Gu X-J, Xiao PG (eds) Medicinal plants. Woodhead Publishing, New York, pp 1–48

Karol KG, McCourt RM, Cimino MT, Delwiche CF (2001) The closest living relatives of land plants. Science 294:2351–2353

Kulkarni AP, Srivastava AA, Harpale PM, Zunjarrao RS (2011) Plant mediated synthesis of silver nanoparticles-tapping the unexploited sources. J Natural Prod Plant Resour 4:100–107

Kulkarni AP, Srivastava AA, Nagalgaon RK, Zunjarrao RS (2012) Phytofabrication of silver nanoparticles from a novel plant source and its application. Int J Biol Pharm Res 3:417–421

Li FW, Nishiyama T, Waller M et al (2020) Anthoceros genomes illuminate the origin of land plants and the unique biology of hornworts. Nat Plants 6:259–272

Ligrone R, Ducket JG, Renzaglia KS (2000) Conducting tissues and phyletic relationships of bryophytes. Phil Trans Royal Soc B: Biol Sci 355:795–813

Ligrone R, Duckett JG, Renzaglia KS (2012) Major transitions in the evolution of early land plants: a bryological perspective. Ann Bot 109:851–871

Liu Y, Johnson MG, Cox CJ, Medina R, Devos N, Vanderpoorten A, Hedenäs L, Bell NE, Shevock JR, Aguero B, Quandt D, Wickett NJ, Shaw AJ, Goffinet B (2019) Resolution of the ordinal phylogeny of mosses using targeted exons from organellar and nuclear genomes. Nat Commun 10:1485

Newton ME (1986) Bryophyte phylogeny in terms of chromosome cytology. J Bryol 14(1):43–57

Nishiyama T, Wolf PG, Kugita M, Sinclair RB, Sugita M, Sugiura C, Wakasugi T, Yamada K, Yoshinaga K, Yamaguchi K, Ueda K, Hasebe M (2004) Chloroplast phylogeny indicates that bryophytes are monophyletic. Mol Biol Evol 21(10):1813–1819

Rice SK (2009) Mosses (bryophytes). In: Likens GE (ed) Encyclopedia of inland waters. Academic Press, New York, pp 88–96

Sathish SS, Vimala A, Kanaga A, Murugan M (2016) Antioxidant and antimicrobial studies on biosynthesized silver nanoparticles using *Bryum medianum* mitt. A bryophyte from Kolli Hills, eastern Ghats of Tamil Nadu, India. J Pharm Sci Res 8(8):704–709

Shanker A (2013) Combined data from chloroplast and mitochondrial genome sequences showed paraphyly of bryophytes. Arch Bryology 171:1–9

Shaw AJ, Szövényi P, Shaw B (2011 Mar) Bryophyte diversity and evolution: windows into the early evolution of land plants. Am J Bot 98(3):352–369

Shaw J, Renzaglia K (2004 Oct) Phylogeny and diversification of bryophytes. Am J Bot 91(10):1557–1581

Söderström L, Hagborg A, von Konrat M, Bartholomew-Began S, Bell D, Briscoe L, Brown E, Cargill DC, Costa DP, Crandall-Stotler BJ, Cooper ED, Dauphin G, Engel JJ, Feldberg K, Glenny D, Gradstein SR, He X, Heinrichs J, Hentschel J, Ilkiu-Borges AL, Katagiri T, Konstantinova NA, Larraín J, Long DG, Nebel M, Pócs T, Puche F, Reiner-Drehwald E, Renner MA, Sass-Gyarmati A, Schäfer-Verwimp A, Moragues JG, Stotler RE, Sukkharak P, Thiers BM, Uribe J, Váňa J, Villarreal JC, Wigginton M, Zhang L, Zhu RL (2016) World checklist of hornworts and liverworts. PhytoKeys 59:1–828

Srivastava AA, Kulkarni AP, Harpale PM, Zunjarrao RS (2011) Plant mediated synthesis of silver nanoparticles using a bryophyte: *Fissidens minutus* and its antimicrobial activity. Int J Eng Sci Technol 3:8342–8347

Steel JB, Wilson JB, Anderson BJ, Lodge RHL, Tangney RS (2004) Are bryophyte communities different from higher-plant communities? Abundance relations. Oikos 104:479–486

Sun S-Q, Wu Y-H, Wang G-X, Zhou J, Yu D, Bing H-J, Luo J (2013) Bryophyte species richness and composition along an altitudinal gradient in Gongga mountain, China. Plos One 8:e58131

Villarreal JC, Cargill DC, Hagborg A, Soderstrom L, Renzaglia KS (2014) A synthesis of hornwort diversity: patterns, causes and future work. Phytotaxa 9:150–166

Vimala A, Sathish SS (2017) Green synthesis, characterization and antibacterial activity of silver nanoparticles from *Bryum salakense* card. A moss plant (bryophyte). World J. pharm. Pharmaceutical Sci 6:1880–1892

Vimala A, Sathish SS, Thamizharasi T, Palani R, Vijayakanth P, Kavitha R (2017) Moss (bryophyte) mediated synthesis and characterization of silver nanoparticles from *Campylopus flexuosus* (Hedw.) bird. J Pharm Sci Res 9:292–297

Check for updates

Biological Synthesis of Nanoparticles: Pteridophytes

9

Abstract

Pteridophytes, also known as "vascular cryptogams," were the first vascular plants that evolved a system of vascular elements. Ferns and lycophytes, believed to be among the oldest lineages of terrestrial plants, are thought to have evolved around 400 million years ago, possibly in the Ordovician or Silurian period when a major divergence event led to separation of bryophytes from the ancestral tracheophytes. The use of pteridophytes for biological production is meager and has been restricted to studies in a few genera. Mostly, fabrication of metal nanoparticles has been carried out using plants of this group which have numerous medicinal properties. However, scant studies available point toward great scope in their utilization in future research programs involving green synthesis of a wide range of nanoparticles.

9.1 Introduction

Pteridophytes, also known as "vascular cryptogams," are characterized by a simple reproductive system that lacks flowers and is devoid of seed production. It includes ferns, horsetails, and lycophytes, although extant ferns alone account for about 90% of the distribution across the world. Pteridophytes do not produce seeds or flowers but have leaves (known as fronds), roots, and sometimes true stems and reproduce via spores. Polyploid species are common in this group, and one pteridophyte, viz., the adder's tongue fern *Ophioglossum reticulatum*, has the highest chromosome number (1440) in the plant kingdom. The gametophyte of pteridophytes is independent of the sporophyte, and detailed studies in this group have shed light on "independent gametophytes," i.e., the gametophyte populations which do not depend on their conspecific sporophytes or in some cases where sporophytes are absent. Pteridophytes were the first group to have evolved a system of vascular

S. Srivastava, A. Bhargava, *Green Nanoparticles: The Future of Nanobiotechnology*, https://doi.org/10.1007/978-981-16-7106-7_9

elements, i.e., xylem and phloem, for the transportation of fluids and thus quite tall as compared to their avascular ancestors. Greater plant height conferred on them an evolutionary advantage with respect to spore dispersal paving the way for evolution of new plants. Around 10,000–15,000 species of pteridophytes are known, and an interesting fact pertains to their distribution which increases from the poles toward the equator.

9.2 Phylogenetic Relationships in Pteridophytes

During the last couple of decades, estimation of fossils and cladistic studies have somewhat solved the complex relationships prevalent among the different plant groups. Ferns and lycophytes are important groups that form a part of the oldest lineages of plants on Earth with their traits like sporophyte branching and indeterminate growth contributing to the rise of vasculature in land plants and establishing a firm ground for further plant radiations. The lycophytes once played a pivotal role in maintaining biodiversity, soil production, and CO_2 sequestration on Earth. These along with ferns and spermatophytes are the surviving members of the ancient lineages and considered to be the descendants of the first tracheophytes (Fig. 9.1). The evolution of ferns and lycophytes as distinct lineages is thought to have taken place around 400 million years ago, possibly in the Ordovician or Silurian period when a major divergence event led to separation of bryophytes from the ancestral tracheophytes. The use of the term “fern allies” needs special mention here. This term refers to a group of plants closely related to ferns but clubbed with mosses by Linnaeus due to phenotypic similarity. Horsetails, whisk ferns, and lycophytes are considered as representatives of the “fern allies.” The common names such pteridophytes like spike mosses (*Selaginella*) and club mosses (*Lycopodium*) amply demonstrate this trend. However, currently the term “fern allies” is restricted to two separate groups (Fig. 9.1):

1. Ferns
2. Lycophytes

The current classification of pteridophytes proposed by the Pteridophyte Phylogeny Group (PPG I 2016) classifies lycophytes into three orders composed of a single family. The Selaginellales and the Isoetales are closely related to each other and together form a sister group with the Lycopodiales (Fig. 9.1).

9.3 Pteridophytes in Nanoparticle Production

Similar to bryophytes, pteridophytes have been scarcely used for biosynthesis of nanomaterials (Table 9.1).

The highly paraphyletic genus *Cheilanthes* comprises about 180 species of rock-dwelling ferns commonly known as “lip ferns.” The leaflet extract of *Cheilanthes*

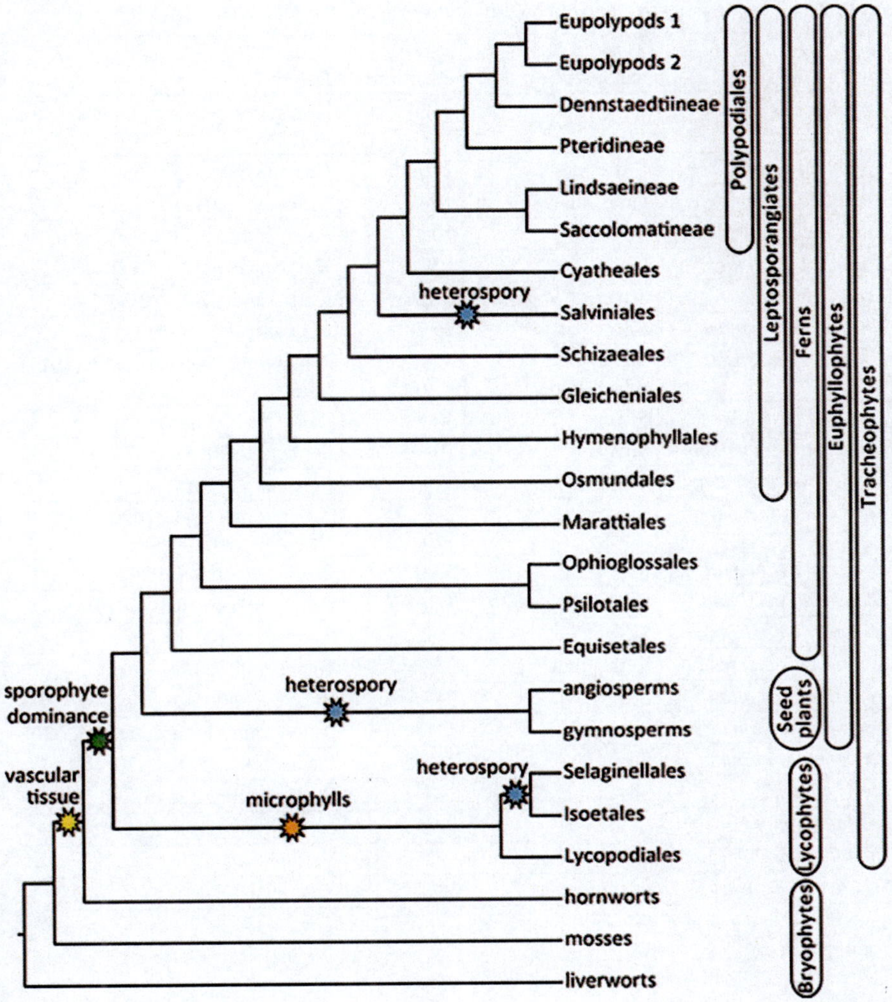

Fig. 9.1 Phylogeny of major groups of land plants (Reprinted with permission from Sessa 2018)

farinose (family: Pteridaceae), a fern of immense medicinal importance, was treated with 1 mM silver nitrate solution by Nalwadea et al. (2013) for the fabrication of silver nanoparticles. The absorption spectra of silver nanoparticles formed had absorption peak at 435 nm. Spherical-shaped polydispersed nanoparticles were observed having size dimensions of 26.58 nm.

Sant et al. (2013) assessed the potential of an aqueous extract of *Adiantum philippense* L. (family: Adiantaceae) fronds for the biosynthesis of AuNPs and AgNPs using tetrachloroauric acid (0.01 M) and silver nitrate (0.1 M). This creeping or semi-erect plant, known by the common name "walking maidenhair fern," is found throughout the southern hemisphere especially in Southeast Asia. Gold nanoparticles obtained were anisotropic and polycrystalline, while silver

Table 9.1 Nanoparticles synthesized by different species of pteridophytes

Plant group	Type of nanoparticle	Size (nm)	References
Actiniopteris radiata	Silver	4–30	Koteswaramma and Varalakshmi (2017)
Adiantum lunulatum	Silver	28	Chatterjee et al. (2019)
Adiantum philippense	Gold, silver	10–18	Sant et al. (2013)
Athyrium filix-femina	Silver	10–94	Soare and Sutan (2018)
Azolla microphylla	Gold	17–40	Kunjiappan et al. (2014)
Azolla pinnata	Silver	6.5	Korbekandi et al. (2014)
Cheilanthes farinosa	Silver	26.58	Nalwadea et al. (2013)
Cyathea nilgirensis	Silver Silver oxide	45–74 <100	Antonysamy Johnson et al. (2017) Pradheesh et al. (2020)
Dicranopteris linearis	Silver	40–60	Rajaganesh et al. (2016)
Diplazium esculentum	Silver	10–45	Paul et al. (2015)
Dryopteris cochleata	Silver	35	Singh et al. (2020)
Equisetum arvense	Silicon Silver	2–3 170.5*	Adinarayana et al. (2020) Das et al. (2020)
Equisetum telmateia	Silica	30	Boromand et al. (2020)
Gleichenia pectinata	Silver	7.51	Femi-Adepoju et al. (2019)
Microsorum pteropus	Silver	nr	Chick et al. (2020)
Nephrolepis biserrata	Ruthenium	25.96	Gupta et al. (2019)
Nephrolepis exaltata	Silver	24.76	Bhor et al. (2014)
Odontosoria chinensis	Silver	22.3–48.2	Antonysamy Johnson et al. (2020)
Pteridium aquilinum	Silver	35–65	Panneerselvam et al. (2016)
Pteris tripartita	Silver	22–41	Baskaran et al. (2016)
Salvinia minima	Lead	17.2	Castro-Longoria et al. (2014)
Salvinia molesta	Silver	1–35	Verma et al. (2016)
Selaginella myosurus	Silver	58.81	Kedi et al. (2018)

[a]d.nm
nr not reported.

nanoparticles were anisotropic and monocrystalline in nature. Particle size of AuNPs and AgNPs using TEM brought out the particle size ranging between 10 and 18 nm (Fig. 9.2). Elemental gold (8.95%) and silver (5.04%) were detected using characterization techniques like EDX and XRD. It was concluded that the presence of several phytochemicals like carbohydrates, phenols, flavonoids, and terpenoids might have played a major role in the bioreduction process.

Castro-Longoria et al. (2014) successfully developed an eco-friendly method for the synthesis of lead nanoparticles from *Salvinia minima* (family: Salviniaceae), a free floating fern. This non-rooted aquatic fern grows wild in South America and Mesoamerica and is referred to as "common *Salvinia*." The leaves and roots of the fern were treated with 80 mM lead nitrate [$Pb(NO_3)_2$] for 12 h. After the incubation time, lyophilized biomass were explored for the presence of lead using different

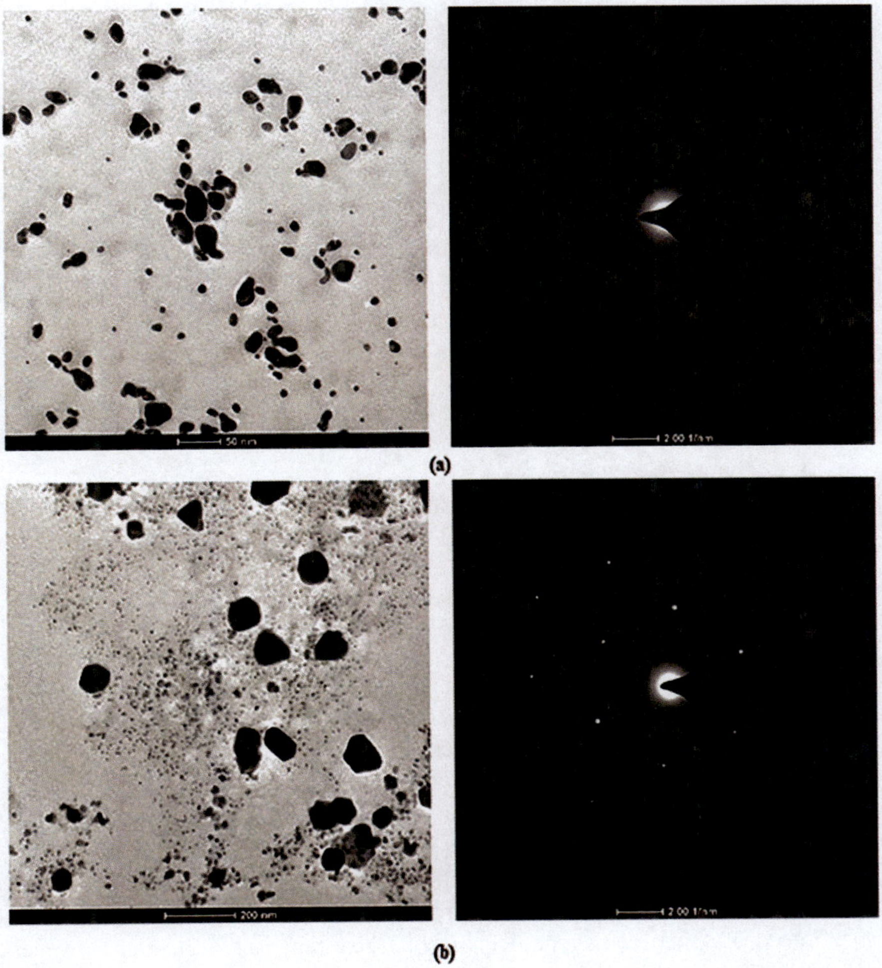

Fig. 9.2 (**a**) TEM image (left) and SAED pattern (right) of gold nanoparticles synthesized using the walking maidenhair fern *Adiantum philippense*. (**b**) TEM image (left) and SAED pattern (right) of silver nanoparticles synthesized using *Adiantum philippense* (Reprinted from Sant et al. 2013)

characterization techniques which confirmed the biosynthesis of lead nanoparticles by the fern. *S. minima* accumulated lead in the form of nanostructures not only in the cell wall of leaflike fronds and rootlike modified fronds but also in the cytoplasm. Nanoparticles found within the cell wall were mostly quasi-spherical having diameter of around 17.2 nm, while the nanostructures near the cell membrane were predominantly elongated and 11.1 nm in width. Cellulose, lignin, and pectin seemed to be the prominent constituents that played a pivotal role in the reduction of lead ions. The leaf extract of another aquatic fern, *S. molesta*, was explored for the green synthesis of AgNPs by Verma et al. (2016). *S. molesta* is a fern native to South America and commonly known as "kariba weed" and "giant *Salvinia*." This free

Fig. 9.3 Image of *Salvinia molesta* (Source: Wikipedia)

floating fern does not attach on the substratum and grows rapidly in still water like lakes, ponds, streams, ditches, marshes, and rivers. The fern forms dense mats over the surface (Fig. 9.3). Freshly harvested leaves of the fern were chopped and boiled in deionized water and the filtrate exposed to 5 mM of $AgNO_3$ solution. A color change pointed toward the formation of nanoparticles and necessitated further studies. The optical properties of silver nanoparticles in the colloidal mixture measured through UV-visible spectroscopy showed a sharp SPR band at 413 nm suggesting toward the formation of silver nanoparticles. HRTEM images established the presence of spherical nanoparticles having size range of 1–35 nm. However, most of the nanoparticles were in the range of 5–15 nm having mean size of 12.46 nm. Atomic force microscopy revealed that both individual and agglomerated AgNPs were responsible for the presence of even surface morphology. XRD analysis pointed toward the crystalline nature of AgNPs. Thus, the study demonstrated that the aqueous leaf extract of *S. molesta* can be efficiently utilized for the production of AgNPs using a single-step process.

Paul et al. (2015) carried out fabrication of AgNPs using *Diplazium esculentum* (family: Athyriaceae), a wild edible rhizomatous fern. The leaf powder of the vegetable fern was treated with 10^{-3} M $AgNO_3$ solution. The XRD pattern of the synthesized nanoparticles exhibited five diffraction peaks at $2\theta = 38.09°$, 44.22°, 64.59°, 77.43°, and 81.41° corresponding to (111), (200), (220), (311), and (222) planes of the face-centered cubic silver. The mean size calculated using the Debye-Scherrer formula was about 9.71 nm, while that obtained using TEM was around 10–45 nm. TEM images also confirmed that the nanostructures were polydispersed and had varied shapes ranging from spherical to oval and triangular (Fig. 9.4).

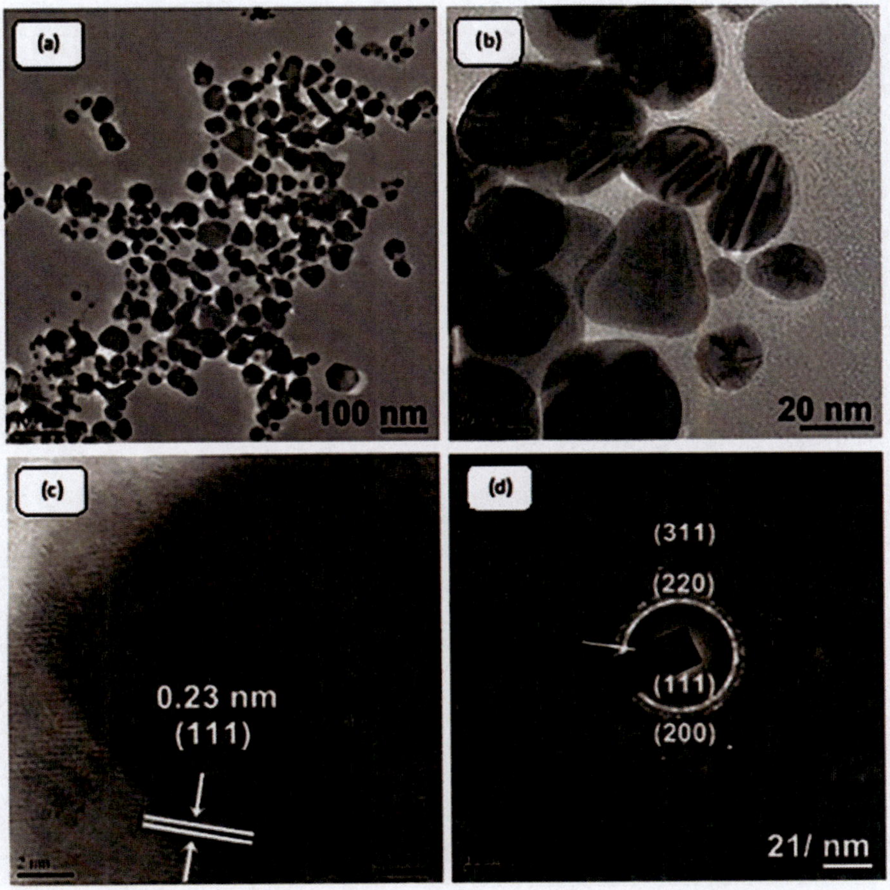

Fig. 9.4 TEM (**a**, **b**), HRTEM image (**c**), and ED pattern (**d**) of silver nanoparticles synthesized using *Diplazium esculentum* (Reprinted with permission from Paul et al. 2015)

Electron diffraction (ED) also confirmed the polycrystalline nature of the biologically synthesized nanoparticles.

Endangered ferns have also been explored for the biosynthesis of nanoparticles. Baskaran et al. (2016) synthesized silver nanoparticles using the leaf extract of *Pteris tripartite* (family: Pteridaceae) and obtained hexagonal-, spherical-, and rod-shaped nanoentities of 22–41 nm dimensions.

Among the pteridophytes, the aquatic fern *Azolla* (family: Salviniaceae) has been at the forefront of nanobiotechnology research for biosynthesis of nanoparticles. Many species of the genus have been used for the biofabrication of gold and silver nanoparticles (Table 9.1). Kunjiappan et al. (2014) produced AuNPs utilizing fine powder prepared from dried plants of *Azolla microphylla*, the fast-growing aquatic fern. The filtered methanolic extract of *A. microphylla* was incubated with 1 mM aqueous solution of hydrogen tetrachloroaurate(III) hydrate ($HAuCl_4.3H_2O$). The

Table 9.2 Antioxidant activity of AuNPs against standard drugs (Reprinted with permission from Kunjiappan et al. 2014)

Name of assay	Free radical scavenging power
$\%DPPH_{sc}$	(85.75 ± 0.439)%
FRAP	(872.03 ± 5.8) μm of $FeSO_4$/mg
$\%H_2O_{2sc}$	(82.13 ± 0.30)%
$\%NO_{sc}$	(75.97 ± 0.31)%

electron microscopy studies revealed the presence of spherical and rectangular AuNPs of 17–40 nm dimensions. An intense signal for elemental gold was displayed during EDX analysis along with an optical peak at 2.15 keV indicating absorption of phytochemicals on the nanostructure wall and resulting stability conferred on them. HRTEM images showed nanoparticles of diverse shapes and having size range of 3–20 nm and mean size of 8.3 nm. The AuNPs also exhibited intense antioxidant activity calculated through DPPH scavenging assay, FRAP activity, and nitric oxide and H_2O_2 scavenging activity (Table 9.2). Jha and Prasad (2016) successfully carried out fabrication of AuNPs using *Azolla* sp. The ethanolic extract of the fern was treated with 0.25 M $HAuCl_4$ solution. Nanoparticle characterization was carried out using X-ray, UV-visible spectroscopy, SEM, and TEM. The spherical nanoparticles having size range of 5–17 nm exhibited surface plasmon resonance at 538 nm in the UV-visible spectra. The authors emphasized on the role of various phytochemicals like phenolics, tannins, anthraquinone glycosides, and carbohydrates in the reduction of the gold salt.

Green synthesis of silver nanostructures was accomplished by Panneerselvam et al. (2016) using *Pteridium aquilinum* L. Kuhn (family: Dennstaedtiaceae), a cosmopolitan species having worldwide distribution and also known as bracken fern. The leaves of this fern, which are a rich source of diverse phytochemicals, are used as vegetable, as analgesic and for preparing alcoholic beverages. The leaf extract of bracken fern was prepared by boiling finely cut leaves in double-distilled water and the filtrate treated with 1 mM $AgNO_3$ solution. Scanning of the absorption maxima at a wavelength of 200–800 nm gave maximum value at 420 nm. SEM analysis demonstrated that the nanoparticles were mostly spherical in shape with 35–65 nm dimensions and were capped with organic material that conferred stability to the nanoparticles.

Rajaganesh et al. (2016) accomplished biofabrication of silver nanoparticles using the “Old World forked fern,” viz., *Dicranopteris linearis* (family: Gleicheniaceae) (Fig. 9.5). An extract was prepared by boiling finely cut leaves of *D. linearis* in double-distilled water, and after filtration, the filtrate was treated with 1 mM $AgNO_3$ at ambient temperature. SEM images confirmed the role of *D. linearis* in the biosynthesis of spherical AgNPs of about 40–60 nm in diameter which aggregated into larger size due to the occurrence of different phytochemicals which played a major role in the process. The EDX spectrum obtained also exhibited intense Ag signal which confirmed the production of AgNPs. The results obtained through SEM were also corroborated by particle size distribution analysis through DLS wherein the particle diameter ranged from about 10 to 110 nm and mean particle size of about 80 nm.

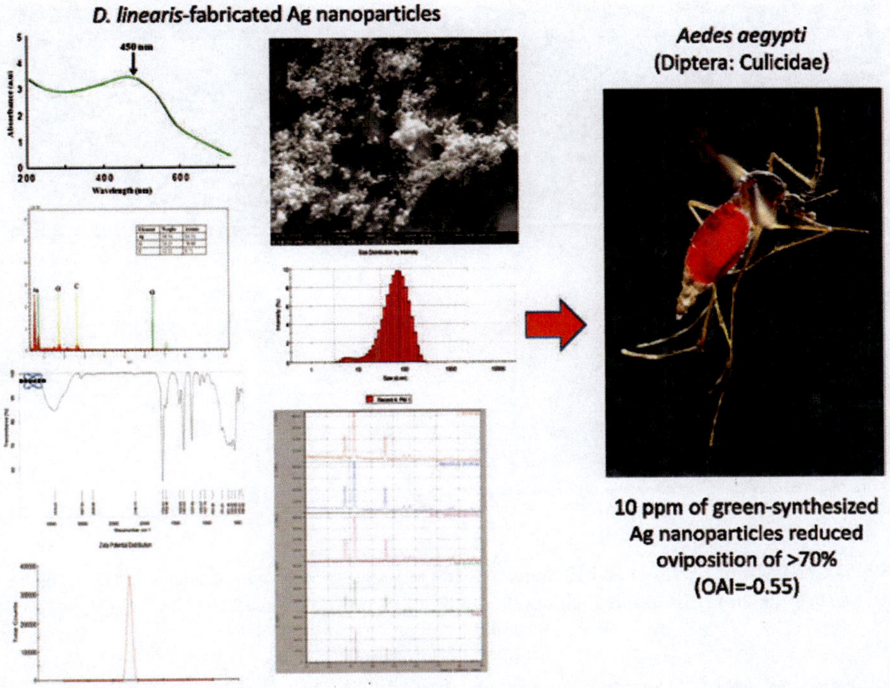

Fig. 9.5 Production of silver nanoparticles using the Old World forked fern *Dicranopteris linearis* (Reprinted with permission from Rajaganesh et al. 2016)

The methanolic extract of *Marsilea quadrifolia* (family: Marsileaceae), the "European waterclover" or the "four-leaf clover," was used for biogenic production of AuNPs by Chowdhury et al. (2017). This pteridophyte has several active constituents that confer antibacterial, cytotoxic, and antioxidant properties besides reducing blood sugar and cholesterol levels. The lyophilized *M. quadrifolia* powder obtained by mixing fern powder with 99% methanol was treated with 0.001 M $HAuCl_4$ solution. The color change from the original light yellow to dark pink indicated the biosynthesis of gold nanoparticles. The nature, stability, and morphological characteristics were analyzed by XRD, thermogravimetric differential thermal analyses (TG-DTA), SEM, TEM, and selected area electron diffraction (SAED). The fern leaf extract served as an efficient reducing and capping agent during the synthesis process and produced spherical gold nanoparticles ranging from 17 to 40 nm (average size: 14 nm) (Fig. 9.6). TEM images showed that the biologically synthesized AuNPs were spherical and triangular in shape with diameter ranging between 3 and 19 nm and a mean of 20 nm.

Fabrication of AgNPs using the extract of *Selaginella myosurus* (family: Selaginellaceae) was successfully accomplished by Kedi et al. (2018). This terrestrial, perennial, pteridophyte commonly known as "tunda" by local communities in Africa is known for its numerous medicinal benefits. The freshly prepared extract of the plant was combined with 0.001 M $AgNO_3$ solution, and the mixture was

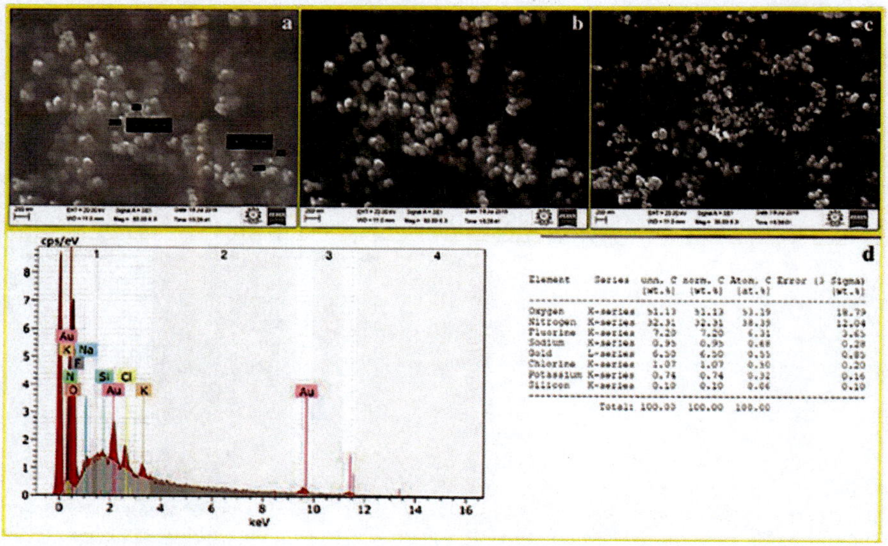

Fig. 9.6 SEM images (**a–c**) and EDX pattern (**d**) of biogenic gold nanoparticles obtained using European waterclover (Reprinted with permission from Chowdhury et al. 2017)

incubated at ambient temperature in an atmosphere devoid of light. The optimum conditions for biosynthesis of AgNPs were assessed using six pH gradients, five extract quantities, and five different concentrations of $AgNO_3$. It was observed that the nanoparticles were most efficiently synthesized at pH 12. The production of nanoparticles increased in proportion to the extract quantity as well as with increasing silver nitrate concentration. The average crystalline particle size using XRD was calculated as 33.7 and 44.2 nm for silver and silver chloride, respectively. HRTEM images showed the predominant presence of spherical nanoparticles having a mean size of 58.81 nm, a fact also confirmed by SAED analysis.

Femi-Adepoju et al. (2019) utilized the neglected, terrestrial fern *Gleichenia pectinata* (family: Gleicheniaceae) for the production of AgNPs using 0.005 M $AgNO_3$ solution. A single, intense SPR band at 460 nm pointed toward the polydisperse nature of the nanostructures. The intensity of the absorbance rose with concomitant increase in the reaction time, the maximum value being observed at 105 minutes indicating that it took less than 2 hours for completion of the bioreduction process. The SEM analysis revealed the presence spherical-shaped nanoparticles as aggregates, while the TEM analysis established the presence of spherical and polydispersed silver nanoparticles having a mean of 7.51 ± 2.88 nm. The XRD patterns also confirmed the sharp and clear peaks of AgNPs indicating the crystalline nature of the biosynthesized entities. The fern-derived AgNPs also exhibited potent antibacterial (*P. aeruginosa*, *E. coli*, and *K. pneumoniae*) and antifungal activity (*C. albicans*).

Ruthenium nanoparticles were biosynthesized via the green method by Gupta et al. (2019) using *Nephrolepis biserrata* (family: Nephrolepidaceae). The methanolic extract of the fern leaves was treated with 2.4 mM methanolic solution of ruthenium chloride ($RuCl_3.xH_2O$). The reaction mixture showed a broad peak at 391 nm and a narrow one at 524 nm during UV-visible spectroscopy. TEM images exhibited the presence of hexagonal, amorphous, and monodispersed nanostructures having average size of 25.96 nm. .

Biofabrication of silver nanoparticles was accomplished by Antonysamy Johnson et al. (2020) using *Odontosoria chinensis* (family: Lindsaeaceae), commonly known as "lace fern." Aqueous extract prepared from mature sporophytes of the fern was added to 0.001 M $AgNO_3$ with different ratios, viz., 0.5:9.5, 1:9, 1.5:8.5, and 2:8. After change in color of the reaction mixture, silver nanoparticles were characterized by UV-vis spectrometry, FTIR, XRD, and SEM-EDAX. It was surprising that complete reduction of silver nanoparticles occurred within 10 min. The EDX spectra of AgNPs synthesized by lace fern exhibited a strong peak for silver 46.228°. The biologically synthesized nanoparticles were also successfully tested for antidiabetic and anti-inflammatory activity.

The methanolic extract of *Microsorum pteropus* (family: Polypodiaceae), the semiaquatic epiphytic fern commonly known as "Java fern," was utilized by Chick et al. (2020) for the biosynthesis of AgNPs. The production of AgNPs was confirmed by color change of the solution from yellow to brown, but strangely the electron microscopic studies or particle size estimation was not carried out to conclusively prove the presence of nanoparticles.

Das et al. (2020) successfully carried out biosynthesis of Ag nanoparticles using leaf and stem extract of *Equisetum arvense* (family: Equisetaceae) (Fig. 9.7), commonly known as "field horsetail" or "common horsetail." This herbaceous fern, native to Europe and considered as weed in many countries, is known to possess antioxidant, anti-inflammatory, antibacterial, antifungal, vasorelaxant, neuroprotective, cardioprotective, and antiproliferative properties. A change in the color of the reaction mixture to brownish red suggests reduction of the silver salt and biofabrication of nanoparticles. The UV-vis spectra showed excitation of surface plasmons in the range of 440 and 558 nm, with maximum absorbance at 484 nm. SEM images revealed the shape of the synthesized nanoparticles as nearly spherical, while EDX demonstrated a sharp peak at 3 keV, thus confirming the existence of AgNPs. The particle size distribution brought out the hydrodynamic diameter of silver nanoparticles to be around 170.5 (d.nm).

The rhizome extract of Indian male fern, Dryopteris cochleata (fFamily: Dryopteridaceae), wasere used for the fabrication of AgNPs by Singh et al. (2020), and the newly synthesized entities were characterized with the help of different characterization techniques. Spherical silver nanoparticles of about 35 nm were observed through SEM, while XRD pattern revealed the presence of three sharp peaks in the spectrum of 2ω value ranging from 20 to 80 °C. The bioactivity of the biologically produced nanoparticles was tested against *Naja naja* snake venom. The results pointed toward the utility of silver nanoparticles obtained from *D. cochleata* as antidote against snakebite.

Fig. 9.7 Image of *Equisetum arvense* (Source: Wikipedia)

9.4 Future Prospects

Like bryophytes, pteridophyte is also an underused plant group with respect to their utilization in the biosynthesis of nanoparticles. However, the presence of several active phytochemical compounds in this group points toward the tremendous scope it offers in this upcoming field.

References

Adinarayana TVS, Mishra A, Singhal I, Ram Koti Reddy DV (2020) Facile green synthesis of silicon nanoparticles from *Equisetum arvense* for fluorescence based detection of Fe(iii) ions. Nanoscale Adv 2:4125–4132

Antonysamy Johnson MA, Santhanam A, Thangaiah S, Narayanan J (2017) Green synthesis of silver nanoparticles using *Cyathea nilgirensis* Holttum and their cytotoxic and phytotoxic potentials. Part Sci Technol 36:578–582

Antonysamy Johnson MA, Shibila T, Amutha S, Menezes IRA, da Costa JGM, Sampaio NFL, Coutinho HDM (2020) Synthesis of silver nanoparticles using *Odontosoria chinensis* (L.) J. Sm. and evaluation of their biological potentials. Pharmaceuticals 13(4):66

Asgarpanah J, Roohi E (2012) Phytochemistry and pharmacological properties of *Equisetum arvense* L. J Med Plant Res 6:3689–3693

Baskaran X, Geo Vigila AV, Parimelazhagan T, Muralidhara-Rao D, Zhang S (2016) Biosynthesis, characterization, and evaluation of bioactivities of leaf extract-mediated biocompatible silver nanoparticles from an early tracheophyte, *Pteris tripartita* Sw. Int J Nanomed 11:5789–5806

Bhor G, Maskare S, Hinge S, Singh L, Nalwade A (2014) Synthesis of silver nanoparticles by using leaflet extract of *Nephrolepis exaltata* L. and evaluation of antibacterial activity against human and plant pathogenic bacteria. Asian J Pharmaceut Technol Innova 02:7

Boromand N, Behbahani M, Dini G (2020) Green synthesis of silica nanoparticles from the horsetail (*Equisetum telmateia* Ehrh.) plant. New Cell Mole Biotechnol 10(39):49–60

Carneiro DM, Jardim TV, Araújo YCL, Arantes AC, Sousa AC (2019) Equisetum arvense: new evidences supports medical use in daily clinic. Pharmacogn Rev 13(26):50–58

Castro-Longoria E, Trejo-Guillén K, Vilchis-Nestor AR, Avalos-Borja M, Andrade-Canto SB, Leal-Alvarado DA, Santamaría JM (2014) Biosynthesis of lead nanoparticles by the aquatic water fern, *Salvinia minima* baker, when exposed to high lead concentration. Colloids Surf B Biointerfaces 114:277–283

Chatterjee A, Khatua S, Acharya K, Sarkar J (2019) A green approach for the synthesis of antimicrobial bio-surfactant silver nanoparticles by using a fern. Dig J Nanomater Biostruct 14:479–490

Chick CN, Misawa-Suzuki T, Suzuki Y, Usuki T (2020) Preparation and antioxidant study of silver nanoparticles of *Microsorum pteropus* methanol extract. Bioorg Med Chem Lett 30:127526

Chowdhury A, Kunjiappan S, Bhattacharjee C, Somasundaram B, Panneerselvam T (2017) Biogenic synthesis of *Marsilea quadrifolia* gold nanoparticles: a study of improved glucose utilization efficiency on 3T3-L1 adipocytes. In Vitro Cell Dev Biol Anim 53(6):483–493

Das G, Patra JK, Shin HS (2020) Biosynthesis, and potential effect of fern mediated biocompatible silver nanoparticles by cytotoxicity, antidiabetic, antioxidant and antibacterial, studies. Korean J Couns Psychother 114:111011

Femi-Adepoju AG, Dada AO, Otun KO, Adepoju AO, Fatoba OP (2019) Green synthesis of silver nanoparticles using terrestrial fern (*Gleichenia pectinata* (Willd.) C. Presl.): characterization and antimicrobial studies. Heliyon 5:e01543

Gini T, Jothi GJ (2015) In vitro screening of antibacterial and antifungal activity of *Marsilea quadrifolia* (Marsileaceae) Linn. Extract. Am J Phytomed Clin Therap 3:313–329

Gupta PK, Ranganath KVS, Dubey NK, Mishra L (2019) Green synthesis, characterization and biological activity of synthesized ruthenium nanoparticles using fishtail fern, sago palm, rosy periwinkle and holy basil. Curr Sci 117:1308–1317

Jha AK, Prasad K (2016) Aquatic fern (*Azolla* sp.) assisted synthesis of gold nanoparticles. Int J Nanosci 15:1650008

Kedi PBE, Meva FE, Kotsedi L, Nguemfo EL, Zangueu CB, Ntoumba AA, Mohamed HEA, Dongmo AB, Maaza M (2018) Eco-friendly synthesis, characterization, in vitro and in vivo anti-inflammatory activity of silver nanoparticle-mediated *Selaginella myosurus* aqueous extract. Int J Nanomedicine 13:8537–8548

Korbekandi H, Chitsazi MR, Asghari G, Najafi RB, Badii A, Iravani S (2014) Green biosynthesis of silver nanoparticles using *Azolla pinnata* whole plant hydroalcoholic extract. Green Process Synth 3(5):365–373

Koteswaramma JK, Varalakshmi S (2017) Biological synthesis of silver nanopartical from aqueous extract of *Actiniopteris radiata* and evaluation of their antimicrobial activity. Int J Pharm Bio Sci 8(1):121–125

Kunjiappan S, Chowdhury R, Bhattacharjee C (2014) A green chemistry approach for the synthesis and characterization of bioactive gold nanoparticles using *Azolla microphylla* methanol extract. Front Mat Sci 8:123–135

Lovis JD (1977) Evolutionary patterns and processes in ferns. Adv Bot Res 4:229–440

Moran RC (2004) A natural history of ferns. Timber Press, Inc., Portland

Nalwadea AR, Badheb MN, Pawalec CB, Hinge SB (2013) Rapid biosynthesis of silver nanoparticles using fern leaflet extract and evaluation of their antibacterial activity. Int J Biol Technol 4(2):12–18

Nitta JH, Ebihara A (2019) Virtual issue: ecology and evolution of pteridophytes in the era of molecular genetics. J Plant Res 132:719–721

Panneerselvam C, Murugan K, Roni M, Aziz AT, Suresh U, Rajaganesh R, Madhiyazhagan P, Subramaniam J, Dinesh D, Nicoletti M, Higuchi A, Alarfaj AA, Munusamy MA, Kumar S, Desneux N, Benelli G (2016) Fern-synthesized nanoparticles in the fight against malaria: LC/MS analysis of *Pteridium aquilinum* leaf extract and biosynthesis of silver nanoparticles with high mosquitocidal and antiplasmodial activity. Parasitol Res 115(3):997–1013

Paul B, Bhuyan B, Purkayastha DD, Dhar SS (2015) Green synthesis of silver nanoparticles using dried biomass of *Diplazium esculentum* (Retz.) sw. And studies of their photocatalytic and anticoagulative activities. J Mol Liq 212:813–817

PPG 1 (2016) A community-derived classification for extant lycophytes and ferns. J Syst Evol 54: 563–603

Pradheesh G, Suresh S, Suresh J, Alexramani V (2020) Antimicrobial and anticancer activity studies on green synthesized silver oxide nanoparticles from the medicinal plant *Cyathea nilgiriensis* Holttum. Int J Pharmaceut Investig 10(2):146–150

Rajaganesh R, Murugan K, Panneerselvam C, Jayashanthini S, Aziz AT, Roni M, Suresh U, Trivedi S, Rehman H, Higuchi A, Nicoletti M, Benelli G (2016) Fern-synthesized silver nanocrystals: towards a new class of mosquito oviposition deterrents? Res Vet Sci 109:40–51

Rani D, Khare PB, Dantu PK (2010) In vitro antibacterial and antifungal properties of aqueous and non-aqueous frond extracts of *Psilotum nudum*, *Nephrolepis biserrata* and *Nephrolepis cordifolia*. Indian J Pharm Sci 72(6):818–822

Ripa FA, Nahar L, Haque M, Islam MM (2009) Antibacterial, cytotoxic and antioxidant activity of crude extract of *Marsilea quadrifolia*. Euro J Sci Res 33:123–129

Sandhu NS, Kaur S, Chopra D (2010) Equisetum arvense: pharmacology and phytochemistry- a review. Asian J Pharm Clin Res 3(3):146–150

Sant DG, Gujarathi TR, Harne SR, Ghosh S, Kitture R, Kale S, Chopade BA, Pardesi KR (2013) Adiantum philippense L. frond assisted rapid green synthesis of gold and silver nanoparticles. J Nanopart 5:182320

Sessa EB (2018) Evolution and classification of ferns and lycophytes. In: Fernández H (ed) Current advances in fern research. Springer, Cham, pp 179–200

Singh P, Yasir M, Khare R, Shrivastava R (2020) Green synthesis of silver nanoparticles using Indian male fern (*Dryopteris cochleata*), operational parameters, characterization and bioactivity on *Naja naja* venom neutralization. Toxicol Res (Camb) 9(5):706–713

Soare LC, Sutan AN (2018) Current trends in pteridophyte extracts: from plant to nanoparticles. In: Fernández H (ed) Current advances in fern research. Springer, Cham

Verma DK, Hasan SH, Banik RM (2016) Photo-catalyzed and phyto-mediated rapid green synthesis of silver nanoparticles using herbal extract of *Salvinia molesta* and its antimicrobial efficacy. J Photochem Photobiol B Biol 155:51–59

Zahan R, Ripa FA, Alam MB, Haque MA, Mosaddik M, Nahar L (2011) Hypoglycemic and in vitro antioxidant effects of methanolic extract of *Marsilea quadrifolia* plant. Pharm J 3:86–92

Check for updates

Biological Synthesis of Nanoparticles: Gymnosperms

10

Abstract

Gymnosperms, also termed as Acrogymnospermae, are an assemblage of seed-bearing vascular plants that contain "naked" seeds. Although the early evolutionary history of seed plants is a bit tricky and confusing due to inconclusive fossil evidences, recent molecular phylogenetic studies place extant gymnosperms in a monophyletic group allied to the Magnoliophyta. All major groups of gymnosperms have found usage in the nanofabrication nanoparticles with majority of the studies carried out in the last 10 years. Among the gymnosperms, conifers have been the popular choice in the eco-friendly synthesis of nanoparticles primarily due to the structural and chemical diversity of natural compounds present in different members of this group. *Ginkgo*, *Cycas*, and *Ephedra* have also yielded nanoparticles, while *Gnetum* has also been utilized to some extent.

10.1 Introduction

Gymnosperms, also termed as Acrogymnospermae, are a group of seed-bearing vascular plants in which the seeds are "naked" since they do not develop in an ovary. These are woody shrubs, trees, or lianas that are found around the world but form dominant vegetation in temperate and arctic regions. This group, like angiosperms, is different from seedless plants in not requiring water as a medium for pollination since the male gamete moves to the female gamete through air (some members are entomophilous as well). Also, there is a major reduction in size and complexity of the gametophyte in gymnosperms. Of the five lineages of seed plants known, the cycads, conifers, *Ginkgo*, and gnetophytes form the gymnosperms, while the fifth comprise the flowering plants and are termed as Angiospermae or

S. Srivastava, A. Bhargava, *Green Nanoparticles: The Future of Nanobiotechnology*,
https://doi.org/10.1007/978-981-16-7106-7_10

Magnoliophyta. The gymnosperm database lists 947 species of gymnosperms as provided below:

- *Ginkgo* (1 spp.)
- Gnetophytes (68 spp.)
- Cycads (289 spp.)
- Conifers (589 spp.)

Thus, in spite of morphological, ecological, and taxonomic diversity, the living gymnosperms are meager in terms of number of species as compared to angiosperms which have estimated to have about 369,000 species. This is due to the fact that although the gymnosperms were widespread and dominated the Earth until the mid-Cretaceous, their diversity and distribution has reduced significantly since the Mesozoic. The spread of angiosperms during the late Cretaceous and major extinction events due to extreme climatic conditions over the Cenozoic led to significant reduction in gymnosperm diversity. Recent data from molecular phylogenetics and evolutionary researches have indicated that gymnosperms can be considered as a sister group to the angiosperms. Gymnosperms bear important information related to evolution of seed plants and represent a pivotal link in the evolution of many gene families and biological pathways.

10.2 Phylogenetic Studies in Gymnosperms

Although the early evolutionary history of spermatophytes is a bit tricky and confusing due to inconclusive fossil evidences, recent molecular phylogenetic studies place extant gymnosperms in a monophyletic group allied to the Magnoliophyta. Earlier researchers, Hill (2005), were of the view that spermatophytes arose in the Late Devonian about 385–359 million years ago. However, more accepted view traces the origin of gymnosperms to the Late Carboniferous period when they replaced the lycopsid rainforests of the tropical region. A molecular study using nuclear and organelle data performed using 305 nuclear genes and 47 plastid genes from 14 species of seed plants made the following inferences (Fig. 10.1):

1. Living gymnosperms were monophyletic.
2. Gnetophyta were haphazardly placed within the Coniferophyta.
3. Cycadophyta and Ginkgophyta formed a clade that showed strong affinities to other extant gymnosperms.

10.3 Gymnosperms in Green Synthesis of Nanoparticles

Gymnosperm has recently attracted the attention of nanobiotechnologists as a promising plant group for green synthesis of metal nanoparticles. All major groups of gymnosperms mentioned previously have found usage in the nanofabrication

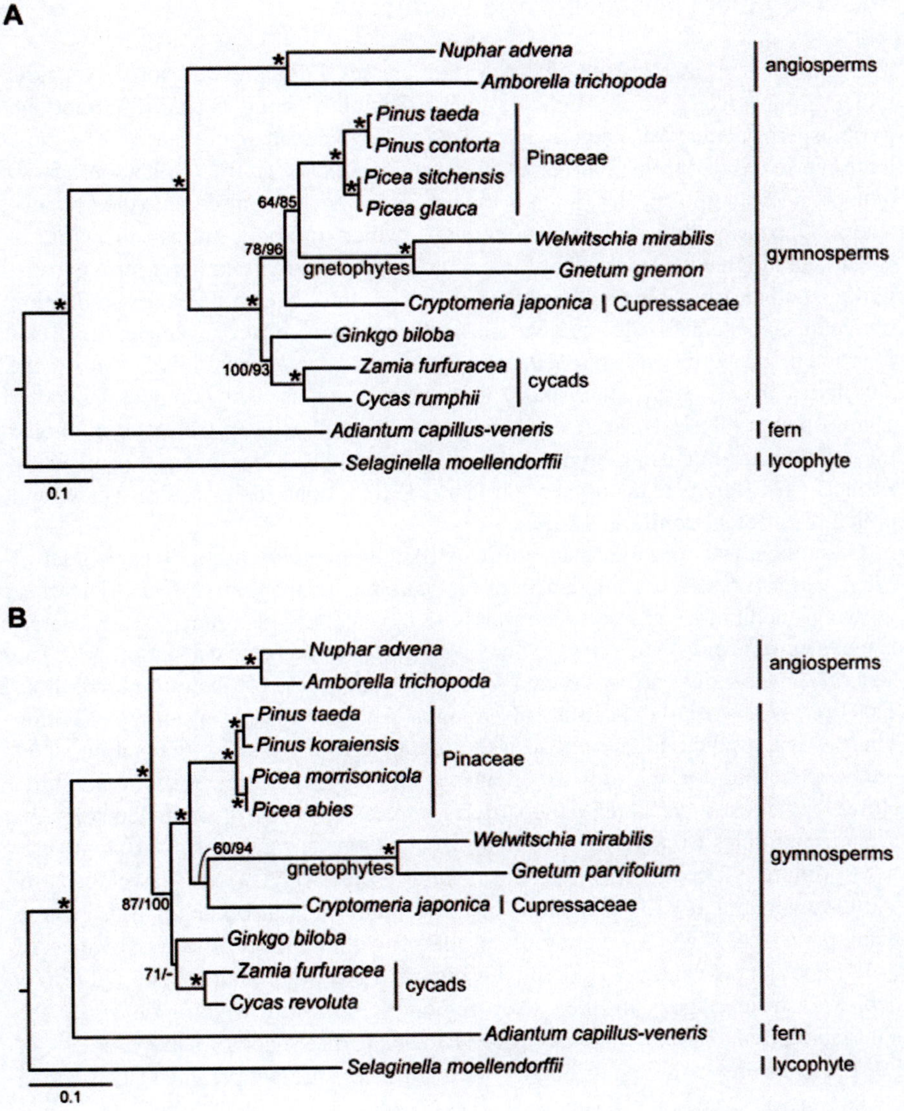

Fig. 10.1 Phylogenetic analysis and construction of species tree using (**a**) nuclear genes and (**b**) plastid genes (Reprinted from Xi et al. 2013)

nanoparticles with majority of the studies carried out in the last 10 years. Among the gymnosperms, conifers have been maximally used for nanofabrication which is attributed to the enormous structural and phytochemical diversity present in different members of this group.

10.3.1 Conifers in Nanoparticle Production

The conifer is an assemblage of two former classes Coniferopsida and Taxopsida. Other groups like Cycads, Ephedras, and Gnetums, along with the monotypic gymnosperm genus *Welwitschia* and *Ginkgo*, the maidenhair tree native to China, have been excluded from conifers. Conifers comprise of 8 families with about 65–70 genera and 600–630 species. Such is the predominance of conifers that the Gymnosperm Database (https://www.conifers.org/) which provides information for all species and higher-ranked taxa of the gymnosperms separates conifers from cycads, ginkgo, and the gnetophytes. A number of conifers have been utilized for production of metal and metal oxide nanoparticles which greatly differ in shape, size, and functional properties (Table 10.1). However, some genera like *Cupressus*, *Juniperus*, *Pinus*, *Taxus*, and *Thuja* have been more used as compared to other gymnosperms (Table 10.1). Among the metals, silver has been the metal of choice for nanofabrication using gymnosperms. Nanosynthesis has been carried using extracts of different plant organs like bark, leaves, cone, callus, gum, berry, and pollen of different coniferous species.

Green synthesis of silver nanoparticles (AgNPs) employing the bark extract of *Pinus eldarica* was accomplished by Iravani and Zolfaghari (2013). This fast growing, medium-sized tree is commonly termed as “Eldarica pine,” “Eldar pine,” “Quetta pine,” and “Afghan pine” having light gray or brownish gray bark. The aqueous pine bark extract was mixed with 1, 2, 4, and 6 mM of silver nitrate solution. The UV-vis absorption spectrum of colloidal Ag showed λ_{max} at around 430 nm. There was a gradual enhancement in the production of AgNPs on elevating silver nitrate concentration as well as on increasing the pH of the reaction mixture. However, the size of silver nanoparticles decreased on increase in temperature. The nanoparticles were mostly spherical having diameter ranging from 10 to 40 nm.

A simple process for biological synthesis of AgNPs was developed by Velmurugan et al. (2013) using cone of *Pinus thunbergii*, also known as “Japanese black pine.” The AgNPs were rapidly produced within a short time at pH values of 8–10 although pH values of 6, 7, and 8 were also a part of the study. The EDS results depicted strong signal in the silver region at 412 nm which confirmed the nanofabrication process. Triangular- to hexagonal-shaped nanoparticles of 20–100 nm were confirmed through TEM analysis at different pH. Seven diffraction peaks were observed in XRD analysis at 38.6°, 44.2°, 46.2°, 65.2°, 68.1°, 78.2°, and 85.2°, and the average particle size was calculated to be about 35 nm.

Zheng et al. (2013) synthesized platinum (PtNPs) by mixing Na_2PtCl_4 with *Cacumen platycladi* (family: Cupressaceae) extract as the reducing agent. Platinum nanoparticles (size: 2.4 ± 0.8 nm) were obtained after incubation of about 24 h under high temperature (90 °C). Even low metal concentration of 0.0005 M was suitable for production of the nanomaterials. Reducing sugars, protein, and flavonoids present in the plant extract served as effective reducing agents and aided in the formation of the nanoentities.

The pioneer research illustrating the biological synthesis of AgNPs using the needles of *Pinus roxburghii*, popularly known as “longleaf Indian pine,” was

Table 10.1 Use of conifer extracts in the biosynthesis of metallic nanoparticles (Modified and reprinted from Bhardwaj et al. (2020))

Type of conifer	Family	Plant part used	Type of metallic nanoparticle	Reducing agent	Reaction time	Reaction temperature	Shape	Size (nm)	References
Abies spectabilis	Pinaceae	NS	Gold	Polyphenols	24 h	29 °C	Sphere	20–200	Wu et al. (2019)
Araucaria heterophylla	Araucariaceae	Gum	Silver	Gum extract	NS	NS	ND	<25	Samrot et al. (2019)
Cupressus goveniana	Cupressaceae	Leaf/needle	Silver	Carboxyl and hydroxyl and amine groups	60 min	80 °C	Sphere	67–200	Hernández et al. (2015)
Cupressus sempervirens	Cupressaceae	Leaf/needle	Silver	Carbonyl and hydroxyl groups	12 h	RT	Sphere, hexahedral, oval, and triangle	10–80	Ebrahiminezhad et al. (2017)
Cupressus torulosa	Cupressaceae	Leaf/needle	Silver	Leaf extract	24 h	NS	Sphere	NS	Rajput et al. (2016)
Juniperus chinensis	Cupressaceae	Leaf/needle	Silver	Proteins	60 min	100 °C	NS	18–25	Al-Dhafri et al. (2019)
Juniperus communis	Cupressaceae	Berry/cone	Gold	Carbonyl, carboxyl, and hydroxyl groups	24 h	RT	Sphere and triangle	10–50	Prashanth et al. (2011); Mariychuk et al. (2020)
Juniperus procera	Cupressaceae	Leaves	Silver	Phenolic acids, chlorogenic acid, flavonoids, caffeoylquinic acids	24 h	RT	Spherical, cubical	30–90	Ibrahim et al. (2019)
Picea abies L.	Pinaceae	Bark	Silver	Phenolic compounds	3 h	60 °C	Sphere and polygonal	44	Tanase et al. (2019)

(continued)

Table 10.1 (continued)

Type of conifer	Family	Plant part used	Type of metallic nanoparticle	Reducing agent	Reaction time	Reaction temperature	Shape	Size (nm)	References
Pinus densiflora	Pinaceae	Cone	Silver	Hydroxyl and carbonyl groups	15 min	RT	Oval and triangular	40–70	Velmurugan et al. (2015)
Pinus eldarica	Pinaceae	Bark	Silver	Bark extract	ND	RT	Sphere	10–40	Iravani and Zolfaghari (2013)
Pinus eldarica	Pinaceae	Leaf/needle	Iron oxide	Leaf extract	30 min	RT	Sphere	8–34	Kheshtzar et al. (2019)
Pinus kesiya	Pinaceae	Pollen	Gold	Pollen extract	60 min	RT	ND	ND	Fernando et al. (2017)
Pinus merkusii	Pinaceae	Cone	Silver	Phenolic hydroxyl groups	60 min	60 °C	Sphere	8–23	Azkiya et al. (2018)
Pinus merkusii	Pinaceae	Cone/flower	Copper	Phenolic groups	NS	NS	ND	20–35	Masruri et al. (2018)
Pinus wallichiana	Pinaceae	Stem	Silver	Hydroxyl groups	30 min	55 °C	Sphere	10–30	Khan et al. (2020)
Taxus baccata	Taxaceae	Leaf/needle	Silver	Proteins and terpenoids	NS	10 and 30 °C	Triangular and hexagonal	75.1	Kajani et al. (2014)
Taxus baccata	Taxaceae	Leaf/needle	Gold	Needle extract	24 h	25 °C	Sphere, semisphere, hexagonal, and triangle	<20	Kajani et al. (2016)
Taxus baccata	Taxaceae	Leaf/needle	Zinc oxide	Proteins	NS	NS	Hexagonal	2–27.64	Sarli and Ghasemi (2020)
Taxus yunnanensis	Taxaceae	Callus	Silver	Callus extract	120 min	NS	Sphere	6.4–27.2	Xia et al. (2016)

Thuja orientalis	Cupressaceae	Leaf/needle	Gold	Phytochemicals	10 min	RT	Sphere	5–94	Noruzi et al. (2012)
Thuja occidentalis	Cupressaceae	Leaf/needle	Silver	Leaf extract	NS	NS	ND	7–14	Barua et al. (2013)
Thuja occidentalis	Cupressaceae	NS	Silver	Carbonyl groups	10 min	27 °C	Sphere	122.8	Das et al. (2013)
Thuja occidentalis	Cupressaceae	Leaf/needle	Silver	Hydroxyl and carbonyl groups	3–4 h	RT	ND	41.48	Bhor et al. (2016)
Thuja occidentalis	Cupressaceae	Leaf/needle	Silver	Leaf extract	NS	RT	ND	10–15	Barua et al. (2017)
Thuja occidentalis	Cupressaceae	Leaf/needle	Silver	Carbonyl, carboxyl, aliphatic, and aromatic amine groups	35–40 min	60 °C	Sphere	<30	Kanawaria et al. (2018)
Thuja occidentalis	Cupressaceae	Leaf/needle	Silver	Leaf extract	3–4 h	RT	ND	ND	Riat et al. (2018)
Torreya nucifera	Taxaceae	Leaf/needle	Silver	Proteins	24 h	20 °C	Sphere	10–125	Kalpana et al. (2019)

NS not specified, *RT* room temperature, *ND* not determined.

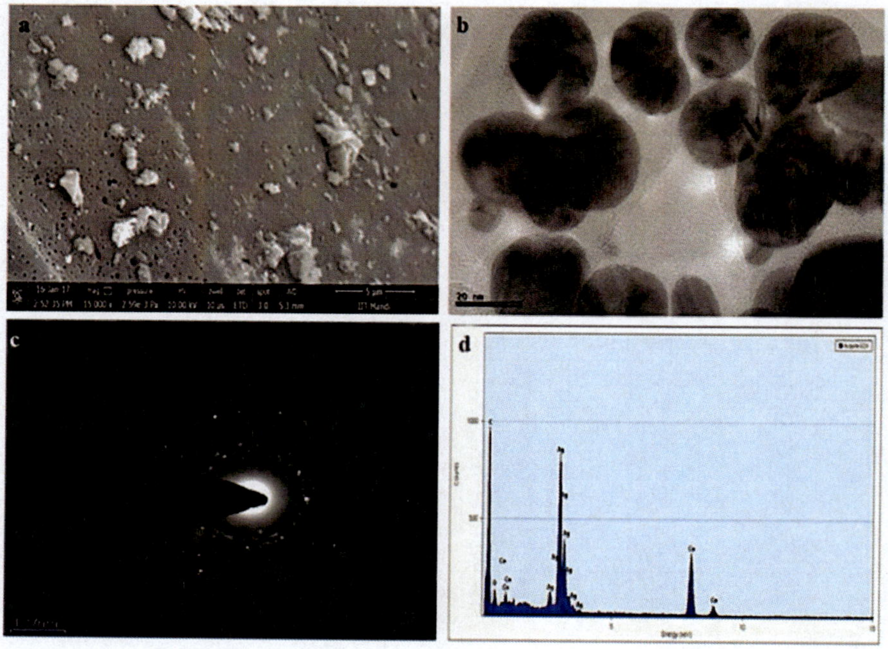

Fig. 10.2 (**a**) FESEM image of AgNPs synthesized by *Pinus roxburghii*. (**b**) TEM images revealing the morphology of AgNPs. (**c**) SAED pattern of AgNPs. (**d**) EDS image showing the elemental composition (Reprinted with permission from Kumari et al. 2020)

accomplished by Kumari et al. (2020) wherein the phytochemicals present in the needles served as efficient capping and reducing agents. The butanol extract (2 mg/ml) prepared by maceration of dried needles was added to 0.1 M $AgNO_3$ solution. The characterization of nanoparticles by UV-vis spectra showed a peak at 459 nm that corresponded to the surface plasmon resonance of AgNPs. The crystallinity of the newly synthesized nanostructures analyzed through XRD showed intense peaks at 38.16°, 44.30°, 64.54°, and 77.40° in the 2θ range confirming the presence of silver as nanocrystals. HRTEM showed that the biofabricated AgNPs were spherical having a mean size of ~80 nm (Fig. 10.2). The FESEM also pointed out toward the spherical nature of silver nanoparticles. The EDS pattern exhibited higher counts at 3 keV, which confirmed the presence of AgNPs.

Apart from whole plant extract and needles that have been used in most of the cases, a number of other plant parts like bark, cones, pollen, and gum have also been utilized for nanoparticle synthesis. The bark extract of *Amentotaxus assamica*, a threatened species of conifer in family Taxaceae, was used for the production of AgNPs under the influence of sunlight irradiation by Bharali et al. (2019). The UV-vis spectrum showed a broad peak at 472 nm, while the XRD analysis pointed toward the crystalline nature of the biologically synthesized nanoparticles. SEM images showed that the AgNPs were spherical in shape. Tanase et al. (2019) utilized the bark extract of spruce (*Picea abies* L.) for biosynthesis of silver nanoparticles.

The AgNPs obtained with silver acetate ($AgC_2H_3O_2$) at different pH were nanometric-, spherical-, and rarely polygonal-shaped having size ranging from 10 to 350 nm and average size of 165.54 ± 8.19 and 44.02 ± 0.31 nm at pH 4 and 9, respectively. The silver nanoparticles obtained using silver nitrate were nanometric and isolately distributed, with average size of 75.91 ± 2.77 and 148.64 ± 11.41 nm at pH 4 and 9, respectively.

Cupressus sempervirens L. is a medium-sized evergreen conifer known by various names as "Italian cypress," "Persian cypress," or "pencil pine." The pollen extract of the conifer was used for the fabrication of AgNPs through an environmentally friendly approach by Turunc et al. (2021). *Cupressus sempervirens* L. (CSPE) pollen extract was mixed with 1 mM $AgNO_3$ solution and stirred vigorously at room temperature. A color change of reaction mixture pointed toward the biosynthesis of AgNPs possibly due to reduction of Ag^+ to Ag^0. The synthesis of nanoparticles was also confirmed by the presence of a characteristic peak located at 440 nm in the UV-vis spectra. The effect of some parameters like amount of pollen extract, temperature during the reduction process, and reaction time was assessed for optimization of reaction conditions for nanofabrication. The magnitude of the absorption peak increased on rise in the reaction time with an insignificant increase witnessed beyond 150 minutes that was specified as the optimum reaction time. SEM analysis showed the presence of spherical AgNPs with orbicule structures on the pollen surface, while the TEM images showed the presence of spherical, oval, and triangular nanoparticles. The size distribution graph obtained by DLS showed the size variation of newly formed AgNPs ranging from 11.70 to 24.36 nm with about 65% of particles within the range of 15.69–21.04 nm.

10.3.2 Nanoparticle Synthesis from Other Gymnosperms

Many gymnosperms other than conifers have also been explored for the eco-friendly synthesis of nanoparticles (Table 10.2). The use of *Ginkgo*, *Cycas*, and *Ephedra* in nanoparticle production needs special mention here, although *Gnetum* has also been utilized to some extent.

10.3.2.1 Ginkgo

Ginkgo biloba Linn. (family: Ginkgoaceae) is known by different names like "duck foot tree," "maidenhair tree," and "silver apricot" (Fig. 10.3). This large tree with fan-shaped leaves is thought to have originated around 200 million years ago. It is native to China and with a large population still existing in Zhejiang province of Central China along with scattered distribution in patches in the central Himalayas. *G. biloba* is termed as a "living fossil" because it is the only remaining representative of a perished botanical family (the Ginkgoaceae). The plant is medicinally very important and displays a broad spectrum of pharmacological activities in human disorders like cardiovascular diseases, cognitive dysfunction, depression, kidney injury, age-related dementia, Alzheimer's disease, and macular degeneration. These medicinal properties of *G. biloba* are attributed to the presence of several

Table 10.2 List of gymnosperms (barring conifers) used for the biosynthesis of nanoparticles

Gymnosperm	Family	Plant part used	Type of nanoparticle	Size (nm)	Activity	References
Cycas sp.	Cycadaceae	Leaf	Silver	2–6	nr	Jha and Prasad (2010)
Cycas circinalis	Cycadaceae	Leaf	Silver	13–51	nr	Johnson and Prabu (2015)
Cycas pschannae	Cycadaceae	Plant	Zinc oxide	108–454	Antioxidant and anticancer	Sudha et al. (2020)
Ephedra intermedia	Ephedraceae	Stem	Silver	10–36	nr	Ebrahiminezhad et al. (2016)
Ephedra sinica	Ephedraceae	Stem	Gold	35.04	Anti-neuroinflammatory	Park et al. (2019)
Ginkgo biloba	Ginkgoaceae	Leaf Leaf Leaf Leaf Leaf	Silver Silver Copper Gold Gold	10–16 20–90 15–20 10–40 18.95	Antibacterial Anticancerous catalytic nr degradation of azo dyes	Ren et al. (2016), Xu et al. (2020) Nasrollahzadeh and Mohammad (2015), Zha et al. (2017), Liu et al. (2020)
Gnetum africanum	Gnetaceae	Leaf	Silver	9.00	nr	Ketebu and Gede (2020)
Gnetum bucholzianum	Gnetaceae	Leaf	Silver	67.40	Antibacterial	Meva et al. (2017)
Thuja occidentalis	Cupressaceae		Silver	122.8	Antiproliferative	Das et al. (2013)

nr not reported.

Fig. 10.3 (**a**) *Ginkgo biloba* tree. (**b**, **d**) Phenotype of the normal green leaf. (**c**, **e**) Phenotype of the mutant leaf (Reprinted from Li et al. 2018)

pharmacologically active phytochemicals, primarily the terpenoids like ginkgolides A, B, C, J, M, and bilobalide and flavonoids like quercetin, kaempferol, and isorhamnetin (Fig. 10.4).

The maidenhair tree has been extensively utilized for the eco-friendly synthesis of nanoparticles especially silver (Fig. 10.5). A simple, eco-friendly protocol was developed by Nasrollahzadeh and Mohammad (2015) for the synthesis of copper nanoparticles using *G. biloba* leaf extract. The aqueous leaf extract was mixed with 5 mM of cupric chloride solution ($CuCl_2.2H_2O$). Within 15 min, a change in the color of the reaction mixture pointed toward the possible formation of nanoparticles. The FT-IR spectrum of the crude extract showed numerous peaks indicating the presence of phytoconstituents like flavonoid and phenolics which had a possible role to play in the nanosynthesis process. The FTIR showed significant differences in the shape and location of signals after addition of cupric chloride solution which showed the presence of a strong association between the salt and different functional groups of the phytochemicals that aided in nanofabrication. The morphology and size of the nanoparticles were confirmed by TEM analysis which indicated the presence of spherical nanoparticles having diameter ranging from 15 to 20 nm.

Zha et al. (2017) adopted a green method for synthesis of AuNPs by the reduction of $HAuCl_4$ solution with the help of *Ginkgo biloba* leaf extract. The surface plasmon resonance peak observed at 545.5 nm confirmed the synthesis of spherical-shaped AuNPs. The size distribution carried out by DLS analysis showed that gold nanoparticles had a narrow distribution range and had a mean size of 130 nm. The SEM images depicted the presence of spherical AuNPs with the size range of 10–30 nm, while TEM analysis showed nanoparticles of 10–40 nm in size. The particle size decreased, while their concentration increased with increase in temperature.

Major constituents in standard *Ginkgo biloba* extract EGb761

Percentage in EGb761 (%)	Bioactive content: Terpenoids (6%)	Bioactive content: Flavonoids (24%)	Others (70%)
	Ginkgolide A Ginkgolide B Ginkgolide C Bilobalide	Quercetin Kaempferol Tamarixetin	

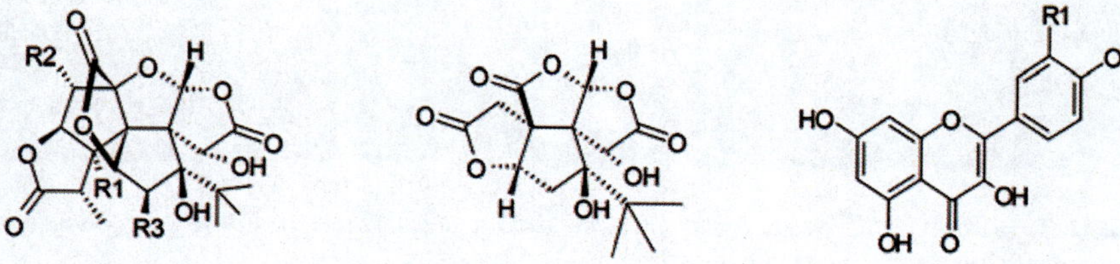

R1	R2	R3	Ginkgolide
OH	H	H	A
OH	OH	H	B
OH	OH	OH	C

Bilobalide

R1	
OH	Quercetin
H	Kaempferol
OCH_3	Tamarixetin

Fig. 10.4 Terpenoids and flavonoids of *Ginkgo biloba* (Reprinted with permission from Li et al. 2018)

Huang et al. (2020) successfully carried out biosynthesis of AgNPs having small diameter and narrow particle size distribution via the green route using *G. biloba* leaf extract. The experiment was performed under different conditions, viz., the ratio of reactants, reaction time, pH, and temperature. The formation of spherical nanoparticles was confirmed by the strong SPR band at 400 nm. It was observed that an increase in the amount of leaf led to profuse synthesis of silver nanoparticles but having small size. Also the yield of nanoparticles exhibited positive association with temperature since the absorption bands increased on elevation of temperature from 40 to 80 °C. The experiment also showed strong correlation between pH and the biosynthesis of nanoparticles as the absorption peaks of silver nanoparticles became sharper and increased on enhancement of pH from 5 to 9. The TEM analysis pointed toward the presence of face-centered and cubic crystalline structure of the newly synthesized AgNPs. The nanoparticle diameter estimated using DLS was around 14.14 ± 4.44 nm with the distribution range of 8.7–21.3 nm. The nanoparticles effectively degraded azo dyes, exhibited noteworthy antibacterial activity against both Gram-negative and Gram-positive bacteria, and exhibited promise for use as fluorescent probes for heavy metal detection.

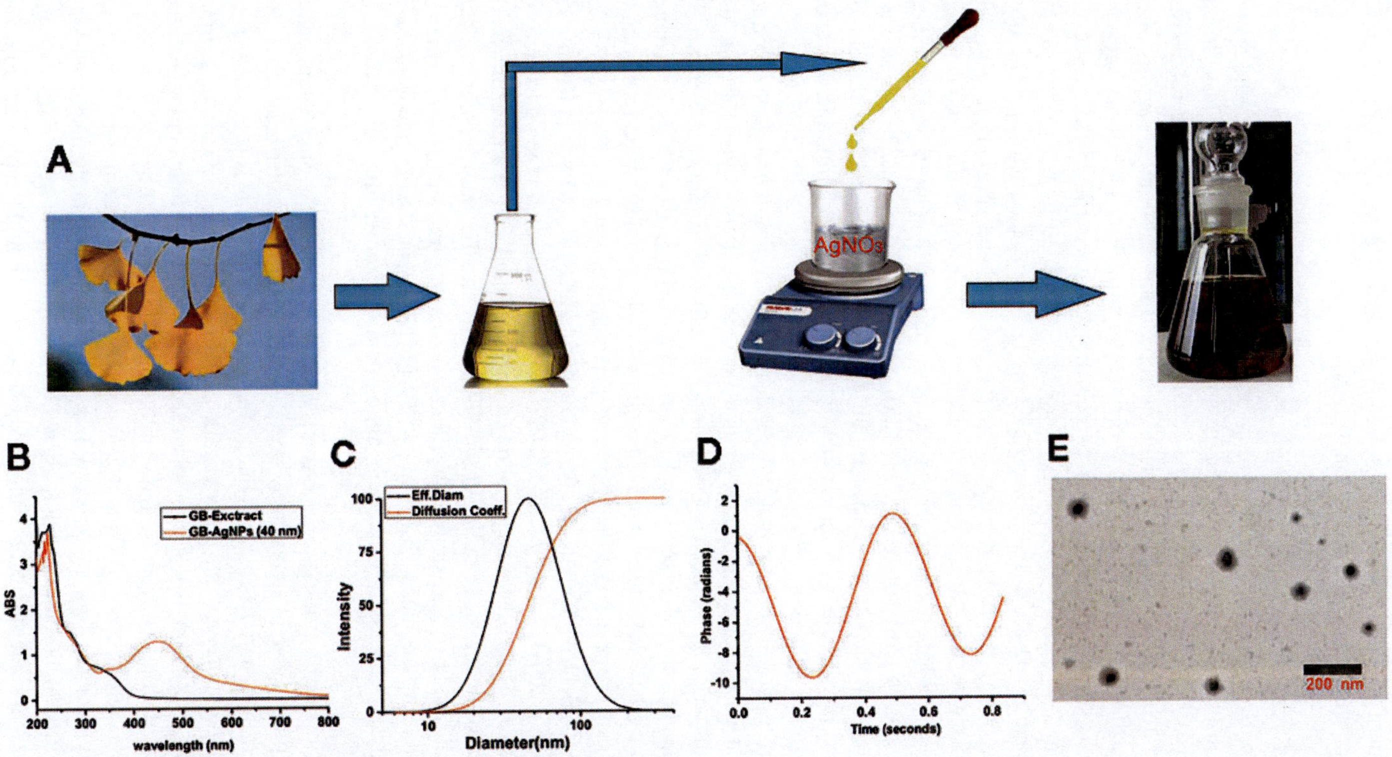

Fig. 10.5 Synthesis of AgNPs using leaf extracts of *G. biloba*. (**a**) Step silver nanoparticle (AgNP) synthesis, (**b**) UV-vis spectra image, (**c**) DLS image, (**d**) Zeta potential images, and (**e**) TEM images (Reprinted from Xu et al. 2020)

G. biloba extract has also been successfully used as a bioreductant for the preparation of the nonmetallic nanostructure graphene: a single layer of carbon hexagons arranged in a two-dimensional sheet.

10.3.2.2 Cycads

Cycads are slow-growing ancient group of plants that have a stout and woody (ligneous) trunk with a crown of large, stiff, pinnately compound leaves. This group exhibits superficial resemblance to palms due to similarities in foliage and plant structure. Cycads are mostly found in tropical and subtropical regions of the world. Apart from having numerous medicinal properties, cycads contain the toxic compounds cycasin and neocycasin, which are unique toxins present in cycad species.

The ethanolic extract of the leaves of *Cycas* sp. was treated with 0.25 M silver nitrate solution for production of AgNPs by Jha and Prasad (2010). The nanoparticles were roughly spherical with diameters ranging from 2 to 6 nm. The XRD analyses indicated that silver nanoparticles had a face-centered cubic unit cell having the sets of lattice planes (111), (200), (220), (311), and (222). The authors were of the view that the presence of different phytochemicals like flavonoids, polyphenols, phytochelatins, glutathiones, and metallothioneins in the leaves had a major role in the reduction process.

Sudha et al. (2020) used powdered dry leaves of *Cycas pschannae* to produce zinc oxide nanoparticles. The leaves of this species are eaten as vegetable and also secrete gum having numerous medicinal benefits. The aqueous extract of *C. pschannae* leaves was mixed with 1 M zinc acetate ($Zn(CH_3COO)_2$). The change in the color of the reaction mixture from yellow to white indicated the formation of zinc oxide nanoparticles. The major intense peaks are seen at 31.70° (100), 34.20° (002), 36.32° (101), 47.45° (012), 56.58° (110), 62.72° (013), 67.96° (112), 69.35° (201), 72.58° (604), and 77.09° (202) which confirmed the presence of nanostructures. The SEM images showed irregular and rod-shaped zinc oxide nanostructures having size ranging from 177 to 249 nm. The morphology of ZnO nanorods analyzed by TEM showed the nanorods to be rodlike-shaped with a size of 108–454 nm and the mean size being ~250 nm. The zinc oxide nanorods were further analyzed successfully for their antioxidant, photocatalytic, and anticancer activity.

Apart from *Ginkgo* and *Cycas*, the genus *Ephedra* which is the only extant genus in family Ephedraceae is increasingly becoming popular among the researchers for production of nanoparticles. This is due to the fact that the genus is a rich source of medicinally important phytochemicals like ephedrine, pseudoephedrine, ephedroxane, cyclopropyl-A amino acids, chlorogenic acid, caffeic acid, quercetin, isoquercitrin, flavones, flavanols, and carboxylic acids (Fig. 10.6).

Biological synthesis of AgNPs was accomplished by Ebrahiminezhad et al. (2016) using *Ephedra intermedia* stem extract. This detailed study also shed light on the effect of metal concentration as well as the amount of stem extract on the reduction process. The sugars present in the extract were thought to play a major role in the synthesis process by serving as reducing and capping agents. Characterization

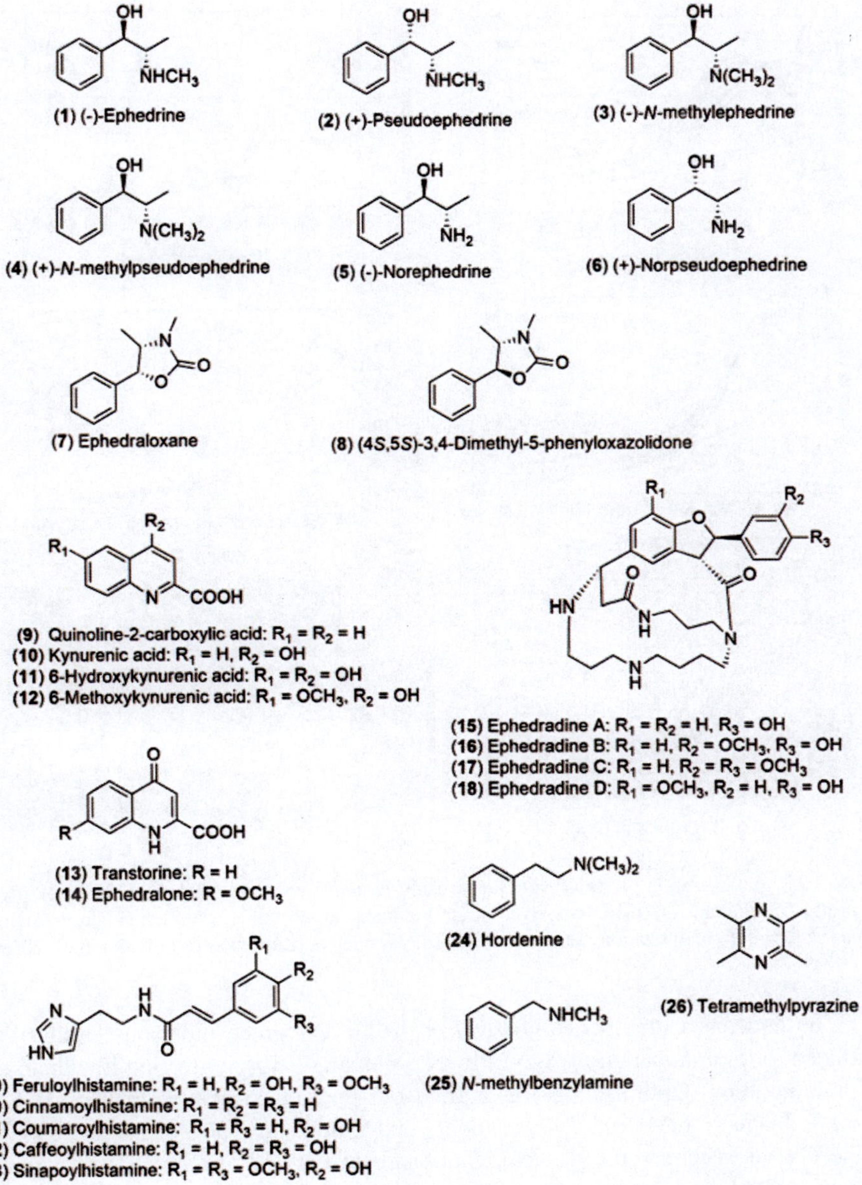

Fig. 10.6 Structures of alkaloids from *Ephedra* species (Reprinted from González-Juárez et al. 2020)

of the particles showed that they were spherical, had a crystal structure, and had diameters ranging from 10 to 36 nm, the average diameter calculated as 24 nm. Zeta potential was around −30 mV which lay between the acceptable limit for colloid structures.

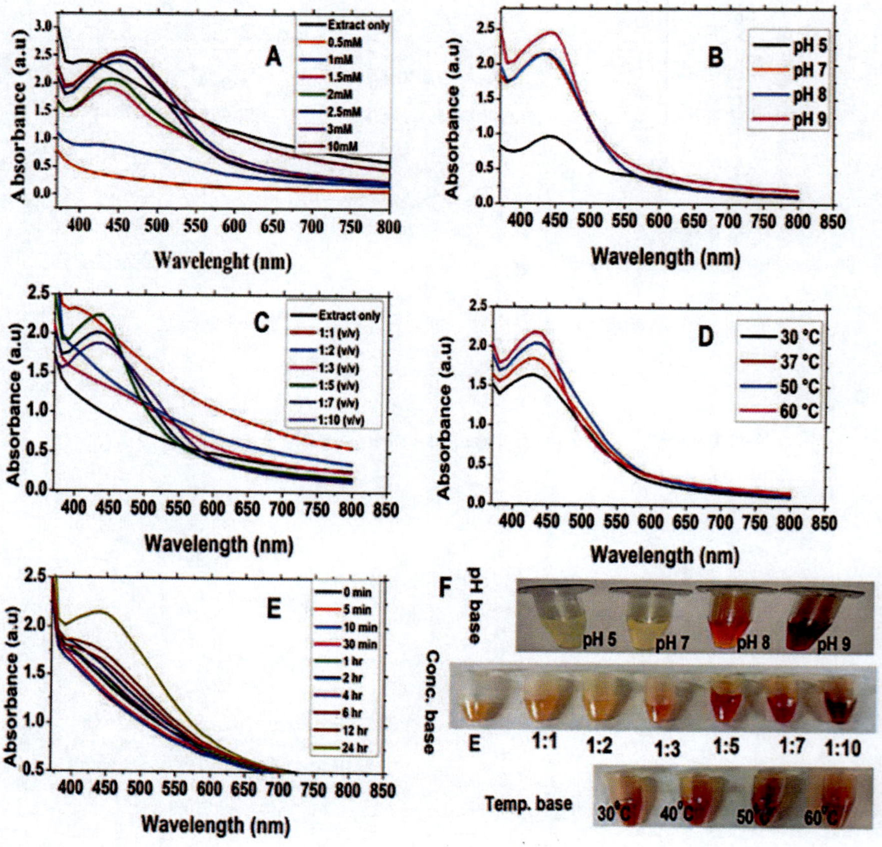

Fig. 10.7 Optimization of AgNPs produced using *Ephedra procera*: (**a**) concentration of silver nitrate, (**b**) different pH, (**c**) ratio of silver nitrate and plant extract, (**d**) effect of temperature, (**e**) time intervals, and (**f**) optimization images for different parameters (Reprinted from Nasar et al. 2019)

The aqueous extract of *Ephedra procera*, an evergreen dioecious shrub, was utilized for biofabrication of AgNPs by Nasar et al. (2019) by treating it with silver nitrate solution. Optimization was carried out for salt concentration (0.5, 1, 1.5, 2, 2.5, 3, and 10 mM), pH (5, 7, 8, and 9), and temperature (30, 37, 50, and 60 °C). The UV-vis spectra showed maximum absorbance at 458 nm for 10 mM $AgNO_3$ (Fig. 10.7). In pH optimization, an increase in pH led to change in the color of reaction mixture from yellow to brown, thus indicating the synthesis of nanostructures. The UV-vis spectra showed maximum SPR of 444 nm at high pH, while for temperature the recorded absorbance intensities increased with increase in temperature. Time optimization studies showed that the absorbance gradually increased with passage of time. The SEM micrograph showed that the AgNPs were 20.4 nm in diameter and roughly spherical and majority of them in aggregated form. The XRD analysis showed the mean crystallite size of the AgNPs as 17.2 nm.

Park et al. (2019) successfully synthesized gold nanoparticles from *Ephedra sinica*, an evergreen shrub native to Central Asia and having the common names "Chinese ephedra" or "Ma Huang." This plant contains several bioactive components like alkaloids, viz., ephedrine, pseudoephedrine (isoephedrine), nor-pseudoephedrine (cathine), norephedrine, methyl ephedrine, methyl pseudo-ephedrine, tannins, and other constituents, including quinoline and 6-hydroxy kynurenic acid which enables the plant to exhibit a wide range of bioactivities. In fact the diverse pharmacological properties of the genus *Ephedra* are attributed to the content of ephedrine-type alkaloids. Chloroauric acid ($HAuCl_4 \cdot 3H_2O$) solution (1 mM) was added to the stem extract of Chinese ephedra with continuous stirring and left at room temperature for some time. The newly synthesized AuNPs exhibited maximum absorption at 538 nm in UV-visible spectroscopy. The diameter of the gold nanoparticles assessed through DLS was about 57.6 ± 3.07 nm, while the zeta potential value of -24.6 ± 0.84 mV was recorded. HR-TEM showed that the AuNPs were spherical having a mean diameter of 35.04 ± 4.02 nm. Crystalline structure of the AuNPs was confirmed by XRD, while the presence of elemental gold was confirmed by energy-dispersive X-ray spectroscopy.

10.3.2.3 Gnetophytes

The systematic position of gnetophytes has been the most debated topics in seed plant phylogeny. Members of this group show numerous similarities with flowering plants such as water-conducting cells similar to xylem vessels, double fertilization, and reticulate leaf venation due to which they were proposed to be the sister group of angiosperms. This has been termed as the Anthophyte hypothesis. However, molecular phylogenetic data and deeper insights into the developmental pathways have led to rejection of this hypothesis. The current evidence and consensus indicates that gnetophytes show close affinity to conifers rather than to angiosperms.

AgNPs of 67.4 nm in size were synthesized by Meva et al. (2017) via the green method using *Gnetum bucholzianum* (Meva et al. 2017). An aqueous extract of fresh leaves of *G. bucholzianum* Engl. (Gnetaceae) was mixed with 1 mM $AgNO_3$ solution and kept at room temperature in the dark. The mean crystalline size of the nanostructures was calculated to be about 67.4 nm.

10.4 Future Prospects

The available literature at our disposal shows that although gymnosperms have been sufficiently utilized for nanoparticle production, these studies are limited mostly to conifers. The use of other gymnosperm groups has been scanty. More studies are needed to tap the potential of the lesser utilized groups like cycads and gnetophytes. Gnetophytes, due to their proximity to angiosperms, should be explored in a much more planned way.

References

Abourashed EA, El-Alfy AT, Khan IA, Walker L (2003) Ephedra in perspective—a current review. Phytother Res 17:703–712

Al-Dhafri K, Ching CL, Philip K (2019) Phyto-synthesis of silver nanoparticles and its bioactivity response towards nosocomial bacterial pathogens. Biocatal Agric Biotechnol 18:101075

Azkiya NI, Masruri M, Ulfa SM (2018) Green synthesis of silver nanoparticles using extract of *Pinus merkusii* Jungh & De Vriese cone flower. IOP Conf Series: Mater Sci Eng 299:12070

Barua S, Banerjee PP, Sadhu A, Sengupta A, Chatterjee S, Sarkar S, Barman S, Chattopadhyay A, Bhattacharya S, Mondal NC (2017) Silver nanoparticles as antibacterial and anticancer materials against human breast, cervical and oral cancer cells. J Nanosci Nanotechnol 17: 968–976

Barua S, Konwarh R, Bhattacharya SS, Das P, Devi KSP, Maiti TK, Mandal M, Karak N (2013) Non-hazardous anticancerous and antibacterial colloidal 'green' silver nanoparticles. Colloids Surf B Biointerfaces 105:37–42

Bharali P, Das S, Bhandari N, Das AK, Kalta MC (2019) Sunlight induced biosynthesis of silver nanoparticle from the bark extract of *Amentotaxus assamica* D.K. Ferguson and its antibacterial activity against *Escherichia coli* and *Staphylococcus aureus*. IET Nanobiotechnol 13(1):18–22

Bhardwaj K, Dhanjal DS, Sharma A, Nepovimova E, Kalia A, Thakur S, Bhardwaj S, Chopra C, Singh R, Verma R, Kumar D, Bhardwaj P, Kuča K (2020) Conifer-derived metallic nanoparticles: green synthesis and biological applications. Int J Mol Sci 21(23):9028

Bhor GL, Kharate S, Nikam S, Kulkarni VD (2016) Synthesis of silver nanoparticles using *Thuja* leaf extract. Res J Mater Sci 4:4–6

Das S, Das J, Samadder A, Bhattacharyya SS, Das D, Khuda-Bukhsh AR (2013) Biosynthesized silver nanoparticles by ethanolic extracts of *Phytolacca decandra, Gelsemium sempervirens, Hydrastis canadensis* and *Thuja occidentalis* induce differential cytotoxicity through G2/M arrest in A375 cells. Colloids Surf B Biointerfaces 101:325–336

De La Torre AR, Piot A, Liu B, Wilhite B, Weiss M, Porth I (2019) Functional and morphological evolution in gymnosperms: a portrait of implicated gene families. Evol Appl 13(1):210–227

Dehkordi NV, Kachouie MA, Pirbalouti AG, Malekpoor F, Rabei M (2015) Total phenolic content, antioxidant and antibacterial activities of the extract of *Ephedra procera* fisch. et mey. Acta Pol Pharm 72:341

Ebrahiminezhad A, Taghizadeh S, Berenjian A, Heidaryan Naeini F, Ghasemi Y (2016) Green synthesis of silver nanoparticles capped with natural carbohydrates using *Ephedra intermedia*. Nanosci Nanotechnol-Asia 6:1–9

Ebrahiminezhad A, Taghizadeh S, Ghasemi Y (2017) Green synthesis of silver nanoparticles using Mediterranean cypress (*Cupressus sempervirens*) leaf extract. Am J Biochem Biotechnol 13:1–6

Fernando SID, Judan-Cruz KG, De Guia ACM (2017) Biologically synthesized gold nanoparticles (Aunp) using pine (*Pinus kesiya*) pollen extract show antifungal activity against *Candida albicans*. Int J Agric Technol 13:2615–2622

Forest F, Moat J, Baloch E, Brummitt NA, Bachman SP, Ickert-Bond S, Hollingsworth PM, Liston A, Little DP, Mathews S, Rai H, Rydin C, Stevenson DW, Thomas P, Buerki S (2018) Gymnosperms on the EDGE. Sci Rep 8(1):6053

Gernandt D, Willyard A, Syring J, Liston A (2011) The conifers (Pinophyta). Genet Genom Breed Conifers 6:29–67

González-Juárez DE, Escobedo-Moratilla A, Flores J, Hidalgo-Figueroa S, Martínez-Tagüeña N, Morales-Jiménez J, Muñiz-Ramírez A, Pastor-Palacios G, Pérez-Miranda S, Ramírez-Hernández A, Trujillo J, Bautista E (2020) A review of the *ephedra* genus: distribution, ecology, ethnobotany, phytochemistry and pharmacological properties. Molecules 25:3283

Gurunathan S, Han JW, Park JH, Eppakayala V, Kim JH (2014) Ginkgo biloba: a natural reducing agent for the synthesis of cytocompatible graphene. Int J Nanomedicine 9:363–377

Hajibabaei M, Xia J, Drouin G (2006) Seed plant phylogeny: gnetophytes are derived conifers and a sister group to Pinaceae. Mol Phylogenet Evol 40(1):208–217

Hernández LG, Islas DA, Guerrero MUF, Ortega PAR, Lechuga LG (2015) Use of extract of *Cupressus goveniana* for synthesis and stabilization of nanoparticles silver. In: Proceedings of the TMS 2015 144th annual meeting & exhibition. Springer, Cham, pp 1105–1112

Hill K (2005) Diversity and evolution of gymnosperms. In: Henry RJ (ed) Plant diversity and evolution. CABI, Cambridge, pp 25–44

Huang L, Sun Y, Mahmud S, Liu H (2020) Biological and environmental applications of silver nanoparticles synthesized using the aqueous extract of *Ginkgo biloba* leaf. J Inorg Organomet Polym 30:1653–1668

Ibrahim EH, Kilany M, Ghramh HA, Khan KA, Islam SU (2019) Cellular proliferation/cytotoxicity and antimicrobial potentials of green synthesized silver nanoparticles (AgNPs) using *Juniperus procera*. Saudi J Biol Sci 26:1689–1694

Iravani S, Zolfaghari B (2013) Green synthesis of silver nanoparticles using *Pinus eldarica* bark extract. Biomed Res Int 2013:639725

Jha AK, Prasad K (2010) Green synthesis of silver nanoparticles using *Cycas* leaf. Int J Green Nanotechnol Phys Chem 1:110–117

Johnson I, Prabu HJ (2015) Green synthesis and characterization of silver nanoparticles by leaf extracts of *Cycas circinalis*, *Ficus amplissima*, *Commelina benghalensis* and *Lippia nodiflora.* Int Nano Lett 5(1):43–51

Kajani AA, Bordbar A-K, Esfahani SHZ, Khosropour AR, Razmjou A (2014) Green synthesis of anisotropic silver nanoparticles with potent anticancer activity using *Taxus baccata* extract. RSC Adv 4:61394–61403

Kajani AA, Bordbar A-K, Esfahani SHZ, Razmjou A (2016) Gold nanoparticles as potent anticancer agent: green synthesis, characterization, and in vitro study. RSC Adv 6:63973–63983

Kalpana D, Han JH, Park WS, Lee SM, Wahab R, Lee YS (2019) Green biosynthesis of silver nanoparticles using *Torreya nucifera* and their antibacterial activity. Arab J Chem 12:1722–1732

Kanawaria SK, Sankhla A, Jatav PK, Yadav RS, Verma KS, Velraj P, Kachhwaha S, Kothari SL (2018) Rapid biosynthesis and characterization of silver nanoparticles: an assessment of antibacterial and antimycotic activity. Appl Phys A Mater Sci Process 124:320

Ketebu O, Gede EE (2020) Green synthesis of silver nanoparticles using mixture of *Gnetum africanum* and *Heinsia crinita* leaf extracts. Int J Academic Eng Res 4:17–23

Khan N, Khan I, Nadhman A, Azam S, Ullah I, Ahmad F, Khan HA (2020) Pinus wallichiana synthesized silver nanoparticles as biomedical agents: in-vitro and in-vivo approach. Green Chem Lett Rev 13:69–82

Kheshtzar R, Berenjian A, Taghizadeh S-M, Ghasemi Y, Asad AG, Ebrahiminezhad A (2019) Optimization of reaction parameters for the green synthesis of zero valent iron nanoparticles using pine tree needles. Green Process Synth 8:846–855

Kopaczyk JM, Warguła J, Jelonek T (2020) The variability of terpenes in conifers under developmental and environmental stimuli. Environ Exp Bot 180:104197

Kumari R, Saini AK, Kumar A, Saini RV (2020) Apoptosis induction in lung and prostate cancer cells through silver nanoparticles synthesized from *Pinus roxburghii* bioactive fraction. J Biol Inorg Chem 25(1):23–37

Lal M, Chandraker SK, Shukla R (2020) Antimicrobial properties of selected plants used in traditional Chinese medicine. In: Functional and preservative properties of phytochemicals. Academic Press, London, pp 119–143

Li L, Stanton JD, Tolson AH, Luo Y, Wang H (2009) Bioactive terpenoids and flavonoids from Ginkgo biloba extract induce the expression of hepatic drug-metabolizing enzymes through pregnane X receptor, constitutive androstane receptor, and aryl hydrocarbon receptor-mediated pathways. Pharm Res 26:872

Li WX, Yang SB, Lu ZG, He ZC, Ye YL, Zhao BB, Wang L, Jin B (2018) Cytological, physiological, and transcriptomic analyses of golden leaf coloration in *Ginkgo biloba* L. Hortic Res 5:12

Liu Y, Huang L, Mahmud S, Liu H (2020) Gold nanoparticles biosynthesized using *Ginkgo biloba* leaf aqueous extract for the decolorization of azo-dyes and fluorescent detection of Cr(VI). J Clust Sci 31:549–560

Lu Y, Ran J-H, Guo D-M, Yang Z-Y, Wang X-Q (2014) Phylogeny and divergence times of gymnosperms inferred from single-copy nuclear genes. PLoS One 9(9):e107679

Mariychuk R, Fejer J, Porubska J, Grishchenko LM, Lisnyak VV (2020) Green synthesis and characterization of gold triangular nanoprisms using extract of *Juniperus communis* L. Appl Nanosci 10:2835–2841

Masruri M, Pangestin DN, Ulfa SM, Riyanto S, Srihardyastutie A, Rahman MF (2018) A potent *Staphylococcus aureus* growth inhibitor of a dried flower extract of *Pinus merkusii* Jungh & De Vriese and copper nanoparticle. IOP Conf Series: Mater Sci Eng 299:12072

Mathews S (2009) Phylogenetic relationships among seed plants: persistent questions and the limits of molecular data. Am J Bot 96:228–236

McLoughlin S (2021) Gymnosperms. In: Alderton D, Elias SA (eds) Encyclopedia of geology. Elsevier, Amsterdam, pp 476–500

Meva FE, Ebongue CO, Fannang VS, Segnou ML, Ntoumba AA, Kedi PBE, Loudang REN, Wanlao AY, Mang RE, Mpondo MEA (2017) Natural substances for the synthesis of silver nanoparticles against *Escherichia coli*: the case of *Megaphrynium macrostachyum* (Marantaceae), *Corchorus olitorus* (Tiliaceae), *Ricinodendron heudelotii* (Euphorbiaceae), *Gnetum bucholzianum* (Gnetaceae), and *Ipomoea batatas* (Convolvulaceae). J Nanomater 2017:6834726

Nasar MQ, Khalil AT, Ali M, Shah M, Ayaz M, Shinwari ZK (2019) Phytochemical analysis, Ephedra procera C. A. Mey. mediated green synthesis of silver nanoparticles, their cytotoxic and antimicrobial potentials. Medicina (Kaunas) 55(7):369

Nasrollahzadeh M, Mohammad Sajadi S (2015) Green synthesis of copper nanoparticles using *Ginkgo biloba* L. leaf extract and their catalytic activity for the Huisgen [3+2] cycloaddition of azides and alkynes at room temperature. J Colloid Interface Sci 457:141–147

Noruzi M, Zare D, Davoodi D (2012) A rapid biosynthesis route for the preparation of gold nanoparticles by aqueous extract of cypress leaves at room temperature. Spectrochim Acta Part A: Mol Biomol Spectros 94:84–88

Park SY, Yi EH, Kim Y, Park G (2019) Anti-neuroinflammatory effects of *Ephedra sinica* Stapf extract-capped gold nanoparticles in microglia. Int J Nanomedicine 14:2861–2877

Pennisi E (2009) On the origin of flowering plants. Science 324(5923):28–31

Phillips GC, Gladfelter HJ (1991) Eldarica pine, afghan pine (*Pinus eldarica* Medw.). In: Bajaj YPS (ed) Trees III. Biotechnology in agriculture and forestry, vol 16. Springer, Berlin, Heidelberg, pp 269–287

Pires ND, Dolan L (2012) Morphological evolution in land plants: new designs with old genes. Philos Trans Royal Soc B: Biol Sci 367(1588):508–518

Prashanth S, Menaka I, Muthezhilan R, Sharma NK (2011) Synthesis of plant-mediated silver nanoparticles using medicinal plant extract and evaluation of its antimicrobial activities. Int J Eng Sci Technol 3:6235–6250

Rajput K, Bhatt A, Agrawal PK (2016) Plant mediated biosynthesis, characterization and application of silver nanoparticles by leaves extract of *Cupressus torulosa*. Int J Adv Res 4:1199–1207

Ren YY, Yang H, Wang T, Wang C (2016) Green synthesis and antimicrobial activity of monodisperse silver nanoparticles synthesized using *Ginkgo biloba* leaf extract. Phys Lett A 380:3773–3777

Riat AK, Geyi D, Rafi M, Kaur G (2018) Efficacy of *Thuja occidentalis* plant mediated synthesis of silver nanoparticles against *Culex quinquefasciatus* larvae. Res J Pharm Technol 11:4981

Samrot AV, Saipriya C, Angalene JLA, Roshini SM, Cypriyana PJJ, Saigeetha S, Raji P, Kumar SS (2019) Evaluation of nanotoxicity of *Araucaria heterophylla* gum derived green synthesized silver nanoparticles on *Eudrilus eugeniae* and *Danio rerio*. J Clust Sci 30:1017–1024

Sarli S, Ghasemi N (2020) Optimization of biosynthesized Zn nanoparticles by poisonous *Taxus baccata* leaves extract and evaluation of their effect on the bacterias and MCF-7 cancer cells. Eurasian Chem Commun 2:302–318

Sudha KG, Ali S, Karunakaran G, Kowsalya M, Kolesnikov E, Rajeshkumar MP (2020) Eco-friendly synthesis of ZnO nanorods using *Cycas pschannae* plant extract with excellent photocatalytic, antioxidant, and anticancer nanomedicine for lung cancer treatment. Appl Organomet Chem 34:e5511

Tanase C, Berta L, Coman N-A, Roşca I, Man A, Toma F, Mocan A, Nicolescu A, Jakab-Farkas L, Biró D (2019) Antibacterial and antioxidant potential of silver nanoparticles biosynthesized using the spruce bark extract. Nanomedicine 9:1541

Turunc E, Kahraman O, Binzet R (2021) Green synthesis of silver nanoparticles using pollen extract: characterization, assessment of their electrochemical and antioxidant activities. Anal Biochem 621:114123

Velmurugan P, Lee SM, Iydroose M, Lee KJ, Oh BT (2013) Pine cone-mediated green synthesis of silver nanoparticles and their antibacterial activity against agricultural pathogens. Appl Microbiol Biotechnol 97(1):361–368

Velmurugan P, Park J-H, Lee S-M, Jang JS, Lee K-J, Han S-S, Lee S-H, Cho M, Oh B-T (2015) Synthesis and characterization of nanosilver with antibacterial properties using *Pinus densiflora* young cone extract. J Photochem Photobiol B Biol 147:63–68

Wang X-Q, Ran J-H (2014) Evolution and biogeography of gymnosperms. Mol Phylogenet Evol 75:24–40

Wu T, Duan X, Hu C, Wu C, Chen X, Huang J, Liu J, Cui S (2019) Synthesis and characterization of gold nanoparticles from *Abies spectabilis* extract and its anticancer activity on bladder cancer T24 cells. Artif Cells Nanomed Biotechnol 47:512–523

Xi Z, Rest JS, Davis CC (2013) Phylogenomics and coalescent analyses resolve extant seed plant relationships. PLoS One 8(11):e80870

Xia QH, Ma YJ, Wang JW (2016) Biosynthesis of silver nanoparticles using *Taxus yunnanensis* callus and their antibacterial activity and cytotoxicity in human cancer cells. Nanomedicine 6: 160

Xu Z, Feng Q, Wang M, Zhao H, Lin Y, Zhou S (2020) Green biosynthesized silver nanoparticles with aqueous extracts of *Ginkgo biloba* induce apoptosis *via* mitochondrial pathway in cervical cancer cells. Front Oncol 10:575415

Zha J, Dong C, Wang X, Zhang X, Xiao X, Yang X (2017) Green synthesis and characterization of monodisperse gold nanoparticles using *Ginkgo biloba* leaf extract. Optik 144:511–521

Zheng B, Kong T, Jing X, Odoom-Wubah T, Li X, Sun D, Lu F, Zheng Y, Huang J, Li Q (2013) Plant-mediated synthesis of platinum nanoparticles and its bioreductive mechanism. J Colloid Interface Sci 396:138–145

Biological Synthesis of Nanoparticles: Angiosperms: Monocotyledons

11

Abstract

Monocots form a monophyletic group of grass and grasslike spermatophytes comprising of about 85,000 species arranged in 77 families. This group is considered more evolved than dicotyledons. This chapter sheds light on the role of monocotyledonous plants in the eco-friendly synthesis of nanoparticles.

11.1 Introduction

Monocots form a monophyletic group of angiosperms whose members contain seeds having one embryonic leaf or cotyledon. This group comprises of about 85,000 species arranged in 77 families with the family Orchidaceae having the largest number of around 28,000 species. Monocots lack a lateral meristem and are therefore largely herbaceous, but secondary growth is reported in some forms termed as "anomalous secondary growth." Some monocotyledonous species like palms, screwpines, bamboos, and agaves achieve great heights. Monocots differ from dicots in having a single cotyledon, parallel leaf venation, scattered vascular bundles, and uniaperturate pollen and due to absence of intrafascicular cambium.

11.2 Origin of Monocotyledonous Plants

The prephyletic classifications considered monocots to be primitive to dicots and placed them between plants other than angiosperms and dicots. However, with the advent of phyletic thinking and the development of Ranalian theory, monocots were considered as more advanced to dicots.

S. Srivastava, A. Bhargava, *Green Nanoparticles: The Future of Nanobiotechnology*,
https://doi.org/10.1007/978-981-16-7106-7_11

Table 11.1 Some theories regarding the origin of monocotyledonous plants

Botanist	Year	Possible progenitor
Hegelmaier	1878	Abortion hypothesis: one cotyledon of dicot failed to develop
Engler	1892	Isoetales, Ophioglossales, and few other pteridophytes
Hallier	1914	Lardizabalaceae
Cheadle	1953	Vesselless dicots like Nymphaeales and Magnoliales
Hutchinson	1959	Ranunculaceae
Takhtajan	1961	Extinct vesselless herb similar to modern Nymphaeales
Stebbins	1974	*Drimys* like ancestor having vesselless wood and short internodes
Burger	1977	Dicot ancestors similar to Chloranthaceae of Piperales

Arthur John Cronquist, the famous American botanist, in his monumental work *The Evolution and Classification of Flowering Plants* (1988) divided angiosperms into two classes:

1. Liliopsida (monocotyledons)
2. Magnoliopsida (dicotyledons)

However, subsequent studies based on ribulose 1,5-bisphosphate carboxylase (rbcL) and 18S rRNA genes indicated that the dicots should be segregated into basal dicots (magnoliids) and eudicots. It was also speculated that the monocots originated from one of the basal magnoliids. Phylogenetic analysis based upon the chloroplast rbcL gene sequences divides monocotyledons into two major groups:

1. "Commelinoid" monocotyledons: cell walls contain ferulic acid and glucuronoarabinoxylans as the major noncellulosic polysaccharides. Grasses, prominent members of this group, contain another polysaccharide, the (1 3, 1 4)-glucans.
2. "Non-commelinoid" monocotyledons: cell walls contain pectic polysaccharides along with xyloglucans as the major noncellulosic polysaccharides and do not contain ferulic acid. The cell wall composition and the metabolic products are similar to that of dicotyledons.

Several views have been put forward with respect to the origin of monocots that have been enumerated in Table 11.1. An interesting study performed by Givnish et al. (2018) presented the first plastome phylogeny of monocot families. The study talked about the development of the "monocot syndrome" that included the following characters:

1. Narrow leaves
2. Parallel venation
3. Clasping leaf bases
4. Loss of secondary thickening
5. Evolution of scattered vascular bundles

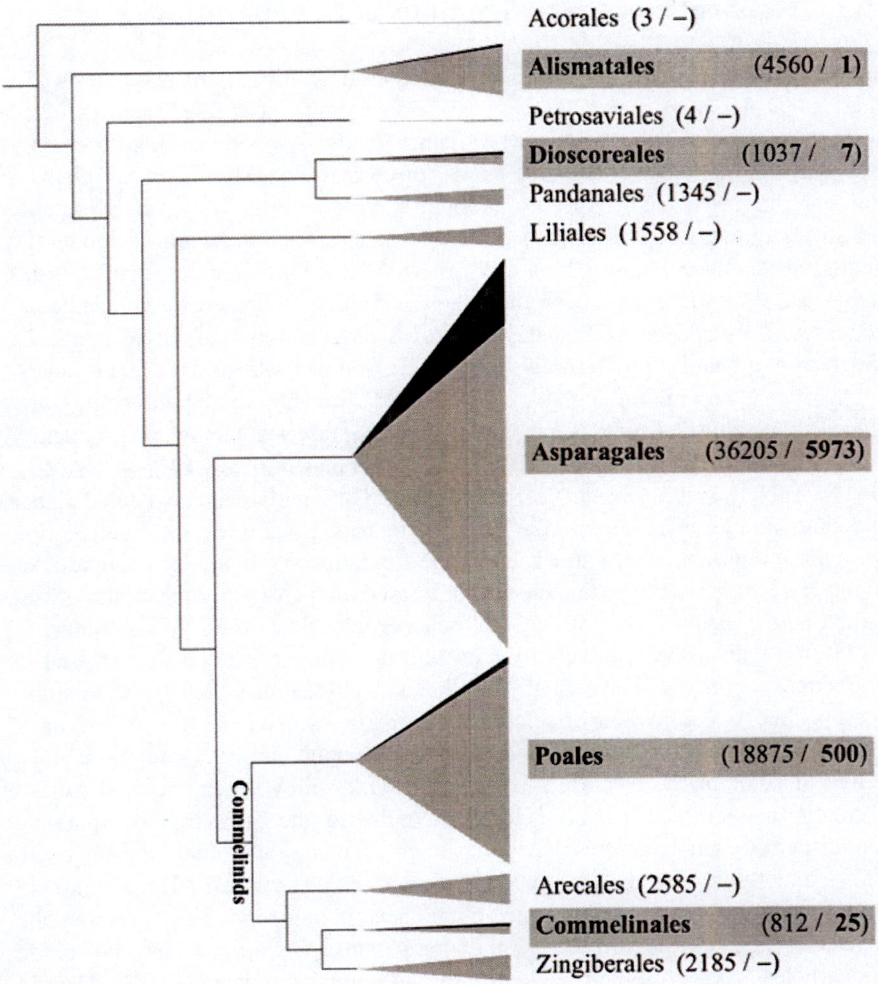

Fig. 11.1 Phylogeny of the monocot orders as per angiosperm phylogeny group (APG 2009, 2016) (Reprinted with permission from Nyffeler and Eggli 2020)

6. Sympodial growth
7. Shallow primary roots
8. Adventitious rooting

The study put forward a novel model for the evolution of monocots from submerged aquatic ancestors, which finds support in the studies comprising of evolution of several characters in aquatic Hydatellaceae (Nymphaeales). The recent classification of monocotyledons as proposed in APG (2009, 2016) recognizes 11 orders in monocots (Fig. 11.1).

11.3 Monocots for Green Synthesis of Nanoparticles

Monocotyledonous plants have been efficiently utilized for the green synthesis of a wide range of nanoparticles (Table 11.2). Some monocot species have gathered increased attention in this respect. *Aloe vera* (family: Asphodelaceae), commonly known as "Indian aloe," "true aloe," and "medicinal aloe," has been the plant of choice in this respect. This species, although having its origin in Arabia, Somalia, and Sudan, now grows wild in dry and arid climates around the world including the Mediterranean region, India, China, and eastern Africa. This "wonder plant," besides having use in cosmetic and food industries, is widely cultivated for its medicinal properties. This species is a rich source of a large number of phytochemicals, namely, polyphenols, phytosterols, fatty acids, indoles, alkaloids, organic acids, glycoproteins, anthraquinones, saccharides, dicarboxylic acids, and some low-molecular-weight substances. Chandran et al. (2006) undertook biogenic production of gold nanotriangles and spherical silver nanoparticles utilizing 10^{-3} M $HAuCl_4$, 10^{-2} M $AgNO_3$, and *A. vera* leaf extract. Spherical, triangular, and hexagonal planar gold nanoparticles were observed in the TEM images. It was observed that addition of small amounts of plant extract led to the formation of triangular nanoparticles having large edge lengths, while the size decreased on increase in the amount of leaf extract in the reaction mixture. Gold nanoparticles had a wide size range of 50–350 nm which was regulated by adjusting the amount of plant extract used in the reduction process. Fardsadegh and Jafarizadeh-Malmiri (2019) successfully accomplished biofabrication of selenium nanoparticles using *A. vera* leaf extract. The leaf extract was treated with 10 mM sodium selenite (Na_2SeO_3) solution, and a change in color of the reaction mixture from dark yellow to red pointed toward successful nanofabrication. TEM images confirmed the formation of spherical monodispersed nanostructures having average diameter of 50 nm. Primo et al. (2020) undertook biogenic production of zinc oxide nanoparticles by mixing *A. vera* gel broth extract with zinc nitrate hexahydrate and then calcining the suspension in a muffle furnace at high temperature for 60 m. The zinc oxide nanoparticles appeared to be pseudospheres and nonuniform hexagonal in shape in SEM images, while the size determination through XRD patterns and Scherrer's equation showed the average size as 43.3 nm.

Syafiuddin et al. (2017) utilized the biochemical components present in the aqueous extracts of two monocot species, viz., *Cyperus rotundus* and *Eleusine indica*, to synthesize silver nanoparticles from 0.15 M silver nitrate ($AgNO_3$) solution. *C. rotundus*, also known as "purple nutsedge," "Java grass," or "nut grass," is a smooth, erect, and perennial weed of the sedge family, Cyperaceae. Although this species has been labeled as one of the worst weed, it has been extensively utilized in traditional medicine systems around the world for remediation of different ailments. *E. indica*, an annual tufted grass commonly known as "Indian goose grass" or "crowfoot grass" (family: Poaceae), is an agricultural and environmental weed which is considered a "serious weed" in about 42 countries. This grass is native to Africa and parts of temperate and tropical Asia and is closely related to finger millet, viz., *E. coracana*. *E. indica* is used in conventional medicine systems

Table 11.2 List of nanoparticles synthesized by monocotyledonous species

Scientific name	Plant family	Type of nanoparticle	Size (nm)	Biological activity	References
Agave americana	Agavaceae	Silver	30–150	Antifungal, hemagglutination	Ahmad et al. (2016)
Agave potatorum	Agavaceae	Gold	14–22	nr	Moreno-Luna et al. (2019)
Allium cepa	Amaryllidaceae	Gold Silver Zinc oxide	~100 33.67 Nr	Cytotoxicity Antibacterial Anti-inflammatory	Parida et al. (2011) Saxena et al. (2010) Wu et al. (2019)
Allium sativum	Amaryllidaceae	Copper Iron Silver Zinc	14.62–22.80 60.30–82.63 13.13–22.69 99.34–134.57	Antimicrobial, Antioxidant Antimicrobial, Antioxidant Antimicrobial, Antioxidant Antimicrobial, Antioxidant	El-Refai et al. (2018) El-Refai et al. (2018) El-Refai et al. (2018) El-Refai et al. (2018)
Aloe barbadensis	Asphodelaceae	Zinc oxide	25–40	nr	Gunalan et al. (2012)
Aloe vera	Asphodelaceae	Copper oxide Silver	20 15.2	Antibacterial nr	Vijay Kumar et al. (2015) Chandran et al. (2006)
Alpinia calcarata	Zingiberaceae	Silver	27.2	Antibacterial, antioxidant	Khandel et al. (2018)
Alpinia katsumadai	Zingiberaceae	Silver	12.6	Antioxidant, cytotoxicity	He et al. (2017)
Alpinia nigra	Zingiberaceae	Silver	6	Antimicrobial, photocatalytic	Baruah et al. (2019)
Alpinia officinarum	Zingiberaceae	Silver	100	Antioxidant, nephroprotective	Zhang et al. (2019)
Ananas comosus	Bromeliaceae	Silver ZnO	– 30–57	Antibacterial, antidiabetic Antibacterial	Das et al. (2019) Ahmad et al. (2016)
Arisaema flavum	Araceae	Silver	12–20	Antibacterial, photocatalytic	Rahman et al. (2019)
Bambusa chung	Poaceae	Gold	28.8	nr	Jia et al. (2015)
Chrysopogon zizanioides	Poaceae	Gold	20–50	nr	Arunachalam and Annamalai (2013)
Curcuma longa	Zingiberaceae	Silver	1–50	Bactericidal	Sathishkumar et al. (2010)
Cymbopogon citratus	Poaceae	Silver Gold Cerium oxide	10–33 20–50 10–40	Antibacterial Insecticidal Nr	Ajayi and Afolayan (2017) Murugan et al. (2015) Maensiri et al. (2014)

(continued)

Table 11.2 (continued)

Scientific name	Plant family	Type of nanoparticle	Size (nm)	Biological activity	References
		Copper Copper oxide	2.90 11.4–14.5	Nr Antibacterial, antibiofilm	Brumbaugh et al. (2014) Cherian et al. (2020)
Cynodon dactylon	Poaceae	Silver	30–50	Nr	Supraja et al. (2013)
Dioscorea batatas	Dioscoreaceae	Silver	Nr	Antimicrobial	Nagajyothi and Lee (2011)
Dioscorea bulbifera	Dioscoreaceae	Silver	8–20	Antimicrobial	Ghosh et al. (2012)
Gloriosa superba	Colchicaceae	Copper oxide Cerium oxide Ruthenium	5–10 5 25–90	Antibacterial Antibacterial Antibacterial	Naika et al. (2015) Arumugam et al. (2015) Gopinath et al. (2014)
Musa acuminata	Musaceae	Gold Silver Zinc oxide	10.1–15.6 12.6–15.7 30–80	Antibacterial, anticancer Antibacterial, anticancer Photocatalytic	Valsalam et al. (2019) Valsalam et al. (2019) Abdullah et al. (2020)
Musa ornata	Musaceae	Iron	43.69	Antibacterial	Saranya et al. (2017)
Musa paradisiaca	Musaceae	Silver	Nr	Antibacterial, antifungal	Bankar et al. (2010)
Musa sapientum	Musaceae	Iron	Nr	Anticorrosive	Ituen et al. (2019)
Myriostachya wightiana	Poaceae	Silver	15–65	Antibacterial, insecticidal	Vadlapudi and Amanchy (2017)
Oryza sp.	Poaceae	Silver	36–107	Antifungal	Adak et al. (2021)
Plumbago zeylanica	Plumbaginaceae	Gold Silver	20–30 60	Antimicrobial, antibiofilm, antimicrobial, antibiofilm	Salunkhe et al. (2014) Salunkhe et al. (2014)
Sorghum sp.	Poaceae	Iron Palladium Platinum Silver	50 3.4 3.34 10	Biodegradation Catalytic Catalytic Biodegradation	Njagi et al. (2011) Hu et al. (2019) Hu et al. (2019) Njagi et al. (2011)
Triticum aestivum	Poaceae	Copper oxide Silver	5–40 5.06	Catalytic Catalytic	Waghmode et al. (2013) Buazar et al. (2019)

Typha latifolia	Typhaceae	Zinc oxide	nr	Photocatalytic	Kumar et al. (2019)
Zingiber officinale	Zingiberaceae	Copper	14.92–27.41	Antimicrobial, antioxidant	El-Refai et al. (2018)
		Gold	5–15	Blood compatibility	Kumar et al. (2011)
		Iron	14.08–21.57	Antimicrobial, antioxidant	El-Refai et al. (2018)
		Silver	20–51	Anticancer	Venkatadri et al. (2020)
		Zinc oxide	30–50	nr	Anand Raj and Jayalakshmy (2015)

nr not reported.

as a diuretic, antihelminthic, hepatoprotective, and remedy against venereal infections and dermatoses. UV-vis spectra of the silver nanoparticles using *C. rotundus* and *E. indica* had maximum peaks at 446 and 524 which indicated their spherical morphology. The FESEM showed that the nanoparticles synthesized using *C. rotundus* and *E. indica* had size dimensions of 20.5 ± 9.6 and 55.0 ± 24.1, respectively. These nanoparticles also showed potent antibacterial activity in different bacterial species.

Silver and gold nanoparticles were biogenically produced by Valsalam et al. (2019) using *Musa acuminata colla* (wild ancestor of the cultivated banana) flower. The aqueous extract of the flowers was treated with 1.0 mM $AgNO_3$ and $HAuCl_4.3H_2O$ solution at room temperature, and the characterization was carried out using different techniques. A significant change in color of the reaction mixture from brown to red indicated the biosynthesis of AgNPs and AuNPs. Both types of nanomaterials were polydispersed and spherical having size ranging from 12.6 to 15.7 nm and 10.1 to 15.6 nm, respectively. The EDAX profile revealed the presence of an intense signal for elemental silver at ~3 keV and for elemental gold at ~1.5 keV.

Cymbopogon citratus, a tall perennial fast growing aromatic grass belonging to the family Poaceae (common name: Indian lemongrass), has been successfully utilized for the biosynthesis of several metal nanoparticles like silver, gold, copper, and cerium dioxide. The therapeutic effects and nanoparticle production of lemongrass are aided by different phytoconstituents of the plant extract notably long chain hydrocarbons, ketones, esters, essential oil mainly comprising of geranial, citral-α, citral-β, myrcene, flavonoids, isoorientin 2′-*O*-rhamnoside, quercetin, kaempferol, apigenin, and luteolin. A one-pot, green method for the production of copper oxide nanoparticles was developed by Cherian et al. (2020) by utilizing the leaf extract of *Cymbopogon citratus* (as capping and stabilizing agent) and copper sulfate ($CuSO_4.5H_2O$) solution (Fig. 11.2). The copper nanoparticle formation was assessed under diverse concentrations of leaf extract (10, 20, 30, 40, and 50%) and five pH values where the salt concentration was kept constant at 0.25 mM. Significant long-term colloidal dispersion and stability were obtained at 10% leaf extract and pH 10. Spherical, hexagonal, and oval nanoparticles having size of 14.5 ± 2.0 nm were evident in TEM images, while SEM analysis showed slender needle-shaped nanoentities. The study pointed toward an active role of esters like di-propylene glycol diacrylate, α-monoolein, and isooctyl phthalate in the surface functionalization of copper ions. Gold nanoparticles were prepared by Murugan et al. (2015) by treating 10^{-3} M solution of $HAuCl_4$ with an aqueous extract of *C. citratus*. TEM analysis showed the presence of gold nanoparticles of diverse morphologies and size ranging from 20 to 50 nm which were highly effective against malaria and dengue mosquitoes. Another perennial species, *Cymbopogon flexuosus*, popularly known as “Malabar grass” or “Cochin grass,” has been utilized for the biogenic synthesis of zinc oxide nanoparticles by Kumari et al. (2018) that inhibited cell proliferation and exhibited ample amount of cytotoxicity against THP-1 leukemia cells.

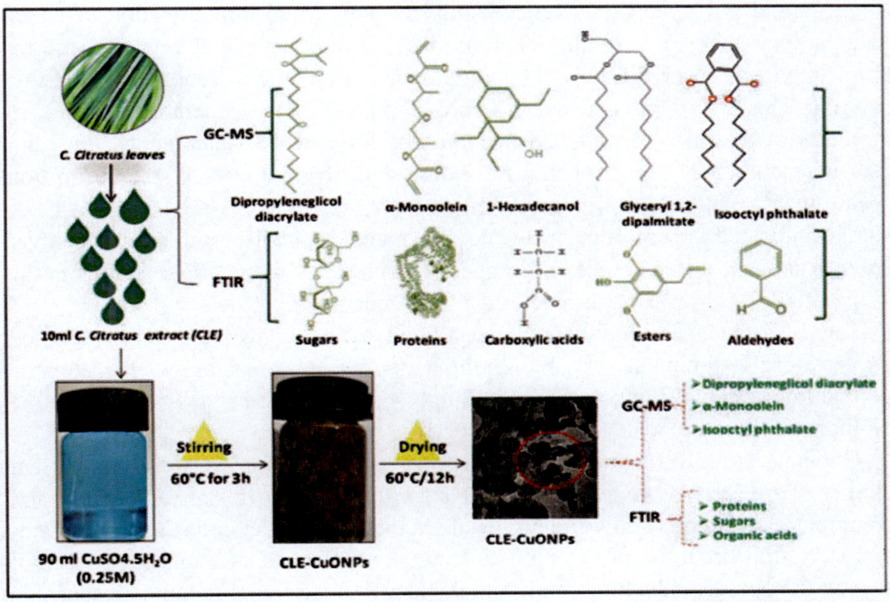

Fig. 11.2 Formation of copper oxide nanoparticles from *Cymbopogon citratus* (Reprinted from Cherian et al. 2020)

Brumbaugh et al. (2014) synthesized non-plasmonic, ultrasmall copper nanoparticles with diameters of 2.90 ± 0.64 nm at room temperature conditions using lemongrass tea as a green reducing agent. Murugan et al. (2015) synthesized gold nanoparticles from lemongrass by treating the aqueous extract of the plant with 10^{-3} M $HAuCl_4$ and incubating for 72 h at 25 °C. Within 30 minutes, the color of the reaction mixture switched from yellow to pinkish red that indicated the synthesis of nanoparticles which was also supported by the presence of an absorption peak at 540 nm. TEM images further confirmed the presence of nanoparticles of diverse shape and size dimensions of 20–50 nm. EDX showed peaks at 2 keV, while studies through Zetasizer showed the primary peak situated between 1 and 40 nm and the size range of 10–110 nm with mean size of 20 nm.

Maensiri et al. (2014) synthesized cerium (IV) oxide (CeO_2) nanoparticles using cerium salt and lemongrass plant extract through a simple cost-effective method. The process initially involved dissolution of cerium (III) acetate hydrate in lemongrass extract with constant agitation at 80 °C. Slightly aggregated cerium oxide nanoparticles measuring 5–30 nm were observed in TEM analysis with their size showing increase on concomitant increase in calcination temperature. Thus calcination at 400 °C had small nanoparticles of 10–15 nm dimensions, high calcination temperatures (500 °C) gave rise to nanoentities of around 15–20 nm, and the highest calcination temperatures of 600 °C exhibited maximum particle size of ~20–30 nm. Interestingly, better crystalline and optical properties were found at calcination temperature of 600 °C as compared to those heated at lower temperatures.

Ajayi and Afolayan (2017) successfully accomplished biofabrication of silver nanoparticles making use of an alkalinized *C. citratus* extract. It was prepared by extracting the powdered leaves of lemongrass in 5% sodium bicarbonate ($NaHCO_3$) solution. This extract was added to 100 mM $AgNO_3$ at room temperature. After 24 h, the color of the mixture switched from yellow to brownish-black suggesting successful reaction between the alkalinized extract and silver nitrate. The UV absorption spectrometric analysis showed absorbance spectra at 435 nm, while SEM and EDX studies indicated the presence of low agglomerated, monodispersed spherical silver nanoparticles of different size and shapes. The presence of spherical AgNPs in the range of 10–33 nm was confirmed in the TEM images.

Family Orchidaceae, the second largest family of flowering plants, is represented by terrestrial, epiphytic, or saprophytic herbs having remarkably characteristic pollination systems. Members of this family have also been investigated for the biofabrication of nanoparticles.

Gopinath et al. (2017) successfully carried out fabrication of AgNPs using leaf extract of the orchid *Anoectochilus elatus*, a terrestrial herb native to India and commonly known as "South Indian Jewel Orchid." This is an endangered monopodial terrestrial orchid with creeping rhizome. The plant is an abundant source of several flavonoids, reducing sugars, and the active bioactive compound kinsenoside that has been explored for treatment against diabetes, hyperliposis, and breast cancer. The aqueous leaf extract of the orchid was treated with 1 mM $AgNO_3$ solution, and the nanoparticles formed were tested for bactericidal activity. Both SEM and TEM images indicated the presence of spherical nanoparticles in the range 10–45 nm having mean size of 20 nm. Potent bactericidal activity was exhibited against several bacterial species like *B. subtilis*, *S. pneumoniae*, *E. coli*, *K. pneumoniae*, *P. aeruginosa*, *S. typhi*, and *S. dysenteriae*. Sarma and Dutta (2019) followed a biogenic approach for the production of AgNPs using fresh leaves of *Rhynchostylis retusa*, popularly known as "foxtail orchid." The aqueous leaf extract was incubated with 0.3 mM $AgNO_3$ that led to formation of star- and cuboidal-shaped nanoparticles mostly having size of 16–180 nm, while some were below the 100 nm range. The AgNPs also exhibited powerful potent antibacterial effect against both forms of bacteria (Gram +ve and –ve).

References

Abdullah FH, Abu Bakar NHH, Abu Bakar M (2020) Low temperature biosynthesis of crystalline zinc oxide nanoparticles from *Musa acuminata* peel extract for visible-light degradation of methylene blue. Optik 206:164279

Adak T, Swain H, Munda S, Mukherjee AK, Yadav MK, Sundaram A, Bag MK, Rath PC (2021) Green silver nanoparticles: synthesis using rice leaf extract, characterization, efficacy, and non-target effects. Environ Sci Pollut Res Int 28(4):4452–4462

Ahmad B, Shireen F, Bashir S, Khan I, Azam S (2016) Green synthesis, characterisation and biological evaluation of AgNPs using Agave americana, Mentha spicata and Mangifera indica aqueous leaves extract. IET Nanobiotechnol 10(5):281–287

Ajayi E, Afolayan A (2017) Green synthesis, characterization and biological activities of silver nanoparticles from alkalinized *Cymbopogon citratus* Stapf. Adv Nat Sci Nanosci Nanotechnol 8:015017

Anand Raj LFA, Jayalakshmy E (2015) Biosynthesis and characterization of zinc oxide nanoparticles using root extract of *Zingiber officinale*. Orient J Chem 2015:31(1)

APG [Angiosperm Phylogeny Group] (2009) An update of the angiosperm phylogeny group classification for the orders and families of flowering plants: APG III. Bot J Linn Soc 161(2):105–121

APG [Angiosperm Phylogeny Group] (2016) An update of the angiosperm phylogeny group classification for the orders and families of flowering plants: APG IV. Bot J Linn Soc 181(1):1–20

Arumugam A, Karthikeyan C, Haja Hameed AS, Gopinath K, Gowri S, Karthika V (2015) Synthesis of cerium oxide nanoparticles using *Gloriosa superba* L. leaf extract and their structural, optical and antibacterial properties. Korean J Couns Psychother 49:408–415

Arunachalam KD, Annamalai SK (2013) Chrysopogon zizanioides aqueous extract mediated synthesis, characterization of crystalline silver and gold nanoparticles for biomedical applications. Int J Nanomed 8:2375–2384

Bankar A, Joshi B, Kumar AR, Zinjarde S (2010) Banana peel extract mediated novel route for the synthesis of silver nanoparticles. Colloids Surf A Physicochem Eng Asp 368:58–63

Baruah D, Yadav RNS, Yadav A, Das AM (2019) Alpinia nigra fruits mediated synthesis of silver nanoparticles and their antimicrobial and photocatalytic activities. J Photochem Photobiol B 201:111649

Brumbaugh AD, Cohen KA, St. Angelo SK (2014) Ultrasmall copper nanoparticles synthesized with a plant tea reducing agent. ACS Sustain Chem Eng 2:1933–1939

Buazar F, Sweidi S, Badri M, Kroushawi F (2019) Biofabrication of highly pure copper oxide nanoparticles using wheat seed extract and their catalytic activity: a mechanistic approach. Green Process Synth 8:691–702

Chandran SP, Chaudhary M, Pasricha R, Ahmad A, Sastry M (2006) Synthesis of gold nanotriangles and silver nanoparticles using *Aloe vera* plant extract. Biotechnol Prog 22:577–583

Chase MW et al (1993) Phylogenetics of seed plants: an analysis of nucleotide sequences from the plastid gene rbcL. Annal Mis Bot Gard 80:528–580

Cherian T, Ali K, Saquib Q, Faisal M, Wahab R, Musarrat J (2020) Cymbopogon citratus functionalized green synthesis of CuO-nanoparticles: novel prospects as antibacterial and antibiofilm agents. Biomol Ther 10(2):169

Cronquist A (1988) The evolution and classification of flowering plants, 2nd edn. New York Botanical Garden, Bronx

Dahlgren RMT, Clifford HT, Yeo PF (1985) Origin of the monocotyledons. In: The families of the monocotyledons. Springer, Berlin, Heidelberg

Das G, Patra JK, Debnath T, Ansari A, Shin HS (2019) Investigation of antioxidant, antibacterial, antidiabetic, and cytotoxicity potential of silver nanoparticles synthesized using the outer peel extract of *Ananas comosus* (L.). PLoS One 14(8):e0220950

El-Refai AA, Ghoniem GA, El-Khateeb AY, Hassaan MM (2018) Eco-friendly synthesis of metal nanoparticles using ginger and garlic extracts as biocompatible novel antioxidant and antimicrobial agents. J Nanostruct Chem 8:71–81

Fardsadegh B, Jafarizadeh-Malmiri H (2019) Aloe vera leaf extract mediated green synthesis of selenium nanoparticles and assessment of their in vitro antimicrobial activity against spoilage fungi and pathogenic bacteria strains. Green Process Synth 8(1):399–407

Ghosh S, Patil S, Ahire M, Kitture R, Kale S, Pardesi K, Cameotra SS, Bellare J, Dhavale DD, Jabgunde A, Chopade BA (2012) Synthesis of silver nanoparticles using *Dioscorea bulbifera* tuber extract and evaluation of its synergistic potential in combination with antimicrobial agents. Int J Nanomed 7:483–496

Givnish TJ, Zuluaga A, Spalink D, Soto Gomez M, Lam VKY, Saarela JM, Sass C, Iles WJD, Lima de Sousa DJ, Leebens-Mack J, Pires JC, Zomlefer WB, Gandolfo MA, Davis JI, Stevenson DW, dePamphilis C, Specht CD, Graham SW, Barrett CF, Ané C (2018) Monocot plastid phylogenomics, timeline, net rates of species diversification, the power of multi-gene analyses, and a functional model for the origin of monocots. Am J Bot 105(11):1–23

Gopinath K, Devi NP, Govindarajan M, Bhakyaraj K, Kumaraguru S, Arumugam A, Alharbi NS, Kadaikunnan S, Benelli G (2017) One-pot green synthesis of silver nanoparticles using the orchid leaf extracts of *Anoectochilus elatus*: growth inhibition activity on seven microbial pathogens. J Clust Sci 28(3):1541–1550

Gopinath K, Karthika V, Gowri S, Senthilkumar V, Kumaresan S, Arumugam A (2014) Antibacterial activity of ruthenium nanoparticles synthesized using *Gloriosa superba* L. leaf extract. J Nanostruct Chem 4:83

Gunalan S, Sivaraj R, Venckatesh R (2012) Green synthesized ZnO nanoparticles against bacterial and fungal pathogens. Prog Nat Sci Mater Int 22(6):693–700

He Y, Wei F, Ma FYZ, Zhang H, Yang Q, Yao B, Huang Z, Li J, Zeng C, Zhang Q (2017) Green synthesis of silver nanoparticles using seed extract of *Alpinia katsumadai*, and their antioxidant, cytotoxicity, and antibacterial activities. RSC Adv 7:39842

Henry RJ, Harris PJ (1997) Molecular distinction between monocotyledons and dicotyledons: more than a simple dichotomy. Plant Mol Biol Rep 15:216–218

Hęś M, Dziedzic K, Górecka D, Jędrusek-Golińska A, Gujska E (2019) Aloe vera (L.) Webb.: natural sources of antioxidants– a review. Plant Foods Hum Nutr 74:255–265

Hu D, Xu H, Yi Z, Chen Z, Ye C, Wu Z, Garces HF, Yan K (2019) Green CO_2-assisted synthesis of mono- and bimetallic pd/pt nanoparticles on porous carbon fabricated from sorghum for highly selective hydrogenation of furfural. ACS Sustain Chem Eng 7:15339–15345

Ituen E, Mkpenie V, Ekemini E (2019) Adsorptive Fe-nanoparticles mediated by *Musa sapientum* peels extract as anticorrosion additive for aqueous oilfield descaling solution. Scientific African 3:e00075

Jia JL, Xu HH, Zhu L, Ye WH, Li DQ (2015) Biosynthesis of gold nanoparticles using novel bamboo (*Bambusa chungii*) leaf extracts. J Nanosci Nanotechnol 15(2):1674–1677

Kamala A, Middha SK, Karigar CS (2018) Plants in traditional medicine with special reference to *Cyperus rotundus* L.: a review. 3 Biotech 8(7):309

Khandel P, Shahi SK, Soni DK, Yadaw RK, Kanwar L (2018) Alpinia calcarata: potential source for the fabrication of bioactive silver nanoparticles. Nano Converge 5:37

Kumar BP, Arthanareeswari M, Devikala S, Sridharan M, Selvi JA, Malini TP (2019) Green synthesis of zinc oxide nanoparticles using *Typha latifolia*. L leaf extract for photocatalytic applications. Mater Today Proc 14:332–337

Kumar KP, Paul W, Sharma CP (2011) Green synthesis of gold nanoparticles with *Zingiber officinale* extract: characterization and blood compatibility. Process Biochem 46:2007–2013

Kumari P, Kumari N, Pratap K, Prasad K (2018) Cymbopogon flexuosus leaves mediated synthesis of ZnO nanoparticles: cytotoxicity assay against THP-1 human leukemia cell line. J Bionanosci 12:683–688

Maensiri S, Labuayai S, Laokul P, Klinkaewnarong J, Swatsitang E (2014) Structure and optical properties of CeO_2 nanoparticles prepared by using lemongrass plant extract solution. Jpn J Appl Phys 53:06JG14

Moreno-Luna FB, Tovar-Corona A, Herrera-Perez JL, Santoyo-Salazar J, Rubio-Rosa E, Vazquez-Cuchillo O (2019) Quick synthesis of gold nanoparticles at low temperature, by using *Agave potatorum* extracts. Mater Lett 235:254–257

Murugan K, Benelli G, Panneerselvam C, Subramaniam J, Jeyalalitha T, Dinesh D, Nicoletti M, Hwang J-S, Suresh U, Madhiyazhagan P (2015) Cymbopogon citratus-synthesized gold nanoparticles boost the predation efficiency of copepod *Mesocyclops aspericornis* against malaria and dengue mosquitoes. Exp Parasitol 153:129–138

Nagajyothi PC, Lee KD (2011) Synthesis of plant mediated silver nanoparticles using *Dioscorea batatas* rhizome extract and evaluation of their antimicrobial activities. J Nanomater 2011: 573429

Naika HR, Lingaraju K, Manjunath K, Kumar D, Nagaraju G, Suresh D, Nagabhushana H (2015) Green synthesis of CuO nanoparticles using *Gloriosa superba* L. extract and their antibacterial activity. J Taibah Univ Sci 9(1):7

Njagi EC, Huang H, Stafford L, Genuino H, Galindo HM, Collins JB, Hoag GE, Suib SL (2011) Biosynthesis of iron and silver nanoparticles at room temperature using aqueous sorghum bran extracts. Langmuir 27(1):264–271

Nyffeler R, Eggli U (2020) Introduction to the classification of monocotyledons. In: Eggli U, Nyffeler R (eds) Monocotyledons. Illustrated handbook of succulent plants. Springer, Berlin, Heidelberg

Parida UK, Bindhani BK, Nayak P (2011) Green synthesis and characterization of gold nanoparticles using onion (*Allium cepa*) extract. World J Nano Sci Eng 1:93–98

Prasad S, Srivastava SK (2021) Oxidative stress and cancer: antioxidative role of Ayurvedic plants. In: Preedy VR, Patel VB (eds) Cancer: oxidative stress and dietary antioxidants. Elsevier, Amsterdam, pp 301–310

Primo JO, Bittencourt C, Acosta S, Sierra-Castillo A, Colomer J-F, Jaerger S, Teixeira VC, Anaissi FJ (2020) Synthesis of zinc oxide nanoparticles by ecofriendly routes: adsorbent for copper removal from wastewater. Front Chem 8:571790

Rahman AU, Khan AU, Yuan Q, Wei Y, Ahmad A, Ullah S, Khan ZUH, Shams S, Tariq M, Ahmad W (2019) Tuber extract of *Arisaema flavum* eco-benignly and effectively synthesize silver nanoparticles: photocatalytic and antibacterial response against multidrug resistant engineered *E. coli* QH4. J Photochem Photobiol B 193:31–38

Sagnia B, Fedeli D, Casetti R, Montesano C, Falcioni G, Colizzi V (2014) Antioxidant and anti-inflammatory activities of extracts from *Cassia alata, Eleusine indica, Eremomastax speciosa, Carica papaya* and *Polyscias fulva* medicinal plants collected in Cameroon. PLoS One 9(8): e103999

Salunke GR, Ghosh S, Santosh Kumar RJ, Khade S, Vashisth P, Kale T, Chopade S, Pruthi V, Kundu G, Bellare JR, Chopade BA (2014) Rapid efficient synthesis and characterization of silver, gold and bimetallic nanoparticles from the medicinal plant *Plumbago zeylanica* and their application in biofilm control. Int J Nanomed 9:2635–2653

Saranya S, Vijayarani K, Pavithra S (2017) Green synthesis of iron nanoparticles using aqueous extract of *Musa ornata* flower sheath against pathogenic bacteria. Indian J Pharm Sci 79(5):688–694

Sarma J, Dutta AM (2019) Green synthesis of silver nanoparticles from *Rhynchostylis retusa* (L.) Blume leaf extract under sunlight. Int J RecentTechnol Eng 8:10421–10424

Sathishkumar M, Sneha K, Yun YS (2010) Immobilization of silver nanoparticles synthesized using *Curcuma longa* tuber powder and extract on cotton cloth for bactericidal activity. Bioresour Technol 101:7958–7965

Saxena A, Tripathi RM, Singh RP (2010) Biological synthesis of silver nanoparticles by using onion (*Allium cepa*) extract and their antibacterial activity. Dig J Nanomater Biostruct 5:427–432

Soltis DE (2000) Angiosperm phylogeny inferred from 18S rDNA, rbcL, and atpB sequences. Bot J Linn Soc 133:381–461

Soltis PS, Soltis DE (1999) Angiosperm phylogeny inferred from multiple genes as a tool for comparative biology. Nature 402:402–404

Supraja S, Mohammed AS, Chakravarthy N, Jayaprakash PA, Sagadevan E, Kasinathan MK, Sindhu S, Arumugam P (2013) Green synthesis of silver nanoparticles from *Cynodon dactylon* leaf extract. Int J Chem Tech 5(1):271–277

Syafiuddin A, Salmiati, Hadibarata T, Beng Hong Kueh A, Razman Salim M (2017) Novel weed-extracted silver nanoparticles and their antibacterial appraisal against a rare bacterium from river and sewage treatment plant. Nanomaterials 8(1):9

Vadlapudi V, Amanchy R (2017) Phytofabrication of silver nanoparticles using *Myriostachya wightiana* as a novel bioresource, and evaluation of their biological activities. Braz Arch Biol Technol 60:e17160329

Valsalam S, Agastian P, Esmail GA, Ghilan AM, Al-Dhabi NA, Arasu MV (2019) Biosynthesis of silver and gold nanoparticles using *Musa acuminata colla* flower and its pharmaceutical activity against bacteria and anticancer efficacy. J Photochem Photobiol B 201:111670

Venkatadri B, Shanparvish E, Rameshkumar MR, Arasu MV, Al-Dhabi NA, Ponnusamy VK, Agastian P (2020) Green synthesis of silver nanoparticles using aqueous rhizome extract of *Zingiber officinale* and *Curcuma longa*: Invitro anti-cancer potential on human colon carcinoma HT-29 cells. Saud J Biol Sci 27(11):2980–2986

Vijay Kumar PPN, Shameem U, Kollu P, Kalyani RL, Pammi SVN (2015) Green synthesis of copper oxide nanoparticles using *Aloe vera* leaf extract and its antibacterial activity against fish bacterial pathogens. BioNanoSci 5:135–139

Waghmode S, Chavan P, Kalyankar V, Dagade S (2013) Synthesis of silver nanoparticles using *Triticum aestivum* and its effect on peroxide catalytic activity and toxicology. J Chem 2013: 265864

Wu F, Chen Y, Li G, Zhu D, Wang L, Wang J (2019) Zinc oxide nanoparticles synthesized from *Allium cepa* prevents UVB radiation mediated inflammation in human epidermal keratinocytes (HaCaT cells). Artif Cells Nanomed Biotechnol 47(1):3548–3558

Zhang Z, Xin G, Zhou G, Li Q, Veeraraghavan VP, Mohan SK, Wang D, Liu F (2019) Green synthesis of silver nanoparticles from *Alpinia officinarum* mitigates cisplatin-induced nephrotoxicity via down-regulating apoptotic pathway in rats. Artif Cells Nanomed Biotechnol 47: 3212–3221

Biological Synthesis of Nanoparticles: Dicotyledons

12

Abstract

Dicots form a paraphyletic group, with basal angiosperms, and are considered more primitive than monocots. Bionanofabrication has been successfully carried out using root, stem, flowers, fruit, seed, tuber, bark, peel, and pulp. Plant extracts of different dicot species have diverse chemical composition which influences the unique characteristics of the nanoparticles. The uses of different plant parts of dicotyledonous plants for nanoparticle production have been exhaustively discussed.

12.1 Introduction

Angiosperms include flowering plants that possess seeds enclosed in a capsule and dominate the terrestrial ecosystems. There are about 300,000 species which are major sources of food and medicine for human beings. Angiosperms have dominated Earth's vegetation since the mid-Cretaceous and are widely distributed along diverse ecological boundaries. Although some researchers are of the view that angiosperms originated in late Palaeozoic or early Mesozoic, credible fossil records indicate their origin to about 140 million years ago in the early Cretaceous period. This was followed by their rapid expansion in the mid-Cretaceous and their domination during the end of the Cretaceous period. Both monophyletic (origin from as single ancestor at a single time) and polyphyletic (origin of dicots and monocots at different time from different primitive stocks) origin of angiosperms have been put forward, but some recent studies have confirmed that angiosperms are monophyletic, while their families are polyphyletic. One interesting fact put forward by Ingrouille (1992) states that angiosperms possibly originated in the tropics since there fossil evidences of early flowering plants exhibited no adaptation to temperate conditions. There seems to be a consensus that angiosperms evolved in West Gondwana, an area

S. Srivastava, A. Bhargava, *Green Nanoparticles: The Future of Nanobiotechnology*,
https://doi.org/10.1007/978-981-16-7106-7_12

equivalent to modern South America and Africa. From this region further spread took place into Laurasia and the rest of Gondwana.

Dicots form a paraphyletic group with basal angiosperms and are considered more primitive than monocots. About 200,000 dicotyledonous species are known to exist in nature. The Angiosperm Phylogeny Group (2009) classified all flowering plants into Magnoliids (4 orders), Monocots (7 orders), Commelinids (5 orders), and Eudicots (36 orders), along with some taxa having uncertain affinity. The chief characteristics of dicotyledonous plants are as below:

1. Tetramerous or pentamerous flowers
2. Three furrows or pores in pollen grains
3. Two cotyledons in seed
4. Vascular bundles in stem arranged in concentric rings
5. Reticulate leaf venation
6. Presence of secondary growth

12.2 Dicotyledonous Plants in Green Synthesis of Nanoparticles

Plant extracts have the unique ability to reduce metal ions and produce nanoparticles of different shapes and sizes. This property, although known for long, was left unutilized. It has only been in the last two decades that studies on the use of plant extracts in the eco-friendly synthesis of nanoparticles have come to the forefront. Biofabrication of nanoparticles has been successfully accomplished using different plant parts of dicotyledonous plants like root, stem, flowers, fruit, seed, tuber, bark, peel, and pulp. Plant extracts may act as reducing as well as stabilizing agents during the biosynthesis of nanomaterials. Plant extracts by virtue of their diverse chemical composition influences the unique characteristics of the nanoparticles. For plant extract-mediated synthesis of nanomaterials, the extract of the plant/specific plant part is treated with an aqueous solution of the metal salt. A remarkable feature of this procedure is that the reaction at most times is carried out at ambient/room temperature and completes within a short duration. The size and shape of the nanoentities is usually regulated by alteration in the metal concentration and the amount of plant extracts in the reaction medium.

12.2.1 Biological Synthesis of Nanoparticles Using Leaves

Leaves are the principal lateral appendages of tracheophytes that perform photosynthesis along with a range of different functions. Leaves are low in calories and fat, but high in dietary fibers, minerals, and phytochemicals like vitamin C, carotenoids, and lutein. Table 12.1 provides information on the use of plant leaves of different dicotyledonous species in nanoparticle production. Various metabolites present in leaf extracts serve as the reducing and stabilizing agents. The plants used include food crops, nonfood crops, and underutilized crops. Even weeds like *Chenopodium*,

Table 12.1 Nanoparticles synthesized using leaves of dicotyledonous plants

Scientific name	Plant family	Type of nanoparticle	Size (nm)	Biological activity	References
Andrographis echioides	Acanthaceae	Ag	68.06	Antibacterial, anticancer	Elangovan et al. (2015)
Borago officinalis	Boraginaceae	Ag	30–80	Antibacterial, anticancer	Singh et al. (2017)
Catharanthus roseus	Apocynaceae	Pd TiO2	38 25–110	Photocatalytic Antiparasitic	Kalaiselvi et al. (2015) Velayutham et al. (2012)
Carica papaya	Caricaceae	Fe	33	–	Latha and Gowri (2014)
Croton sparsiflorus	Euphorbiaceae	Ag	22–52	Antibacterial, antifungal	Kathiravan et al. (2015)
Camellia sinensis	Theaceae	Ni ZnO	43.87–48.76 16	Photocatalytic Antimicrobial	Bibi et al. (2017) Senthilkumar and Sivakumar (2014)
Combretum indicum	Combretaceae	Ag	12	Antibacterial	Birusanti et al. (2019)
Derris trifoliata	Fabaceae	Ag	20	Larvicidal	Kumar et al. (2017)
Eucalyptus	Myrtaceae	Fe	20–80	–	Wang et al. (2014)
Euphrasia officinalis	Orobanchaceae	Ag Au	40.37 49.72	Anticancer, biofilm inhibition Anticancer, biofilm inhibition	Singh et al. (2018) Singh et al. (2018)
Jasminum officinal	Oleaceae	Ag	9.22	Anticancer	Elhawary et al. (2020)
Leucas aspera	Lamiaceae	Ag	25–80	Larvicidal	Suganya et al. (2014)
Malva sylvestris	Malvaceae	CuO	5–30	Antibacterial	Awwad et al. (2015)
Moringa oleifera	Moringaceae	Bi Cu	40.4–57.8 23	Antimicrobial, antioxidant Antimicrobial, antioxidant	Das et al. (2020a) Das et al. (2020b)
Ocimum basilicum	Lamiaceae	Sulfur	23	Antistress	Ragab and Saad-Allah (2020)
Olea purpurea	Oleaceae	Ag	20–25	Antibacterial	Khalil et al. (2014)
Phyllanthus amarus	Phyllanthaceae	CuO	20	Antibacterial	Acharyulu et al. (2014)

(continued)

Table 12.1 (continued)

Scientific name	Plant family	Type of nanoparticle	Size (nm)	Biological activity	References
Punica granatum	Punicaceae	Ag	26.22	Antimicrobial	Swilam and Nematallah (2020)

Parthenium, *Cannabis*, and *Lantana* have also been exploited for green fabrication of nanoparticles. Dwivedi and Gopal (2010) utilized leaf extract of *Chenopodium album* (family: Amaranthaceae), that is one of the top 10 weeds of the world, for the biosynthesis of silver and gold nanoparticles of quasi-spherical shape having size of 10–30 nm. Zahir and Rahuman (2012) successfully carried out eco-friendly synthesis of AgNPs using hexane, chloroform, ethyl acetate, acetone, methanol, and leaf extract of *Euphorbia prostrata* (family: Euphorbiaceae), a weed species native to the Caribbean and commonly known as "ground spurge" or "prostrate Sandmat." Manjamadha and Muthukumar (2016) developed an eco-friendly method of synthesis of spherical AgNPs having mean size of 33.8 nm using *Lantana camara* (family: Verbenaceae) L. leaf extract. Ahsan et al. (2020) accomplished biofabrication of AgNPs having size of 187.87 ± 4.89 nm from *Parthenium hysterophorus* (family: Asteraceae) leaf extract. Similarly, AgNPs have been biologically produced using leaf extract of *Tridax procumbens* (family: Asteraceae) (mean size: 99 nm) by Chopade and Kamble (2021) and *Chenopodium murale* (family: Amaranthaceae) (mean size: 30–50 nm) by Abdel-Aziz et al. (2014). Gold nanoparticles of ~100 nm were fabricated by Jisha et al. (2012) using leaf extract of *Tephrosia purpurea* (family: Fabaceae), an erect, annual, or perennial legume wasteland weed.

Leaves of *Calotropis gigantea* (family: Apocynaceae), a pantropical wild herbaceous weed species commonly known as "crown flower," "giant milkweed," and "shallow wort," have been utilized to produce diverse types of nanoparticles. Copper oxide nanoparticles having mean size of 20–30 nm were synthesized by Sharma et al. (2015) using leaf extract of *C. gigantea* and cupric nitrate salt. Ayodhya and Veerabhadram (2017) synthesized CdS nanoparticles using aqueous extract of milkweed leaves, 1 mM of the colorless salt $Cd(CH_3CO_2)_2$, and 1 mM $Na_2S.9H_2O$. Spherical nanoparticles having mean size of 12 nm were obtained that were photostable and exhibited strong antibacterial and antifungal activity. Tin oxide/stannic oxide (SnO_2) nanoparticles were biosynthesized by Bhosale et al. (2018) using the leaf extract of *C. gigantea* and 50 mM solution of tin chloride. The nanoparticles were spherical having a mean size of 35 nm. Din et al. (2018) developed a biofriendly method for the synthesis of nickel and nickel oxide nanoparticles using hydro-methanolic extract of fresh leaves of *C. gigantea*. Alcohols and halogens played a major role in the reduction process, while amines functioned as the capping agent. The particle size measured using XRD analysis ranged from 20 to 40 nm for nickel nanoparticles and had size dimensions of 60 nm for nickel oxide nanoparticles. Both types of nanoparticles exhibited strong antibacterial activity against *Pseudomonas aeruginosa* that was comparable to that of chloramphenicol. Hii et al. (2018) used leaf extract of giant milkweed for the formation and stabilization of magnesium oxide (MgO) nanoparticles and their further encapsulation using PVP and Eudragit L. The mean size of MgO nanoparticles was 48.38 nm, and the crystallite size was calculated as 8 nm. Similarly zinc oxide nanoparticles of 30–35 nm dimensions were biologically produced using *C. gigantea* leaf extract by Vidya et al. (2013).

12.2.2 Biological Synthesis of Nanoparticles Using Roots

Roots of numerous dicotyledonous plants have been used for nanofabrication. However, Table 12.2 clearly shows that the biological synthesis is largely limited to production of silver and gold nanoparticles. Meager studies have been carried out with other metals and metal oxides. The recent trend involves the use of hairy root culture for nanoparticle production. Hairy roots are differentiated tissue from callus and developed by infection of *Agrobacterium rhizogenes* (family: Rhizobiaceae), a soilborne bacterium present near plant roots which results in the development of the characteristic "hairy root syndrome."

Over the last some few decades, *R. rhizogenes* mediated transformation has facilitated elucidation of physiological processes, biosynthetic pathways, and generation of plant-derived molecules, along with assistance in molecular breeding, improvement in phytoremediation strategies, and production of therapeutic recombinant proteins that find use in pharmacological, cosmetic, and nutraceutical industries. The first use of using hairy root cultures for synthesis of nanostructures was reported by Al-Shalabi in 2010 using hairy roots of tomato (*Solanum lycopersicum*; family: Solanaceae). This eco-friendly method of biosynthesis of CdS quantum dots involved exposing the hairy roots of tomato to 100 μM Cd. Cadmium sulfide nanoparticles with crystal sizes of 2–20 nm were obtained. It was interesting to note that the tomato hairy roots accumulated around 17% of the cadmium added to the culture medium. Of this, only approximately 19% of the accumulated heavy metal was recovered by extraction of the root biomass, and about 1.4% of the accumulated metal was recovered in the form of strongly fluorescing nanocrystals. Another study involving the production of luminescent cadmium sulfide quantum dots (CdSQDs) was carried out by Borovaya et al. (2014) through hairy root culture of *Linaria maroccana* L. (family: Plantaginaceae). The root extract of the plant was incubated with $CdSO_4.H_2O$ and $Na_2S.9H_2O$. The cadmium sulfide nanocrystals exhibited absorption peaks for UV-visible spectra at 362, 398, and 464 nm and luminescent peaks at 425, 462, and 500 nm. TEM images revealed the biologically produced nanoentities were spherical having size of 5.5–6.9 nm (Fig. 12.1). However, numerous nanoparticles having diameter of 2–5 nm were also present in the medium. These nanoparticles were elliptic or spherical in shape without structural defects and crystalline structure.

Biosynthesis of CdSQDs using extracts of the regenerated hairy roots of radish (*Raphanus sativus* L.; family: Solanaceae) was successfully accomplished by Gholami et al. (2020). The hairy root extract of radish was added to 0.025 M

Table 12.2 Nanoparticles synthesized using roots of dicotyledonous plants

Scientific name	Plant family	Type of nanoparticle	Size (nm)	Biological activity	References
Arnebia hispidissima	Boraginaceae	Ag	10–200	Anticancer	Nindawat and Agrawal (2019)
Berberis lyceum	Berberidaceae	Ag	10–100	Antimicrobial	Mehmood et al. (2016)
Berberis vulgaris	Berberidaceae	Ag	30–70	Antibacterial	Behravan et al. (2019)
Bistorta amplexicaulis	Polygonaceae	Ag	50–60	Catalytic	Ahmed et al. (2020)
Catharanthus roseus	Apocynaceae	Ag	35–55	Larvicidal	Rajagopal et al. (2015)
Decalepis hamiltonii	Apocynaceae	Ag	32.5	Antibacterial, antioxidant	Rashmi and Sanjay (2017)
Delphinium denudatum	Ranunculaceae	Ag	⩽85	Antibacterial, larvicidal	Suresh et al. (2014)
Diospyros paniculata	Ebenaceae	Ag	17	Antimicrobial	Rao et al. (2016)
Diospyros sylvatica	Ebenaceae	Ag	10	Antimicrobial	Pethakamsetty et al. (2017)
Erythrina indica	Fabaceae	Ag	20–118	Antibacterial, cytotoxic	Rathi Sre et al. (2015)
Euphorbia condylocarpa	Euphorbiaceae	Pd/Fe3O4,	39	Catalytic	Nasrollahzadeh et al. (2015a)
Morinda citrifolia	Rubiaceae	Ag Au	30–55 12.17	Cytotoxic –	Suman et al. (2013) Suman et al. (2014)
Polygala tenuifolia	Polygalaceae	Au	9.77	Anticoagulant	Jun et al. (2014)
Rheum palmatum	Polygonaceae	Ag	121	Antibacterial	Arokiyaraj et al. (2017)
Withania somnifera	Solanaceae	Ag	–	Antibacterial	Gaurav et al. (2020)
Zingiber officinale	Zingiberaceae	Ag Au	10–20 5–20	Antibacterial Antibacterial	Velmurugan et al. (2014) Velmurugan et al. (2014)

$CdSO_4$ solution, kept in the dark for 3 days, and then 0.5 M Na_2S was added. CdSQDs were spherical and found to be well dispersed and had mean size of 2–7 nm. FTIR analysis showed the active presence of different functional groups on the surface of the newly synthesized CdS quantum dots.

The synthesis of silver nanoparticles using extracts from the hairy root cultures of *Artemisia tilesii* and *Artemisia annua* (family: Asteraceae) was reported for the first time by Kobylinska et al. (2020). The effect of the type of solvent, temperature, flavonoid concentration, and reducing power of the hairy root extracts on the

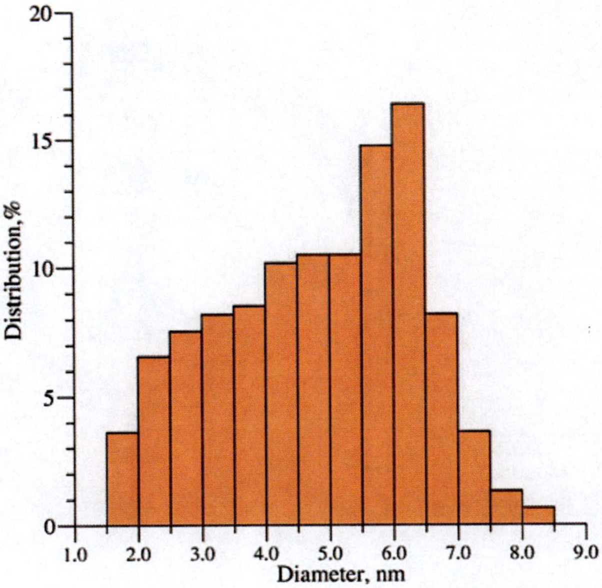

Fig. 12.1 TEM image showing particle size distribution during biosynthesis of cadmium sulfide quantum dots using *Linaria maroccana* (reprinted from Borovaya et al. 2014)

nanoparticle size and yield was reported. Silver nanoparticles of diverse shapes like spherical, oval, and triangular and varied size ranging from 5 to 100 nm were obtained. The TEM analysis showed immense variability in the shapes of the nanoparticles obtained from the extracts using different root clones. The nanoparticles obtained from ethanol extracts were aggregated and irregular and of the dimensions of 10–50 nm. However, when aqueous extract was used, the sizes of the nanoentities decreased to 10–30 nm at 80 °C. Biosynthesized AgNPs showed surface plasmon resonance in the range of 400–450 nm. The antibacterial activity of the AgNPs was assessed against both Gram-positive (*Staphylococcus aureus*) and Gram-negative (*Pseudomonas aeruginosa*, *Escherichia coli*, and *Candida albicans*) bacteria. It was observed that the antibacterial activity of Ag nanoparticles was more effective than $AgNO_3$ solution. It was concluded that hairy root extracts of both *A. annua* and *A. tilesii* can be effectively used for production of AgNPs. Thus, hairy roots can be a convenient and feasible biological system that could be utilized for production of metal-based nanoparticles. However, meager studies have been conducted so far with respect to nanoparticle production using hairy roots.

12.2.3 Biological Synthesis of Nanoparticles Using Fruits

Fruits are rich sources of vitamins, dietary fiber, and a wide range of minerals, vitamins, and numerous phytochemicals having antioxidant properties which are good for human health. The intake of fruits and its products not only alleviates the levels of vital micronutrients and improves health but also significantly reduces the chances of different human disorders. The extensive range of phytochemicals found

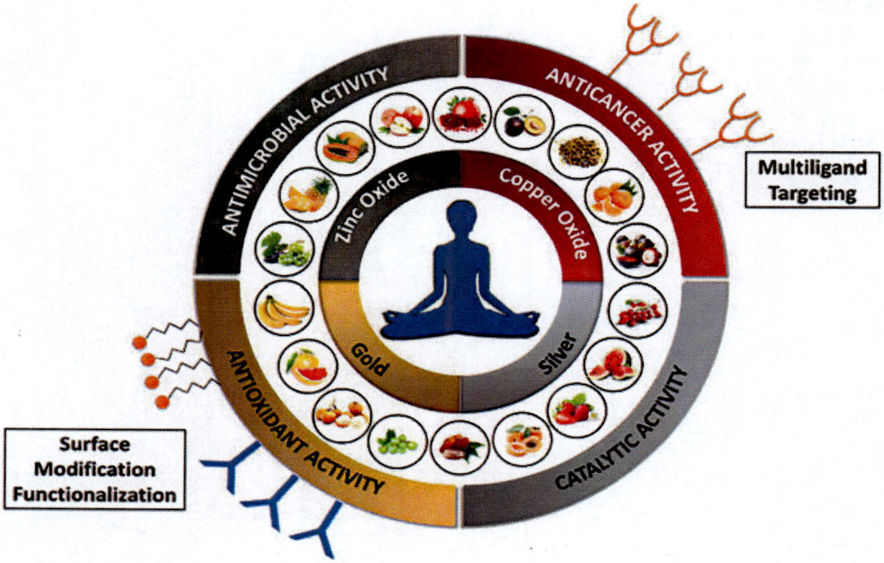

Fig. 12.2 Biological activity of different metal nanoparticles synthesized from flowers (reprinted from Kumar et al. 2020a)

in fruits that confer numerous health benefits include carotenoids, anthocyanins, flavanones, catechins, flavonols, flavones, and isoflavones. Besides preventing diseases ranging from cardiovascular disorders, different types of cancers, urinary tract infections, cataract, and macular degeneration, most of these phytochemicals also act as antioxidants in improving human health. The usage of fruits has been quite successful in the green synthesis of nanomaterials, and the nanoparticles obtained from fruits are reported to possess various biological activities (Fig. 12.2) (Table 12.3).

The method for biosynthesis of nanoparticles from fruits involves preparation of the fruit extract by blending or grinding and thereafter its exposure to metal ion solution. Standardization needs to be done with respect to concentration of fruit extract as well as the salt, pH, temperature, duration of incubation, and the type of solvent used. Figure 12.3 illustrates the mechanism involved in the biofabrication of nanoparticles using fruit extract. Fruits confer numerous advantages in terms of cost, energy efficiency, safety, and being eco-friendly in nanoparticle production. An interesting observation with respect to use of fruits in nanoparticle production is that it has been largely limited to some metal (gold and silver) and metal oxide (copper oxide and zinc oxide) nanoparticles (Table 12.3). Earnest research efforts are needed, and a lot needs to be explored with respect to other nanoparticles.

Table 12.3 Nanoparticles synthesized using fruits of dicotyledonous plants

Scientific name	Plant family	Type of nanoparticle	Size (nm)	Biological activity	References
Artocarpus heterophyllus	Moraceae	Au	20–25	Antibacterial	Basavegowda et al. (2015)
Capparis spinosa	Brassicaceae	Cu_2O	17–41	Antibacterial	Ebrahimi et al. (2017)
Cinnamomum camphora	Lauraceae	Ag	20.3	Antifungal	Huang et al. (2019)
Citrullus lanatus	Cucurbitaceae	Ag	109.97	Antibacterial, anticandidal, antioxidant	Patra et al. (2016)
Cornus mas	Cornaceae	Ag Au	16 19	Immunomodulatory Immunomodulatory	Filip et al. (2019) Filip et al. (2019)
Crataegus pentagyna	Rosaceae	Ag	25–45	Antibacterial	Ebrahimzadeh et al. (2020)
Cucumis melo	Cucurbitaceae	Au	78.11	Antibacterial, anticandidal, antioxidant	Patra and Baek (2016)
Dillenia indica	Dilleniaceae	Ag	40–100	–	Singh et al. (2013)
Emblica officinalis	Phyllanthaceae	Ag MgO	15 27	Antibacterial Antibacterial	Ramesh et al. (2015) Ramanujam et al. (2014)
Ficus carica	Moraceae	FeO	9	–	Aksu Demirezen et al. (2019)
Fragaria ananassa		Cu_2O	10–30	Antifungal, antioxidant	Hemmati et al. (2020)
Garcinia indica	Clusiaceae	Ag	5–30	Antibacterial, antioxidant	Sangaonkar et al. (2018)
Lycopersicon esculentum	Solanaceae	ZnO	5.6–133	Antibacterial	Ogunyemi et al. (2019)
Momordica cymbalaria	Cucurbitaceae	Ag	15.5	Antimicrobial, antioxidant	Swamy et al. (2015)
Prunus persica	Rosaceae	Au	39.90	Antibacterial, anticandidal, antioxidant	Patra and Baek (2016)
Psidium guajava	Myrtaceae	Cu_2O	15–30	Antibacterial	Carloling et al. (2015)
Solanum mammosum	Solanaceae	Ag	10–14	Larvicidal	Pilaquinga et al. (2019)
Solanum nigrum	Solanaceae	Ag	20–30	Biofilm inhibition	Malaikozhundan et al. (2017)
Tamarindus indica	Fabaceae	Ag	6–8	Antibacterial	Jayaprakash et al. (2017)

(continued)

Table 12.3 (continued)

Scientific name	Plant family	Type of nanoparticle	Size (nm)	Biological activity	References
Tribulus terrestris	Zygophyllaceae	Ag	16–28	Antimicrobial	Gopinath et al. (2012)
Ziziphus jujuba	Rhamnaceae	ZnO	29	Photocatalytic	Golmohammadi et al. (2020)

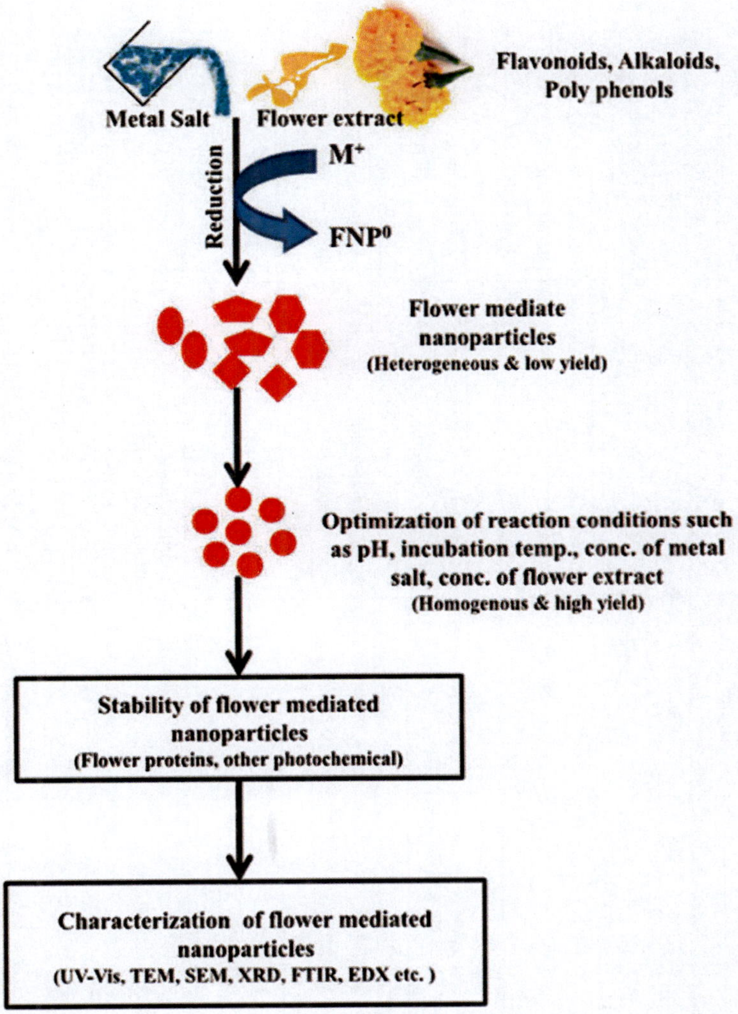

Fig. 12.3 Methodology of nanoparticle production using flower extracts (reprinted from Kumar et al. 2020c)

12.2.4 Biological Synthesis of Nanoparticles Using Seeds

Table 12.4 lists the use of seeds of some dicotyledonous plants in nanoparticle production. Metal and metal oxide nanoparticles of different size ranging from 2 to 162 nm have been successfully synthesized using seed extracts.

Cocoa (*Theobroma cacao* L.), a plant of family Malvaceae, is well known for being the raw material of chocolate. The seed extract of this plant species was used by Nasrollahzadeh et al. (2015b) for eco-friendly synthesis of palladium/CuO nanoparticles utilizing $PdCl_2$ (0.2 mmolar)/$CuCl_2.2H_2O$ (1.0 mmolar). The

Table 12.4 Nanoparticles synthesized using seeds of dicotyledonous plants

Scientific name	Plant family	Type of nanoparticle	Size (nm)	Biological activity	References
Artocarpus hirsutus	Moraceae	Ag	25–40	Antibacterial	Shobana et al. (2020)
Bunium persicum	Apiaceae	Ag	20–50	Catalytic	Rostami-Vartooni et al. (2016)
Camellia oleifera	Theaceae	Fe-Sapogenin	162	Neuroprotection	Yang et al. (2017)
Citrullus colocynth	Cucurbitaceae	FeO	6–15	Antimicrobial	Farouk et al. (2020)
Coffea arabica	Rubiaceae	Ag	20–30	Antibacterial	Dhand et al. (2016)
Croton tiglium	Euphorbiaceae	Ag	Nr	Ameliorative	Aboulthana et al. (2020)
Cuminum cyminum	Apiaceae	Ag	100	Anticancer	Dinparvar et al. (2020)
Dimocarpus longum	Sapindaceae	ZnO	10–100	Photocatalyst	Chankaew et al. (2019)
Hibiscus cannabinus	Malvaceae	Ag	7–11	Anticancer, antimicrobial	Adnan et al. (2020)
Mimusops elengi	Sapotaceae	Ag	12.80–30.48	Antibacterial, antioxidant	Kiran Kumar et al. (2014)
Nigella arvensis	Ranunculaceae	Ag	2–15	Antibacterial, cytotoxicity	Chahardoli et al. (2017)
Peganum harmala	Nitariaceae	Au	52.04	Antibacterial	Moustafa and Alomari (2019)
Pistacia atlantica	Anacardiaceae	Ag	27	Antibacterial	Sadeghi et al. (2015)
Punica granatum	Punicaceae	Cu	40–80	Photocatalysis	Nazar et al. (2018)
Putranjiva roxburghii	Putranjivaceae	Ag	~8	Anticancer	Balkrishna et al. (2020)
Rumex crispus	Polygonaceae	CuO	8–60	Catalytic	Rostami-Vartooni (2017)
Silybum marianum	Asteraceae	Ag	18.12–13.20	Antidiabetic, anti-inflammatory	Shah et al. (2020)
Tectona grandis	Verbenaceae	Ag	10–30	Antimicrobial	Rautela et al. (2019)
Theobroma cacao	Malvaceae	Pd/CuO	40	Catalytic	Nasrollahzadeh et al. (2015b)
Trigonella foenum-graecum	Fabaceae	Ag FeO	~20 ~20	Antibacterial, antioxidant Antibacterial, antioxidant	Deshmukh et al. (2019) Deshmukh et al. (2019)

nanoparticles having dimensions of 40 nm were also evaluated for the catalytic activity under aerobic conditions. Another beverage plant, *Coffea arabica* (family: Rubiaceae), also known as "Arabian coffee," was used by Dhand et al. (2016) to biologically produce AgNPs using the dried roasted bean extract. The silver nanoparticles (size: 20–30 nm) were crystalline and exhibited a cubic, face centered lattice having typical (111), (200), (220), and (311) orientations. Yang et al. (2017) synthesized iron and sapogenin nanostructures using sapogenin extracted from the defatted seeds of oil-seed camellia (*Camellia oleifera*; family: Theaceae) and 0.5 mM $FeCl_3.6H_2O$ solution. Spherical nanoparticles having mean size of 162 nm were obtained which showed potent antioxidative, anti-inflammatory, and analgesic activities. Silver nanoparticles have been produced by Aboulthana et al. (2020) using the seed extract of *Croton tiglium* (family: Euphorbiaceae), a medicinally important plant used as a purgative. The AgNPs exhibited beneficial effect against azoxymethane induced colon cancer by restoring their levels to pretreatment values.

12.2.5 Biological Synthesis of Nanoparticles Using Flowers

Flowers have been used by human beings since time immemorial not only for their esthetic appeal but also for their distinct medicinal properties in the conventional medicine systems. Flowers are the sources of vegetables, spices, honey, dyes, and perfumes and serve as flavoring agent in beer production and help in pollination. The unique chemical properties of flowers especially the presence of large amounts of flavonoids, coumarins, glucosides, anthocyanins, terpenoids, sterol, and xanthones make them ideal candidates for the biofabrication of nanoparticles. Figure 12.4 provides a generalized approach for the green synthesis of nanoparticles using flowers as source material. Flowers have been largely utilized for the synthesis of metal and metal oxide nanostructures. The metal nanoparticles synthesized from flowers have been found to be mostly spherical in shape, while metal oxide nanoparticles were spherical, cuboid, and even flower shaped as in the case of manganese oxide nano-flowers obtained utilizing flower extracts of rosemary (*Rosmarinus officinalis* L.), a fragrant evergreen shrub by Abdallah et al. (2019). The rosemary extract was constantly stirred at 70 °C for 4 h. SEM analysis showed that the manganese nanostructures having uniform distribution over the entire surface had dimensions of less than 20 nm. Although most studies with nanoparticle production using flowers has been done with gold and silver (Table 12.5), production of other metal nanoparticles is gaining momentum due to their unique properties.

The utilization of flowers in the biogenesis of diverse nanostructures is gaining increased attention, but a lot needs to be done with respect to the standardization of protocols, scaling-up of production, stability, and the detailed mechanism behind the process. These aspects need to be looked into greater depth and require extensive research in the coming years.

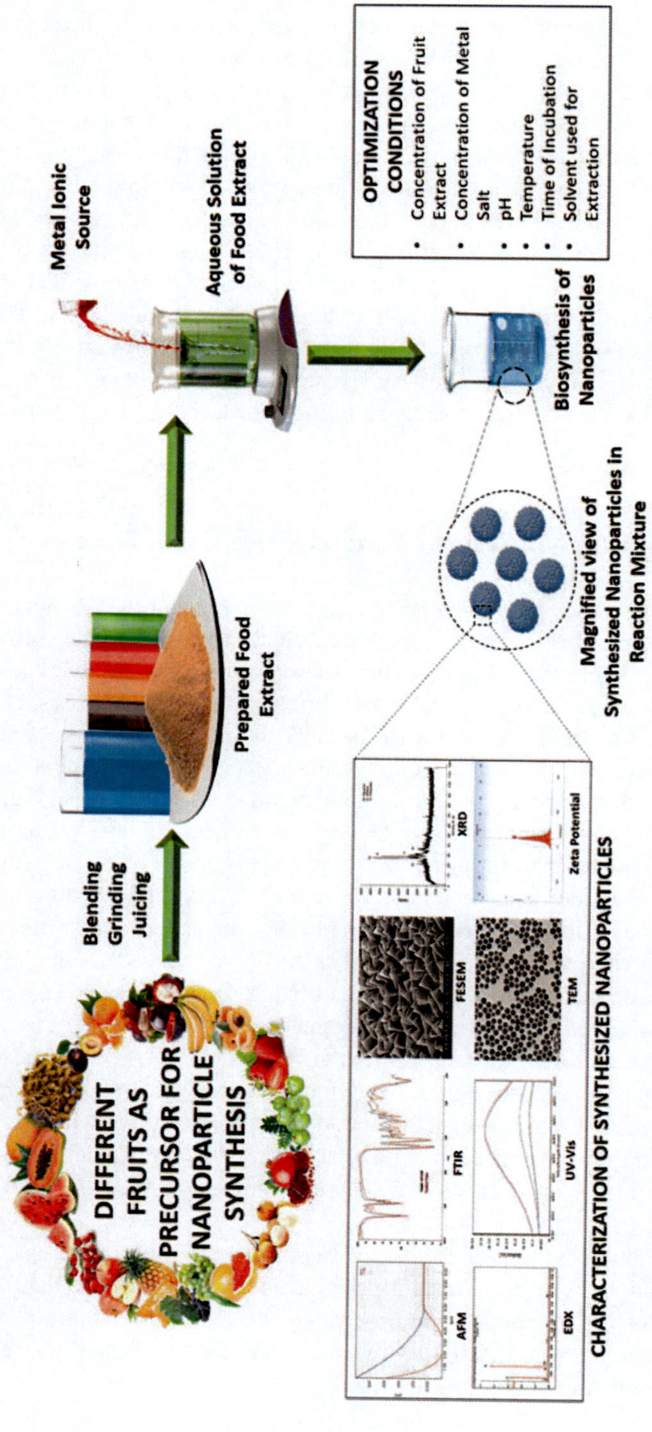

Fig. 12.4 A general mechanism of nanoparticle formation from flower extract. M^+ (metal); FNP^0 (flower nanoparticles) (reprinted from Kumar et al. 2020a)

Table 12.5 Nanoparticles synthesized using flowers of dicotyledonous plants

Scientific name	Plant family	Type of nanoparticle	Size (nm)	Biological activity	References
Achillea biebersteinii	Asteraceae	Ag	12	Antiangiogenic	Baharara et al. (2014)
Bauhinia purpurea	Fabaceae	Ag	20	Antibacterial	Chinnappan et al. (2018)
Caesalpinia pulcherrima	Caesalpiniaceae	Ag	12	Antimicrobial, antioxidant	Moteriya and Chanda (2017)
Cassia alata	Fabaceae	CuO	110–280	–	Jayalakshmi and Yogamoorthi (2014)
Cassia auriculata	Fabaceae	Ag ZnO	10–35 -	Catalytic -	Muthu and Priya (2017) Ramesh et al. (2014)
Clitoria ternatea	Fabaceae	Ag	53–300	Anticancer	Srinivas et al. (2019)
Coccinia grandis	Cucurbitaceae	Cu	18–20	Catalytic	Devi and Aharuzzaman (2018)
Gnidia glauca	Thymelaeaceae	Au	50–150	Chemocatalytic	Ghosh et al. (2012)
Hydrangea paniculata	Hydrangeaceae	Mg	56–107	Antibacterial, antioxidant	Karunakaran et al. (2017)
Lonicera japonica	Caprifoliaceae	Au	10–40	Anticancer	Patil et al. (2019)
Malva sylvestris	Malvaceae	Ag	20–40	Antibacterial	Mahmood et al. (2016)
Matricaria chamomilla	Asteraceae	CuO ZnO	140 49.8–191.0	Antioxidant Antibacterial	Duman et al. (2016) Ogunyemi et al. (2019)
Moringa oleifera	Moringaceae	Ag	8	Antibacterial, sensor	Bindhu et al. (2020)
Nyctanthes arbortristis	Pleaceae	Ag	5–20	Antibacterial	Gogoi et al. (2015)
Piper longum	Piperaceae	Ag	10–42	Antibacterial	Huang et al. (2020)
Rosa chinensis	Rosaceae	Ag	9–20	Antibacterial	Meng and Sun (2016)
Tagetes sp.	Asteraceae	Cd	–	Larvicidal	Hajra et al. (2016)
Tussilago farfara	Asteraceae	Ag Au	13.57 18.20	Antibacterial, anticancer Antibacterial, anticancer	Lee et al. (2019) Lee et al. (2019)
Wedelia urticifolia	Asteraceae	Ag	30	Antibacterial	Rather et al. (2020)

12.2.6 Biological Synthesis of Nanoparticles Using Bark

Bark refers to the tissues outside the vascular cambium in the stem and roots of woody plants. Bark is differentiated into an outer bark comprised of a fat, suberin-rich, dead tissue and the inner bark that is composed of mostly living cells referred to collectively as the phloem. Bark consists primarily of lignin (40%) but also contains various biopolymers, tannins, lignin, suberin, and polysaccharides. Bark provides protection to the inner components of the plant by acting as a barrier to outside interferences, thus preserving the internal system of the tree, allowing continued optimum function. Apart from protective role, bark has other accessory functions like providing mechanical strength, aeration, water storage, microhabitat for organisms, and even photosynthesis as in the green bark of *Acer tegmentosum*. The diversity of bark morphology is also taxonomically important since it can serve as an important identification toll at the family and species level for trees in the tropical region. In nanobiotechnology, bark has been extensively explored in the biosynthesis of metallic nanoparticles by serving as reducing agents (Table 12.6). The production of nanoparticles in this case is aided by the diverse functional groups present in the bark, namely, phenolics, flavonoids, lignin, stilbenes, alkaloids, monoterpenes, sesquiterpenes, triterpenes, saponins, carbohydrates, and proteins.

12.2.7 Biological Synthesis of Nanoparticles Using Other Plant Parts

A number of other plant parts like peels, rhizome, pollen grains and stem, and exudates have been successfully utilized in the biofabrication of nanoparticles (Table 12.7).

Plant gums, the hydrophilic hydrocolloids that ooze out as a result of mechanical injury, have been used for biological synthesis of diverse nanoparticles. Natural gums are more advantageous in producing nanoparticles since the functional groups present in it help in better surface stabilization of nanoparticles. Gums are known to reduce metal ions at a more rapid rate as compared to other biomaterials. Apart from metal nanoparticles produced using plant gum exudates, nonmetallic nanoparticles have also been produced using plant gums. Gallic acid (GA), a natural phenolic compound, was encapsulated using gum arabic into nanoparticles and evaluated for antioxidant and antihypertensive effects by Hassani et al. (2020). The particle size of gallic acid nanoparticles ranged from 33 to 250 nm, while the zeta potential was calculated as -15.2 mV.

Kolenčík et al. (2019) successfully accomplished green synthesis of gold nanoparticles using suspension of mixed flower pollen grains. Mixed pelletized flower pollen grains of different plant species, viz., *Brassica napus* L. (oilseed rape, Brassicaceae) 30%, *Taraxacum officinale* auct. Non Weber (dandelion, Asteraceae) 10%, *Helianthus annuus* L. (sunflower, Asteraceae) 30%, and *Trifolium pretense* L. (red clover, Fabaceae) 30%, were dispersed in redistilled water and treated with Au(III) solution prepared from chloroauric acid ($HAuCl_4$). The green production of AuNPs was confirmed by UV-vis spectra that showed a characteristic

Table 12.6 Nanoparticles synthesized using bark of dicotyledonous plants

Scientific name	Plant family	Type of nanoparticle	Size (nm)	Biological activity	References
Acacia leucophloea	Fabaceae	Ag	17–29	Antibacterial	Murugan et al. (2014)
Afzelia quanzensis	Fabaceae	Ag	10–80	Antibacterial	Moyo et al. (2015)
Albizia chevalieri	Fabaceae	Ag	~30	Antiproliferative, bactericidal	Khan et al. (2018)
Albizia lebbeck	Fabaceae	ZnO	66.25	Antimicrobial, antioxidant	Umar et al. (2018)
Alstonia scholaris	Apocynaceae	Ag	50	Antimicrobial	Shetty et al. (2014)
Artocarpus elasticus	Moraceae	Ag	19.74	–	Abdullah et al. (2015)
Azadirachta indica	Meliaceae	Ag	19.22	Antioxidative	Nahak and Sahu (2010)
Butea monosperma	Fabaceae	Ag	35	Antibacterial, anticancer	Pattanayak et al. (2017)
Cassia fistula	Fabaceae	Au	55.2–98.4	Antidiabetic	Daisy et al. (2012)
Cinnamomum cassia	Lauraceae	Ag	25–55	Antiviral	Fatima et al. (2015)
Cinnamomum verum	Lauraceae	ZnO	–	Antibacterial	Ansari et al. (2020)
Coccinia grandis	Cucurbitaceae	Au	20	Anticataract	Wang et al. (2018)
Cochlospermum gossypium	Bixaxeae	Pd	6.5	Catalytic	Rastogi et al. (2015)
Eucalyptus camaldulensis	Myrtaceae	Ag	468.7	Antiaging	Radwan et al. (2020)
Eucommia ulmoides	Eucommiaceae	Pd	12.6	Catalytic	Duan et al. (2015)
Fagus sylvatica	Fagaceae	Ag	32	Antibacterial, antioxidant	Tanase et al. (2019)
Ficus benghalensis	Moraceae	Ag	85.95	Antimicrobial, antiproliferative	Nayak et al. (2016)
Holarrhena antidysenterica	Apocynaceae	Ag	32	Larvicidal	Kumar et al. (2018)
Melia azedarach	Meliaceae	Ag	5–30	Antimicrobial	Pani et al. (2016)
Prosopis juliflora	Fabaceae	Ag	10–50	Antibacterial, anticancer	Arya et al. (2018)
Senna alata	Fabaceae	Ag	10–30	Antimicrobial	Ontong et al. (2019)
Syzygium alternifolium	Myrtaceae	Ag	44	Antimicrobial	Yugandhar et al. (2015)
Terminalia arjuna	Combretaceae	Ag	30–50	Antibacterial	Ahmed et al. (2017)
Ziziphus xylopyrus	Rhamnaceae	Ag	60–70	–	Babu et al. (2014)

Table 12.7 Nanoparticles synthesized using other parts of dicotyledonous plants

Plant part	Scientific name	Plant family	Type of nanoparticle	Size (nm)	Biological activity	References
Gum						
	Acacia Senegal	Fabaceae	Pd	9.1	Nr	Devi et al. (2011)
	Aegle marmelos	Rutaceae	Gold	2 μm–90 nm	Nr	Subramanian et al. (2016)
	Anogeissus latifolia	Combretaceae	Pd Pt	4.8 3	Antioxidant, catalytic Antioxidant, catalytic	Kora and Rastogi (2018) Kora (2019)
	Buchanania lanzan	Anacardiaceae	Ag	14.74–19.86	Antimicrobial	Siddiqui et al. (2021)
	Cochlospermum gossypium	Bixaceae	Au	4.08–12.73	Antiproliferative	Selvi et al. (2017)
	Cyamopsis tetragonolobus	Fabaceae	Au	~6.5	Biosensing	Pandey et al. (2013)
	Sterculia urens	Malvaceae	Au CuO	12 4.8–7.8	Anticancer drug delivery Antibacterial	Pooja et al. (2015) Thekkae Padil et al. (2013)
	Ferula foetida	Apiaceae	Ag	5.6–8.6	Antimicrobial	Devanesan et al. (2020)
Peel						
	Annona squamosa	Annonaceae	Pd	100	Acaricidal, insecticidal	Roopan et al. (2012)
	Citrus sinensis	Rutaceae	Ag	10–35	Antibacterial	Kaviya et al. (2011)
	Ipomoea batatas	Convolvulaceae	Ag	–	Antidiabetic	Das et al. (2019)
	Pisum sativum	Fabaceae	Ag	10–25	Antidiabetic, antioxidant	Patra et al. (2019)
Rhizome						
	Acorus calamus	Acoraceae	Ag	31.83	Antibacterial, apoptotic	Nakkala et al. (2014)
	Anogeissus latifolia	Combretaceae	Ag Pt	5.7 3	Antibacterial Antioxidant	Kora et al. (2012a) Kora (2019)
	Bergenia ciliata	Saxifragaceae	Ag	35	Antibacterial, antibiofilm	Zia et al. (2018)

	Boswellia serrata	Burseraceae	Ag	7.5	Antibacterial	Kora et al. (2012b)
	Ferula persica	Apiaceae	Au	37.05	Anticancer	Hosseinzadeh et al. (2020)
Stem						
	Euphorbia	Euphorbiaceae	Fe	13–21	–	Shah et al. (2014)
	Cissus quadrangularis	Vitaceae	Ag	42.46	Insecticidal	Santhoshkumar et al. (2012)
	Piper nigrum	Piperaceae	Ag	9–30	Antibacterial	Paulkumar et al. (2014)

peak at 549 nm. Transmission electron micrograph images indicated the presence of gold nanoparticles in the size range of 3–11 nm. About 90% of the nanoparticles were spherical, while the rest were oval and triangular in shape.

12.3 Future Perspectives

Plant extracts offer several advantages over the use of whole plants and other living forms. These are:

1. Simple procedure
2. Cost-effectiveness
3. Scalability
4. Abundance and easy availability of plants
5. Safe to handle
6. Presence of a wide array of metabolites
7. No requirement of elaborate maintenance of cell cultures

Plant-mediated synthesis of nanoparticles is thought to be assisted by various phytochemicals like ketones, terpenoids, amides, flavones, carboxylic acids, and aldehydes which lead to formation of nanoparticles in the reaction mixture. It is also known that the reducing agents from the plant extracts also act as the capping and stabilizing agent, which is an added advantage since there is no requirement for the addition of capping and stabilizing agents. However, harvesting some plant species, especially the rare, threatened, and endangered ones, may pose a risk of reducing plant diversity. Thus novel technologies need to be developed and adopted that ensure isolation of functional biomolecules which are useful in nanofabrication. The future challenge includes more knowledge regarding different biomolecules and the mechanism involved in the biosynthesis of nanoparticles so that considerable improvement is made vis-à-vis utilization of nanoparticles for diverse uses.

References

Abdallah Y, Ogunyemi SO, Abdelazez A, Zhang M, Hong X, Ibrahim E, Hossain A, Fouad H, Li B, Chen J (2019) The green synthesis of MgO nano-flowers using *Rosmarinus officinalis* L. (Rosemary) and the antibacterial activities against *Xanthomonas oryzae* pv. *Oryzae*. Biomed Res Int 2019:5620989

Abdel-Aziz MS, Shaheen MS, El-Nekeety AA, Abdel-Wahhab MA (2014) Antioxidant and antibacterial activity of silver nanoparticles biosynthesized using *Chenopodium murale* leaf extract. J Saudi Chem Soc 18:356–363

Abdullah N, Ahmad M, Kamyar S (2015) Biosynthesis of silver nanoparticles using *Artocarpus elasticus* stem bark extract. Chem Cent J 9:61

Aboulthana WM, Ibrahim NE, Osman NM, Seif MM, Hassan AK, Youssef AM, El-Feky AM, Madboli AA (2020) Evaluation of the biological efficiency of silver nanoparticles biosynthesized using *Croton tiglium* L. seeds extract against azoxymethane induced colon cancer in rats. Asian Pac J Cancer Prev 21(5):1369–1389

Acharyulu NPS, Dubey RS, Swaminadham V, Kalyani RL, Kollu P, Pammi SVN (2014) Green synthesis of CuO nanoparticles using *Phyllanthus amarus* leaf extract and their antibacterial activity against multidrug resistance bacteria. Int J Eng Res Technol 3:639–641

Adnan M, Obyedul Kalam Azad M, Madhusudhan A, Saravanakumar K, Hu X, Wang MH, Ha CD (2020) Simple and cleaner system of silver nanoparticle synthesis using kenaf seed and revealing its anticancer and antimicrobial potential. Nanotechnology 31(26):265101

Ahmed F, Kabir H, Xiong H (2020) Dual colorimetric sensor for Hg^{2+}/Pb^{2+} and an efficient catalyst based on silver nanoparticles mediating by the root extract of *Bistorta amplexicaulis*. Front Chem 8:591958

Ahmed Q, Gupta N, Kumar A, Nimesh S (2017) Antibacterial efficacy of silver nanoparticles synthesized employing *Terminalia arjuna* bark extract. Artif Cells Nanomed Biotechnol 45(6):1–9

Ahsan A, Farooq MA, Ahsan Bajwa A, Parveen A (2020) Green synthesis of silver nanoparticles using *Parthenium hysterophorus*: optimization, characterization and in vitro therapeutic evaluation. Molecules 25(15):3324

Aksu Demirezen D, Yıldız YŞ, Yılmaz Ş, Demirezen YD (2019) Green synthesis and characterization of iron oxide nanoparticles using *Ficus carica* (common fig) dried fruit extract. J Biosci Bioeng 127(2):241–245

Al-Shalabi Z (2010) Production of cadmium sulphide quantum dots in tomato hairy rootcultures. Dissertation, University of New South Wales, Australia

Ansari MA, Murali M, Prasad D, Alzohairy MA, Almatroudi A, Alomary MN, Udayashankar AC, Singh SB, Asiri SMM, Ashwini BS, Gowtham HG, Kalegowda N, Amruthesh KN, Lakshmeesha TR, Niranjana SR (2020) Cinnamomum verum bark extract mediated green synthesis of ZnO nanoparticles and their antibacterial potentiality. Biomol Ther 10(2):336

Arokiyaraj S, Vincent S, Saravanan M, Lee Y, Oh YK, Kim KH (2017) Green synthesis of silver nanoparticles using *Rheum palmatum* root extract and their antibacterial activity against *Staphylococcus aureus* and *Pseudomonas aeruginosa*. Artif Cells Nanomed Biotechnol 45(2):372–379

Arya G, Kumari RM, Gupta N, Kumar A, Chandra R, Nimesh S (2018) Green synthesis of silver nanoparticles using *Prosopis juliflora* bark extract: reaction optimization, antimicrobial and catalytic activities. Artif Cells Nanomed Biotechnol 46(5):985–993

Awwad AM, Albiss BA, Salem NM (2015) Antibacterial activity of synthesized copper oxide nanoparticles using *Malva sylvestris* leaf extract. SMU Med J 2:91–101

Ayodhya D, Veerabhadram G (2017) One-pot green synthesis, characterization, photocatalytic, sensing and antimicrobial studies of *Calotropis gigantea* leaf extract capped CdS NPs. Mat Sci Eng 225:33–44

Babu S, Devadiga A, Shetty KV, Saidutta MB (2014) Synthesis of silver nanoparticles using medicinal *Zizyphus xylopyrus* bark extract. Appl Nanosci 5:755–762

Baharara J, Namvar F, Ramezani T, Hosseini N, Mohamad R (2014) Green synthesis of silver nanoparticles using *Achillea biebersteinii* flower extract and its anti-angiogenic properties in the rat aortic ring model. Molecules 19(4):4624–4634

Baker S, Rakshith D, Kavitha KS, Santosh P, Kavitha HU, Rao Y, Satish S (2013) Plants: emerging as nanofactories towards facile route in synthesis of nanoparticles. Bioimpacts 3(3):111–117

Balkrishna A, Sharma VK, Das SK, Mishra N, Bisht L, Joshi A, Sharma N (2020) Characterization and anti-cancerous effect of *Putranjiva roxburghii* seed extract mediated silver nanoparticles on human colon (HCT-116), pancreatic (PANC-1) and breast (MDA-MB 231) cancer cell lines: a comparative study. Int J Nanomedicine 15:573–585

Baruah D, Yadav RNS, Yadav A, Das AM (2019) Alpinia nigra fruits mediated synthesis of silver nanoparticles and their antimicrobial and photocatalytic activities. J Photochem Photobiol B 201:111649

Basavegowda N, Dhanya Kumar G, Tyliszczak B, Wzorek Z, Sobczak-Kupiec A (2015) One-step synthesis of highly-biocompatible spherical gold nanoparticles using *Artocarpus heterophyllus* lam. (jackfruit) fruit extract and its effect on pathogens. Ann Agric Environ Med 22(1):84–89

Behravan M, Hossein Panahi A, Naghizadeh A, Ziaee M, Mahdavi R, Mirzapour A (2019) Facile green synthesis of silver nanoparticles using *Berberis vulgaris* leaf and root aqueous extract and its antibacterial activity. Int J Biol Macromol 124:148–154

Bhosale TT, Shinde HM, Gavade NL, Babar SB, Gawade VV, Sabale SR, Kamble RJ, Shirke BS, Garadkar KM (2018) Biosynthesis of SnO_2 nanoparticles by aqueous leaf extract of *Calotropis gigantea* for photocatalytic applications. J Mater Sci 8:6826

Bibi I, Kamal S, Ahmed A, Iqbal M, Nouren S, Jilani K, Nazar N, Amir M, Abbas A, Ata S, Majid F (2017) Nickel nanoparticle synthesis using *Camellia sinensis* as reducing and capping agent: growth mechanism and photo-catalytic activity evaluation. Int J Biol Macromol 103:783–790

Bindhu MR, Umadevi M, Esmail GA, Al-Dhabi NA, Arasu MV (2020) Green synthesis and characterization of silver nanoparticles from *Moringa oleifera* flower and assessment of antimicrobial and sensing properties. J Photochem Photobiol B 205:111836

Birusanti AB, Mallavarapu U, Nayakanti D, Espenti CS, Mala S (2019) Sustainable green synthesis of silver nanoparticles by using *Rangoon creeper* leaves extract and their spectral analysis and anti-bacterial studies. IET Nanobiotechnol 13(1):71–76

Borovaya MN, Naumenko AP, Matvieieva NA, Blume YB, Yemets AI (2014) Biosynthesis of luminescent CdS quantum dots using plant hairy root culture. Nanoscale Res Lett 9:686

Burlacu E, Tanase C, Coman NA, Berta L (2019) A review of bark-extract-mediated green synthesis of metallic nanoparticles and their applications. Molecules 24(23):4354

Carloling G, Priyadharshini MN, Vinodhini E, Ranjitham AM, Shanthi P (2015) Biosynthesis of copper nanoparticles using aqueous guava extract-characterisation and study of antibacterial effects. Int J Pharm Biol Sci 5:25–43

Chahardoli A, Karimi N, Fattahi A (2017) Biosynthesis, characterization, antimicrobial and cytotoxic effects of silver nanoparticles using *Nigella arvensis* seed extract. Iran J Pharm Res 16(3):1167–1175

Chang C-C, Chen H-L, Li W-H, Chaw S-M (2004) Dating the monocot? Dicot divergence and the origin of core eudicots using whole chloroplast genomes. J Mol Evol 58(4):424–441

Chankaew C, Tapala W, Grudpan K, Rujiwatra A (2019) Microwave synthesis of ZnO nanoparticles using longan seeds biowaste and their efficiencies in photocatalytic decolorization of organic dyes. Environ Sci Pollut Res Int 26(17):17548–17554

Chinnappan S, Kandasamy S, Arumugam S, Seralathan KK, Thangaswamy S, Muthusamy G (2018) Biomimetic synthesis of silver nanoparticles using flower extract of *Bauhinia purpurea* and its antibacterial activity against clinical pathogens. Environ Sci Pollut Res Int 25(1):963–969

Chopade V, Kamble D (2021) Application of green silver nanoparticles synthesized using leaf extract of *Tridax procumbens* for preparation of clinical antimicrobial bandages. Int J Pharmaceut Investig 11:10–13

Crane PR, Friis EM, Pedersen KR (1995) The origin and early diversification of angiosperms. Nature 374:27–33

Crepet WL (2000) Progress in understanding angiosperm history, success, and relationships: Darwin's abominably "perplexing phenomenon". PNAS 97:12939–12941

Daisy P, Saipriya K (2012) Biochemical analysis of *Cassia fistula* aqueous extract and phytochemically synthesized gold nanoparticles as hypoglycemic treatment for diabetes mellitus. Int J Nanomedicine 7:1189–1202

Das G, Patra JK, Basavegowda N, Vishnuprasad CN, Shin HS (2019) Comparative study on antidiabetic, cytotoxicity, antioxidant and antibacterial properties of biosynthesized silver nanoparticles using outer peels of two varieties of *Ipomoea batatas* (L.) lam. Int J Nanomedicine 14:4741–4754

Das PE, Abu-Yousef IA, Majdalawieh AF, Narasimhan S, Poltronieri P (2020b) Green synthesis of encapsulated copper nanoparticles using a hydroalcoholic extract of *Moringa oleifera* leaves and assessment of their antioxidant and antimicrobial activities. Molecules 25(3):555

Das PE, Majdalawieh AF, Abu-Yousef IA, Narasimhan S, Poltronieri P (2020a) Use of a hydroalcoholic extract of *Moringa oleifera* leaves for the green synthesis of bismuth nanoparticles and evaluation of their anti-microbial and antioxidant activities. Materials 13(4):876

Deshmukh AR, Gupta A, Kim BS (2019) Ultrasound assisted green synthesis of silver and iron oxide nanoparticles using fenugreek seed extract and their enhanced antibacterial and antioxidant activities. Biomed Res Int 2019:1714358

Devanesan S, Ponmurugan K, AlSalhi MS, Al-Dhabi NA (2020) Cytotoxic and antimicrobial efficacy of silver nanoparticles synthesized using a traditional phytoproduct, asafoetida gum. Int J Nanomedicine 15:4351–4362

Devi DK, Pratap SV, Haritha R, Sivudu KS, Radhika P, Sreedhar B (2011) Gum acacia as a facile reducing, stabilizing, and templating agent for palladium nanoparticles. J Appl Polym Sci 121(3):1765–1773

Devi TB, Aharuzzaman M (2018) Removal of perilous nitro compound from aqueous phase using biogenic copper nanoparticles as catalyst. IJCT 25:561–564

Dhand V, Soumya L, Bharadwaj S, Chakra S, Bhatt D, Sreedhar B (2016) Green synthesis of silver nanoparticles using *Coffea arabica* seed extract and its antibacterial activity. Korean J Couns Psychother 58:36–43

Din MI, Nabi AG, Rani A, Aihetasham A, Mukhtar M (2018) Single step green synthesis of stable nickel and nickel oxide nanoparticles from *Calotropis gigantea*: catalytic and antimicrobial potentials. Environ Nanotechnol Monit Manag 9:29–36

Dinparvar S, Bagirova M, Allahverdiyev AM, Abamor ES, Safarov T, Aydogdu M, Aktas D (2020) A nanotechnology-based new approach in the treatment of breast cancer: biosynthesized silver nanoparticles using *Cuminum cyminum* L. seed extract. J Photochem Photobiol B 208:111902

Duan L, Li M, Liu H (2015) Biosynthesised palladium nanoparticles using *Eucommia ulmoides* bark aqueous extract and their catalytic activity. IET Nanobiotechnol 9(6):349–354

Duman F, Ocsoy I, Kup FO (2016) Chamomile flower extract-directed CuO nanoparticle formation for its antioxidant and DNA cleavage properties. Korean J Couns Psychother 60:333–338

Dwivedi AD, Gopal K (2010) Biosynthesis of silver and gold nanoparticles using *Chenopodium album* leaf extract. Colloids Surf A Physicochem Eng Asp 369:27–33

Ebrahimi K, Shiravand S, Mahmoudvand H (2017) Biosynthesis of copper nanoparticles using aqueous extract of *Capparis spinosa* fruit and investigation of its antibacterial activity. Marmara Pharm J 21:866–871

Ebrahimzadeh MA, Naghizadeh A, Amiri O, Shirzadi-Ahodashti M, Mortazavi-Derazkola S (2020) Green and facile synthesis of ag nanoparticles using *Crataegus pentagyna* fruit extract (CP-AgNPs) for organic pollution dyes degradation and antibacterial application. Bioorg Chem 94:103425

Elangovan K, Elumalai D, Anupriya S, Shenbhagaraman R, Kaleena PK, Murugesan K (2015) Phyto mediated biogenic synthesis of silver nanoparticles using leaf extract of *Andrographis echioides* and its bio-efficacy on anticancer and antibacterial activities. J Photochem Photobiol B Biol 151:118–124

Elhawary S, El-Hefnawy H, Mokhtar FA, Sobeh M, Mostafa E, Osman S, El-Raey M (2020) Green synthesis of silver nanoparticles using extract of *Jasminum officinal* L. leaves and evaluation of cytotoxic activity towards bladder (5637) and breast cancer (MCF-7) cell lines. Int J Nanomedicine 15:9771–9781

Farouk F, Abdelmageed M, Azam Ansari M, Azzazy HME (2020) Synthesis of magnetic iron oxide nanoparticles using pulp and seed aqueous extract of *Citrullus colocynth* and evaluation of their antimicrobial activity. Biotechnol Lett 42(2):231–240

Fatima M, Zaidi N-U-S, Amraiz D, Afzal F (2015) In vitro antiviral activity of *Cinnamomum cassia* and its nanoparticles against H7N3 influenza a virus. J Microbiol Biotechnol 26:151–159

Filip GA, Moldovan B, Baldea I, Olteanu D, Suharoschi R, Decea N, Cismaru CM, Gal E, Cenariu M, Clichici S, David L (2019) UV-light mediated green synthesis of silver and gold nanoparticles using cornelian cherry fruit extract and their comparative effects in experimental inflammation. J Photochem Photobiol B 191:26–37

Gaurav I, Singh T, Thakur A, Kumar G, Rathee P, Kumari P, Sweta K (2020) Synthesis, in-vitro and in-silico evaluation of silver nanoparticles with root extract of *Withania somnifera* for antibacterial activity *via* binding of penicillin-binding protein-4. Curr Pharm Biotechnol 21(15):1674–1687

Gholami Z, Dadmehr M, Jelodar NB, Hosseini M, Oroojalian F, Parizi AP (2020) One-pot biosynthesis of CdS quantum dots through *in vitro* regeneration of hairy roots of *Rhaphanus sativus L.* and their apoptosis effect on MCF-7 and AGS cancerous human cell lines. Mater Res Exp 7:015056

Ghosh S, Patil S, Ahire M, Kitture R, Gurav DD, Jabgunde AM, Kale S, Pardesi K, Shinde V, Bellare J, Dhavale DD, Chopade BA (2012) Gnidia glauca flower extract mediated synthesis of gold nanoparticles and evaluation of its chemocatalytic potential. J Nanobiotechnol 10:17

Gogoi N, Babu PJ, Mahanta C, Bora U (2015) Green synthesis and characterization of silver nanoparticles using alcoholic flower extract of *Nyctanthes arbortristis* and in vitro investigation of their antibacterial and cytotoxic activities. Korean J Couns Psychother 46:463–469

Golmohammadi M, Honarmand M, Ghanbari S (2020) A green approach to synthesis of ZnO nanoparticles using jujube fruit extract and their application in photocatalytic degradation of organic dyes. Spectrochim Acta A Mol Biomol Spectrosc 229:117961

Gopinath V, MubarakAli D, Priyadarshini S, Priyadharsshini NM, Thajuddin N, Velusamy P (2012) Biosynthesis of silver nanoparticles from *Tribulus terrestris* and its antimicrobial activity: a novel biological approach. Colloids Surf B Biointerfaces 96:69–74

Gutierrez-Valdes N, Häkkinen ST, Lemasson C, Guillet M, Oksman-Caldentey K-M, Ritala A, Cardon F (2020) Hairy root cultures—a versatile tool with multiple applications. Front Plant Sci 11:33

Hajra A, Dutta S, Mondal NK (2016) Mosquito larvicidal activity of cadmium nanoparticles synthesized from petal extracts of marigold (*Tagetes* sp.) and rose (*Rosa* sp.) flower. J Parasit Dis 40:1519–1527

Häkkinen ST, Oksman-Caldentey K-M (2018) Progress and prospects of hairy root research. In: Srivastava V, Mehrotra S, Mishra S (eds) Hairy roots: an effective tool of plant biotechnology. Springer, Singapore, pp 3–19

Hassani A, Azarian MMS, Ibrahim WN, Hussain SA (2020) Preparation, characterization and therapeutic properties of gum arabic-stabilized gallic acid nanoparticles. Sci Rep 10:17808

Hemmati S, Ahmeda A, Salehabadi Y, Zangeneh A, Zangeneh MM (2020) Synthesis, characterization, and evaluation of cytotoxicity, antioxidant, antifungal, antibacterial, and cutaneous wound healing effects of copper nanoparticles using the aqueous extract of strawberry fruit and L-ascorbic acid. Polyhedron 180:114425

Hii YS, Jaison Jeevanandam J, San Chan YS (2018) Plant mediated green synthesis and nanoencapsulation of MgO nanoparticle from *Calotropis gigantea*: characterisation and kinetic release studies. Inorg Nano-Met Chem 48:620–631

Hosseinzadeh N, Shomali T, Hosseinzadeh S, Raouf Fard F, Pourmontaseri M, Fazeli M (2020) Green synthesis of gold nanoparticles by using Ferula persica Willd. gum essential oil: production, characterization and in vitro anti-cancer effects. J Pharm Pharmacol 72(8):1013–1025

Huang H, Shan K, Liu J, Tao X, Periyasamy S, Durairaj S, Jiang Z, Jacob JA (2020) Synthesis, optimization and characterization of silver nanoparticles using the catkin extract of Piper longum for bactericidal effect against food-borne pathogens via conventional and mathematical approaches. Bioorg Chem 103:104230

Huang W, Xu M, Duan H, Bi Y, Yu H (2019) Inhibition of *fusarium oxysporum* by AgNPs biosynthesised using *Cinnamomum camphora* fruit extract. IET Nanobiotechnol 13(1):42–45

Ingrouille M (1992) Flowers: evolution and diversity. In: diversity and evolution of land plants. Springer, Dordrecht, pp 133–186

Jayalakshmi, Yogamoorthi A (2014) Green synthesis of copper oxide nanoparticles using aqueous extract of flowers of *Cassia alata* and particles characterisation. Int J Nanomat Biostruct 4:66–71

Jayaprakash N, Vijaya JJ, Kaviyarasu K, Kombaiah K, Kennedy LJ, Ramalingam RJ, Munusamy MA, Al-Lohedan HA (2017) Green synthesis of ag nanoparticles using tamarind fruit extract for the antibacterial studies. J Photochem Photobiol B 169:178–185

Jisha ER, Balamurugan G, Edison N, Selvakumar P, Rathiga R (2012) Biogenic synthesis of gold nanoparticles using leaf extract of *Tephrosia purpurea* and study of their antibacterial effect. Int J Pharm Tech Res 4:1323–1331

Jun SH, Kim HS, Koo YK, Park Y, Kim J, Cho S, Park Y (2014) Root extracts of *Polygala tenuifolia* for the green synthesis of gold nanoparticles. J Nanosci Nanotechnol 14(8):6202–6208

Kalaiselvi A, Roopan SM, Madhumitha G, Ramalingam C, Elango G (2015) Synthesis and characterization of palladium nanoparticles using *Catharanthus roseus* leaf extract and its application in the photo-catalytic degradation. Spectrochim Acta Part A 135:116–119

Karunakaran G, Jagathambal M, Venkatesh M, Kumar GS, Kolesnikov E, Dmitry A, Gusev A, Kuznetsov D (2017) Hydrangea paniculata flower extract-mediated green synthesis of MgNPs and Ag-NPs for health care applications. Powder Technol 305:488–494

Kathiravan V, Ravi S, Ashokkumar S, Velmurugan S, Elumalai K, Khatiwada CP (2015) Green synthesis of silver nanoparticles using Croton sparsiflorus morong leaf extract and their antibacterial and antifungal activities. Spectrochim Acta Part A 139:200–205

Kaviya S, Santhanalakshmi J, Viswanathan B, Muthumary J, Srinivasan K (2011) Biosynthesis of silver nanoparticles using *citrus sinensis* peel extract and its antibacterial activity. Spectrochim Acta A Mol Biomol Spectrosc 79:594–598

Khalil MMH, Ismail EH, Baghdady KZE, Mohamed D (2014) Green synthesis of silver nanoparticles using olive leaf extract and its antibacterial activity. Arabian J Chem 7:1131–1139

Khan S, Bello B, Khan J, Anwar Y, Mirza M, Qadri F, Farooq A, Adam IK, Asiri AM, Khan S (2018) Albizia chevalier based ag nanoparticles: anti-proliferation, bactericidal and pollutants degradation performance. J Photochem Photobiol B 182:62–70

Kiran Kumar HA, Mandal BK, Mohan Kumar K, Sb M, Sai Kumar T, Madhiyazhagan P, Ghosh AR (2014) Antimicrobial and antioxidant activities of *Mimusops elengi* seed extract mediated isotropic silver nanoparticles. Spectrochim Acta A Mol Biomol Spectrosc 130:13–18

Kobylinska N, Shakhovsky A, Khainakova O, Klymchuk D, Avdeeva L, Ratushnyak Y, Duplij V, matvieiva N. (2020) 'Hairy' root extracts as source for 'green' synthesis of silver nanoparticles and medical applications. RSC Adv 10:39434–39446

Kolenčík M, Štrba P, Šebesta M, Kratošová G, Urík M, Vávra I, Peikertová P, Polláková N, Šimanský V, Feng H, Qian Y, Ramakanth I (2019) Nanogold biosynthesis mediated by mixed flower pollen grains. J Nanosci Nanotechnol 19(5):2983–2988

Kora AJ (2019) Multifaceted activities of plant gum synthesized platinum nanoparticles: catalytic, peroxidase, PCR enhancing and antioxidant activities. IET Nanobiotechnol 13:602–608

Kora AJ, Beedu SR, Jayaraman A (2012a) Size-controlled green synthesis of silver nanoparticles mediated by gum ghatti (*Anogeissus latifolia*) and its biological activity. Org Med Chem Lett 2(1):17

Kora AJ, Rastogi L (2018) Green synthesis of palladium nanoparticles using gum ghatti (*Anogeissus latifolia*) and its application as an antioxidant and catalyst. Arab J Chem 11: 1097–1106

Kora AJ, Sashidhar RB, Arunachalam J (2012b) Aqueous extract of gum olibanum (*Boswellia serrata*): a reductant and stabilizer for the biosynthesis of antibacterial silvernanoparticles. Process Biochem 47:1516–1520

Kumar Bachheti R, Fikadu A, Bachheti A, Husen A (2020) Biogenic fabrication of nanomaterials from flower-based chemical compounds, characterization and their various applications: a review. Saudi J Biol Sci 27(10):2551–2562

Kumar D, Kumar G, Agrawal V (2018) Green synthesis of silver nanoparticles using *Holarrhena antidysenterica* (L.) wall. Bark extract and their larvicidal activity against dengue and filariasis vectors. Parasitol Res 117(2):377–389

Kumar H, Bhardwaj K, Dhanjal DS, Nepovimova E, Şen F, Regassa H, Singh R, Verma R, Kumar V, Kumar D, Bhatia SK, Kuča K (2020a) Fruit extract mediated green synthesis of metallic nanoparticles: a new avenue in pomology applications. Int J Mol Sci 21(22):8458

Kumar H, Bhardwaj K, Kuča K, Kalia A, Nepovimova E, Verma R, Kumar D (2020c) Flower-based green synthesis of metallic nanoparticles: applications beyond fragrance. Nanomaterials 10(4):766

Kumar VA, Ammani K, Jobina R, Subhaswaraj P, Siddhardha B (2017) Photo-induced and phytomediated synthesis of silver nanoparticles using *Derris trifoliata* leaf extract and its larvicidal activity against *Aedes aegypti*. J Photochem Photobiol B 171:1–8

Latha N, Gowri M (2014) Biosynthesis and characterisation of Fe_3O_4 nanoparticles using *Carica papaya* leaves extract. Int J Sci Res 3:1551–1556

Lee YJ, Song K, Cha SH, Cho S, Kim YS, Park Y (2019) Sesquiterpenoids from *Tussilago farfara* flower bud extract for the eco-friendly synthesis of silver and gold nanoparticles possessing antibacterial and anticancer activities. Nanomaterials 9(6):819

Mahmoodi Esfanddarani H, Abbasi Kajani A, Bordbar AK (2018) Green synthesis of silver nanoparticles using flower extract of *Malva sylvestris* and investigation of their antibacterial activity. IET Nanobiotechnol 12(4):412–416

Malaikozhundan B, Vijayakumar S, Vaseeharan B, Jenifer AA, Chitra P, Prabhu NM, Kannapiran E (2017) Two potential uses for silver nanoparticles coated with *Solanum nigrum* unripe fruit extract: biofilm inhibition and photodegradation of dye effluent. Microb Pathog 111:316–324

Manjamadha VP, Muthukumar K (2016) Ultrasound assisted green synthesis of silver nanoparticles using weed plant. Bioprocess Biosyst Eng 39:401–411

Mehmood A, Murtaza G, Bhatti TM, Kausar R, Ahmed MJ (2016) Biosynthesis, characterization and antimicrobial action of silver nanoparticles from root bark extract of *Berberis lycium* Royle. Pak J Pharm Sci 29(1):131–137

Meng Y, Sun Y (2016) Development of biogenic silver nanoparticle using *Rosa chinensis* flower extract and its antibacterial property. J Nanosci Nanotechnol 16(4):3969–3972

Moteriya P, Chanda S (2017) Synthesis and characterization of silver nanoparticles using *Caesalpinia pulcherrima* flower extract and assessment of their in vitro antimicrobial, antioxidant, cytotoxic, and genotoxic activities. Artif Cells Nanomed Biotechnol 45(8):1556–1567

Moustafa NE, Alomari AA (2019) Green synthesis and bactericidal activities of isotropic and anisotropic spherical gold nanoparticles produced using *Peganum harmala* L leaf and seed extracts. Biotechnol Appl Biochem 66(4):664–672

Moyo M, Gomba M, Nharingo T (2015) Afzelia quanzensis bark extract for green synthesis of silver nanoparticles and study of their antibacterial activity. Int J Ind Chem 6:329–338

Murugan K, Senthilkumar B, Senbagam D, Al-Sohaibani S (2014) Biosynthesis of silver nanoparticles using *acacia leucophloea* extract and their antibacterial activity. Int J Nanomedicine 9:2431–2438

Muthu K, Priya S (2017) Green synthesis, characterization and catalytic activity of silver nanoparticles using *Cassia auriculata* flower extract separated fraction. Spectrochim Acta A Mol Biomol Spectrosc 179:66–72

Nahak G, Sahu R (2010) In vitro antioxidative acitivity of *Azadirachta indica* and *Melia azedarach* leaves by DPPHscavenging assay. Nat Sci 8:22–28

Nakkala JR, Mata R, Gupta AK, Sadras SR (2014) Biological activities of green silver nanoparticles synthesized with *Acorous calamus* rhizome extract. Eur J Med Chem 85:784–794

Nasrollahzadeh M, Sajadi SM, Rostami-Vartooni A, Bagherzadeh M (2015b) Green synthesis of Pd/CuO nanoparticles by *Theobroma cacao* L. seeds extract and their catalytic performance for the reduction of 4-nitrophenol and phosphine-free heck coupling reaction under aerobic conditions. J Colloid Interface Sci 448:106–113

Nasrollahzadeh M, Sajadi SM, Rostami-Vartooni A, Khalaj M (2015a) Green synthesis of Pd/Fe3O4 nanoparticles using *Euphorbia condylocarpa* M. bieb root extract and their catalytic applications as magnetically recoverable and stable recyclable catalysts for the phosphine-free Sonogashira and Suzuki coupling reactions. J Mol Catal 396:31–39

Nayak D, Ashe S, Rauta PR, Kumari M, Nayak B (2016) Bark extract mediated green synthesis of silver nanoparticles: evaluation of antimicrobial activity and antiproliferative response against osteosarcoma. Korean J Couns Psychother 58:44–52

Nazar N, Bibi I, Kamal S, Iqbal M, Nouren S, Jilani K, Umair M, Ata S (2018) Cu nanoparticles synthesis using biological molecule of *P. granatum* seeds extract as reducing and capping agent: growth mechanism and photo-catalytic activity. Int J Biol Macromol 106:1203–1210

Nicholas KJ, Tiffney BH, Knoll AH (1983) Patterns in vascularland plant diversification. Nature 303:614–616

Nindawat S, Agrawal V (2019) Fabrication of silver nanoparticles using *Arnebia hispidissima* (Lehm.) A. DC. Root extract and unravelling their potential biomedical applications. Artif Cells Nanomed Biotechnol 47(1):166–180

Ogunyemi SO, Abdallah Y, Zhang M, Fouad H, Hong X, Ibrahim E, Masum MMI, Hossain A, Mo J, Li B (2019) Green synthesis of zinc oxide nanoparticles using different plant extracts and their antibacterial activity against *Xanthomonas oryzae* pv. *oryzae*. Artif Cells Nanomed Biotechnol 47(1):341–352

Ontong JC, Paosen S, Shankar S, Voravuthikunchai SP (2019) Eco-friendly synthesis of silver nanoparticles using *Senna alata* bark extract and its antimicrobial mechanism through enhancement of bacterial membrane degradation. J Microbiol Methods 165:105692

Pandey S, Goswami GK, Nanda KK (2013) Green synthesis of polysaccharide/gold nanoparticle nanocomposite: an efficient ammonia sensor. Carbohydr Polym 94:229–234

Pani A, Lee JH, Yun SI (2016) Autoclave mediated one-pot-one-minute synthesis of AgNPs and au-ag nanocomposite from *Melia azedarach* bark extract with antimicrobial activity against food pathogens. Chem Cent J 10:15

Patil MP, Bayaraa E, Subedi P, Piad LLA, Tarte NH (2019) Biogenic synthesis, characterization of gold nanoparticles using *Lonicera japonica* and their anticancer activity on HeLa cells. J Drug Deliv Sci Technol 51:83–90

Patil SP (2020) Calotropis gigantea assisted green synthesis of nanomaterials and their applications: a review. Beni-Suef Univ J Basic Appl Sci 9:14

Patra JK, Baek KH (2016) Comparative study of proteasome inhibitory, synergistic antibacterial, synergistic anticandidal, and antioxidant activities of gold nanoparticles biosynthesized using fruit waste materials. Int J Nanomedicine 11:4691–4705

Patra JK, Das G, Baek KH (2016) Phyto-mediated biosynthesis of silver nanoparticles using the rind extract of watermelon (*Citrullus lanatus*) under photo-catalyzed condition and investigation of its antibacterial, anticandidal and antioxidant efficacy. J Photochem Photobiol B 161:200–210

Patra JK, Das G, Shin HS (2019) Facile green biosynthesis of silver nanoparticles using *Pisum sativum* L. outer peel aqueous extract and its antidiabetic, cytotoxicity, antioxidant, and antibacterial activity. Int J Nanomedicine 14:6679–6690

Pattanayak S, Mollick MMR, Maity D, Chakraborty S, Dash SK, Chattopadhyay S, Roy S, Chattopadhyay D, Chakraborty M (2017) Butea monosperma bark extract mediated green synthesis of silver nanoparticles: characterization and biomedical applications. J Saudi Chem Soc 21:673–684

Paulkumar K, Gnanajobitha G, Vanaja M, Rajeshkumar S, Malarkodi C, Pandian K, Annadurai G (2014) Piper nigrum leaf and stem assisted green synthesis of silver nanoparticles and evaluation of its antibacterial activity against agricultural plant pathogens. Sci World J 2014:829894

Pethakamsetty L, Kothapenta K, Nammi HR, Ruddaraju LK, Kollu P, Yoon SG, Pammi SVN (2017) Green synthesis, characterization and antimicrobial activity of silver nanoparticles using methanolic root extracts of *Diospyros sylvatica*. J Environ Sci (China) 55:157–163

Pilaquinga F, Morejón B, Ganchala D, Morey J, Piña N, Debut A, Neira M (2019) Green synthesis of silver nanoparticles using *Solanum mammosum* L. (Solanaceae) fruit extract and their larvicidal activity against *Aedes aegypti* L. (Diptera: Culicidae). PLoS One 14(10):e0224109

Pooja D, Panyaram S, Kulhari H, Reddy B, Rachamalla SS, Sistla R (2015) Natural polysaccharide functionalized gold nanoparticles as biocompatible drug delivery carrier. Int J Biol Macromol 80:48–56

Radwan RA, El-Sherif YA, Salama MM (2020) A novel biochemical study of anti-ageing potential of *Eucalyptus camaldulensis* bark waste standardized extract and silver nanoparticles. Colloids Surf B Biointerfaces 191:111004

Ragab GA, Saad-Allah KM (2020) Green synthesis of sulfur nanoparticles using *Ocimum basilicum* leaves and its prospective effect on manganese-stressed *Helianthus annuus* (L.) seedlings. Ecotoxicol Environ Saf 191:110242

Rajagopal T, Jemimah IA, Ponmanickam P, Ayyanar M (2015) Synthesis of silver nanoparticles using *Catharanthus roseus* root extract and its larvicidal effects. J Environ Biol 36(6):1283–1289

Ramanujam K, Sundrarajan M (2014) Antibacterial effects of biosynthesized MgO nanoparticles using ethanolic fruit extract of *Emblica officinalis*. J Photochem Photobiol B 141:296–300

Ramesh P, Rajendran A, Shisundaram MM (2014) Green synthesis of zinc oxide nanoparticles using flower extract of *Cassia*. J Nanosci Nanotechnol 2:41–45

Ramesh PS, Kokila T, Geetha D (2015) Plant mediated green synthesis and antibacterial activity of silver nanoparticles using *Emblica officinalis* fruit extract. Spectrochim Acta A Mol Biomol Spectrosc 142:339–343

Rao NH, Lakshmidevi N, Pammi SVN, Kollu P, Ganapaty S, Lakshmi P (2016) Green synthesis of silver nanoparticles using methanolic root extracts of *Diospyros paniculata* and their antimicrobial activities. Korean J Couns Psychother 62:553–557

Rashmi V, Sanjay KR (2017) Green synthesis, characterisation and bioactivity of plant-mediated silver nanoparticles using *Decalepis hamiltonii* root extract. IET Nanobiotechnol 11(3):247–254

Rastogi L, Beedu SR, Kora AJ (2015) Facile synthesis of palladium nanocatalyst using gum kondagogu (*Cochlospermum gossypium*): a natural biopolymer. IET Nanobiotechnol 9(6):362–367

Rather MY, Shincy M, Sundarapandian S (2020) Silver nanoparticles synthesis using *Wedelia urticifolia* (Blume) DC. Flower extract: characterization and antibacterial activity evaluation. Microsc Res Tech 83(9):1085–1094

Rathi Sre PR, Reka M, Poovazhagi R, Arul Kumar M, Murugesan K (2015) Antibacterial and cytotoxic effect of biologically synthesized silver nanoparticles using aqueous root extract of *Erythrina indica* lam. Spectrochim Acta A Mol Biomol Spectrosc 135:1137–1144

Rautela A, Rani J, Debnath M (2019) Green synthesis of silver nanoparticles from *Tectona grandis* seeds extract: characterization and mechanism of antimicrobial action on different microorganisms. J Anal Sci Technol 10:5

Roopan SM, Bharathi A, Kumar R, Khanna VG, Prabhakarn A (2012) Acaricidal, insecticidal, and larvicidal efficacy of aqueous extract of *Annona squamosa* L peel as biomaterial for the reduction of palladium salts into nanoparticles. Colloid Surface B 92:209–212

Rostami-Vartooni A (2017) Green synthesis of CuO nanoparticles loaded on the seashell surface using *Rumex crispus* seeds extract and its catalytic applications for reduction of dyes. IET Nanobiotechnol 11(4):349–359

Rostami-Vartooni A, Nasrollahzadeh M, Alizadeh M (2016) Green synthesis of seashell supported silver nanoparticles using *Bunium persicum* seeds extract: application of the particles for catalytic reduction of organic dyes. J Colloid Interface Sci 470:268–275

Sadeghi B, Rostami A, Momeni SS (2015) Facile green synthesis of silver nanoparticles using seed aqueous extract of *Pistacia atlantica* and its antibacterial activity. Spectrochim Acta A Mol Biomol Spectrosc 134:326–332

Sangaonkar GM, Pawar KD (2018) Garcinia indica mediated biogenic synthesis of silver nanoparticles with antibacterial and antioxidant activities. Colloids Surf B Biointerfaces 164: 210–217

Santhoshkumar T, Rahuman AA, Bagavan A, Marimuthu S, Jayaseelan C, Kirthi AV, Kamaraj C, Rajakumar G, Zahir AA, Elango G, Velayutham K, Iyappan M, Siva C, Karthik L, Rao KV (2012) Evaluation of stem aqueous extract and synthesized silver nanoparticles using *Cissus quadrangularis* against *Hippobosca maculata* and *Rhipicephalus* (*Boophilus*) *microplus*. Exp Parasitol 132(2):156–165

Selvi SK, Mahesh J, Sashidhar RB (2017) Anti-proliferative activity of gum kondagogu (*Cochlospermum gossypium*)-gold nanoparticle constructs on B16F10 melanoma cells: an in vitro model. Bioact Carbohydr Diet Fibre 11:38–47

Senthilkumar SR, Sivakumar T (2014) Green tea (*Camellia sinensis*) mediated synthesis of zinc oxide (ZnO) nanoparticles and studies on their antimicrobial activities. Int J Pharm Pharm Sci 6: 461–465

Shah M, Nawaz S, Jan H, Uddin N, Ali A, Anjum S, Giglioli-Guivarc'h N, Hano C, Abbasi BH (2020) Synthesis of bio-mediated silver nanoparticles from *Silybum marianum* and their biological and clinical activities. Korean J Couns Psychother 112:110889

Shah S, Dasgupta S, Chakraborty M, Vadakkekara R, Hajoori M (2014) Green synthesis of iron nanoparticles using plant extracts. Int J Biol Pharm Res 5:549–552

Sharma JK, Akhtar MS, Ameen S, Srivastava P, Singh G (2015) Green synthesis of CuO nanoparticles with leaf extract of *Calotropis gigantea* and its dye-sensitized solar cells applications. J Alloys Compd 632:321–325

Shetty P, Supraja N, Garud M, Prasad TNVKV (2014) Synthesis, characterization and antimicrobial activity of *Alstonia Scholaris* bark-extract-mediated silver nanoparticles. J Nanostructure Chem 4:161–170

Shobana S, Veena S, Sameer SSM, Swarnalakshmi K, Vishal LA (2020) Green synthesis of silver nanoparticles using *Artocarpus hirsutus* seed extract and its antibacterial activity. Curr Pharm Biotechnol 21(10):980–989

Siddiqui MZ, Chowdhury AR, Singh BR, Maurya S, Prasad N (2021) Synthesis, characterization and antimicrobial evaluation of *piyar* gum-induced silver nanoparticles. Natl Acad Sci Lett 44: 203–208

Singh H, Du J, Singh P, Yi TH (2018) Ecofriendly synthesis of silver and gold nanoparticles by *Euphrasia officinalis* leaf extract and its biomedical applications. Artif Cells Nanomed Biotechnol 46(6):1163–1170

Singh H, Du J, Yi TH (2017) Green and rapid synthesis of silver nanoparticles using *Borago officinalis* leaf extract: anticancer and antibacterial activities. Artif Cells Nanomed Biotechnol 45(7):1310–1316

Singh S, Saikia JP, Buragohain AK (2013) A novel 'green' synthesis of colloidal silver nanoparticles (SNP) using *Dillenia indica* fruit extract. Colloids Surf B Biointerfaces 102:83–85

Srinivas BK, Shivamadhu MC, Siddappaji KK, Krishnappa DK, Jayarama S (2019) Angiosuppressive effects of bio-fabricated silver nanoparticles synthesis using *Clitoria ternatea* flower: an in vitro and in vivo approach. J Biol Inorg Chem 24(7):1115–1126

Subramanian SB, Bezawada SR, Dhamodharan R (2016) Green, selective, seedless and one-pot synthesis of triangular Au nanoplates of controlled size using bael gum and mechanistic study. ACS Sustain Chem Eng 4:3830–3839

Suganya G, Karthi S, Shivakumar MS (2014) Larvicidal potential of silver nanoparticles synthesized from *Leucas aspera* leaf extracts against dengue vector *Aedes aegypti*. Parasitol Res 113(5):1673–1679

Suman TY, Radhika Rajasree SR, Kanchana A, Elizabeth SB (2013) Biosynthesis, characterization and cytotoxic effect of plant mediated silver nanoparticles using *Morinda citrifolia* root extract. Colloids Surf B Biointerfaces 106:74–78

Suman TY, Rajasree SR, Ramkumar R, Rajthilak C, Perumal P (2014) The green synthesis of gold nanoparticles using an aqueous root extract of *Morinda citrifolia* L. Spectrochim Acta A Mol Biomol Spectrosc 118:11–16

Suresh G, Gunasekar PH, Kokila D, Prabhu D, Dinesh D, Ravichandran N, Ramesh B, Koodalingam A, Vijaiyan SG (2014) Green synthesis of silver nanoparticles using *Delphinium denudatum* root extract exhibits antibacterial and mosquito larvicidal activities. Spectrochim Acta A Mol Biomol Spectrosc 127:61–66

Swamy MK, Akhtar MS, Mohanty SK, Sinniah UR (2015) Synthesis and characterization of silver nanoparticles using fruit extract of *Momordica cymbalaria* and assessment of their in vitro antimicrobial, antioxidant and cytotoxicity activities. Spectrochim Acta A Mol Biomol Spectrosc 151:939–944

Swilam N, Nematallah KA (2020) Polyphenols profile of pomegranate leaves and their role in green synthesis of silver nanoparticles. Sci Rep 10(1):14851

Tanase C, Berta L, Coman NA, Roşca I, Man A, Toma F, Mocan A, Jakab-Farkas L, Biró D, Mare A (2019) Investigation of in vitro antioxidant and antibacterial potential of silver nanoparticles obtained by biosynthesis using beech bark extract. Antioxidants 8(10):459

Thekkae Padil VV, Černík M (2013) Green synthesis of copper oxide nanoparticles using gum karaya as a biotemplate and their antibacterial application. Int J Nanomedicine 8:889–898

Umar H, Kavaz D, Rizaner N (2018) Biosynthesis of zinc oxide nanoparticles using *Albizia lebbeck* stem bark, and evaluation of its antimicrobial, antioxidant, and cytotoxic activities on human breast cancer cell lines. Int J Nanomedicine 14:87–100

Velayutham K, Rahuman AA, Rajakumar G, Santhoshkumar T, Marimuthu S, Jayaseelan C (2012) Evaluation of *Catharanthus roseus* leaf extract-mediated biosynthesis of titanium dioxide nanoparticles against *Hippobosca maculate* and *Bovicola ovis*. Parasitol Res 111:2329–2337

Velmurugan P, Anbalagan K, Manosathyadevan M, Lee KJ, Cho M, Lee SM, Park JH, Oh SG, Bang KS, Oh BT (2014) Green synthesis of silver and gold nanoparticles using *Zingiber officinale* root extract and antibacterial activity of silver nanoparticles against food pathogens. Bioprocess Biosyst Eng 37(10):1935–1943

Vidya C, Hiremath S, Chandraprabha MN, Antonyraj MAL, Venu GI, Jain A, Kokil BK (2013) Green synthesis of ZnO nanoparticles by *Calotropis gigantea*. Int J Curr Eng Technol 1:118–120

Wang T, Lin J, Chen Z, Megharaj M, Naidu R (2014) Green synthesis of Fe nanoparticles using *Eucalyptus* leaf extracts for treatment of eutrophic wastewater. Sci Total Environ 466–467:210–213

Wang X (2018) The Dawn angiosperms: uncovering the origin of flowering plants. Springer, Cham

Wang Y, Xia R, Hu H, Peng T (2018) Biosynthesis, characterization and cytotoxicity of gold nanoparticles and their loading with N-acetylcarnosine for cataract treatment. J Photochem Photobiol B 187:180–183

Willis KJ, McElwain JC (2002) The evolution of plants. Oxford University Press, New York

Yang Q, Zhao C, Zhao J, Ye Y (2017) Synthesis and neuroprotective effects of the complex nanoparticles of iron and sapogenin isolated from the defatted seeds of *Camellia oleifera*. Pharm Biol 55(1):428–434

Yugandhar P, Haribabu R, Savithramma N (2015) Synthesis, characterization and antimicrobial properties of green-synthesised silver nanoparticles from stem bark extract of *Syzygium alternifolium* (Wt.) Walp. 3 Biotech 5(6):1031–1039

Zahir AA, Rahuman AA (2012) Evaluation of different extracts and synthesised silver nanoparticles from leaves of *Euphorbia prostrata* against *Haemaphysalis bispinosa* and *Hippobosca maculata*. Vet Parasitol 187(3–4):511–520

Zia G, Sadia H, Nazir S, Ejaz K, Ali S, Ihsan-Ul-Haq IT, Khan MAR, Raza A, Andleeb S (2018) In vitro studies on cytotoxic, DNA protecting, antibiofilm and antibacterial effects of biogenic silver nanoparticles prepared with *Bergenia ciliata* rhizome extract. Curr Pharm Biotechnol 19(1):68–78

Biological Synthesis of Nanoparticles: Animal Systems

13

Abstract

There has been a recent interest in the biosynthesis of nanoparticles using animal systems. Several parts of animals and/or their extracts can serve as good reducing agents for the production of nanoparticle. Animal derivatives like wings of insects, honey and propolis from honeybees, spider silk, nests of paper wasp, wasp venom, silk fibroin, and silk sericin from silkworm have increasingly been explored in the last decade. Different components of chicken eggs like yolk, egg white, and egg shells have also shown promising results. Recently a number of studies have been conducted that utilized different species of earthworms in nanofabrication process. This area requires a lot of consistent efforts since most of the members of animal kingdom have not been explored till now. Urgent efforts are needed to develop this area and enrich the scant knowledge available till date for better utilization in green nanobiotechnology.

13.1 Introduction

The kingdom Animalia, also known as Metazoa, includes multicellular eukaryotic organisms that are dependent directly or indirectly on other organisms for nutrition. Animal cells lack the rigid cell walls as found in plants but are held together by structural proteins like collagen. Most of the animals are diploid, with cells organized into tissues each of which performs a specific function. They breathe oxygen and show movement. Of the estimated seven million species of animals inhabiting the Earth, only about 1.5 million have been documented of which around one million are insects. Animalia is monophyletic and is believed to have arisen from a single ancestor that lived 650 million years ago.

Recently there has been a spurt in the biosynthesis of nanoparticles using animal systems. Invertebrates have been largely utilized for this purpose. Several parts of

S. Srivastava, A. Bhargava, *Green Nanoparticles: The Future of Nanobiotechnology*, https://doi.org/10.1007/978-981-16-7106-7_13

animals and their water extracts can serve as good reducing agents for production of nanoparticles. Of the different animal systems, arthropods belonging to the phylum Arthropoda, the largest phylum of all living organisms having ~2–six million species, have been maximally explored for the biogenic synthesis of nanoparticles. This phylum comprises of insects, arachnids, myriapods, and crustaceans that possess segmented bodies, jointed appendages, and cuticle composed of chitin and proteins.

13.2 Insect Wings

Han et al. (2009) performed pioneer studies wherein butterfly wings were taken as natural photonic crystal (PhC) scaffolds for the biofabrication of cadmium sulfide (CdS) nanocrystallites using cadmium dichloride hemipentahydrate ($CdCl_2.2.5H_2O$). *Euploea mulciber* (family: Danaidae) and *Papilio paris* (family: Papilionidae) were the butterflies used for the study. The butterfly wings were activated through an EDTA (ethylenediaminetetraacetic acid)/DMF (dimethylformamide) treatment for production of COO functional groups on their surfaces, which served as reactive scaffolds for the formation and assembly of nano-CdS. The cadmium sulfide layer comprised of cubic phase nanocrystallites (diameter: 6–7 nm) as the building units.

The wings of *Papilio paris* (family: Papilionidae) were used as template for green synthesis of Au/TiO_2 photocatalyst by deposition of gold nanoparticles on titanium dioxide matrix by Chen et al. (2012) using the salts $TiOSO_4$ and $HAuCl_4.4H_2O$. This species commonly known as the "Paris peacock" and a species of swallowtail butterflies are distributed in the Indian subcontinent and Southeast Asia. Au particles were dispersed uniformly across the surface of the biomorphic titanium dioxide. The mean particle size of titanium dioxide nanoparticles was 11 nm and of gold nanoparticles was 7 nm.

Biofabrication of gold nanoparticles (AuNPs) was successfully accomplished by Mu et al. (2013) using wings of three butterflies, viz., Menelaus blue morpho (*Morpho menelaus*; family: Nymphalidae), blue mountain glory (*Papilio ulysses telegonus*; family: Papilionidae), and Wallace's golden birdwing (*Ornithoptera croesus lydius*; family: Papilionidae). For nanoparticle synthesis, wings were put in 0.01% $HAuCl_4$ solution and incubated at 90 °C for 12 h. This was followed by addition of 1% sodium citrate and incubated at 80 °C for 2 h. Gold nanoparticles are formed in situ and grew in the wings and were of different shapes, viz., sphere, triangle, square, and pentagon.

A simple method was developed by Guan et al. (2018) to fabricate hollow metal composite nanospheres using wings of *Euploea mulciber* as a biotemplate upon which hollow silver-gold (Ag-Au) nanospheres attached to form an ordered array of structures. *E. mulciber*, the striped blue crow, is a nymphalid butterfly species found in India and Southeast Asia. The butterfly wings were initially immersed in HNO_3 and thereafter in ethylenediamine for activation and enhancement of the active sites on chitin. Diammine silver (I) hydroxide [$\{Ag(NH_3)_2\}OH$] and chloroauric acid

($HAuCl_4$) were used as silver and gold salts, respectively. The HRTEM images of Ag/Wing, the intermediate product, showed nanoentities of 5–10 nm diameters. After initial formation, the Ag nanoparticles served as seeds for the fabrication of Au-Ag nanoparticle. The distribution of Ag-Au nanospheres on the wings investigated through advanced microscopic techniques showed even distribution of the nanospheres on the "ridge" and "strut."

Khatami et al. (2019) undertook a pioneer study for the biosynthesis of AgNPs making use of the chitin rich wings' extract of *Periplaneta americana* (American cockroach). This cockroach is used as food and medicine in some regions of China. The wing extract was treated with 100 mM silver nitrate solution, and the nanoparticles were characterized using UV-visible spectroscopy, TEM, and XRD. An absorption band at 425 nm in the UV-spectra indicated the presence of AgNPs. This was also confirmed by showing four diffraction peaks at 38, 46, 64, and 78 in the XRD pattern, while TEM images showed the nanoparticle size below 50 nm. The insecticidal activity explored using the nanoparticles exhibited significant effect of these biologically synthesized AgNPs at 100 μg/ml concentrations on *Aphis gossypii*, an agricultural pest known by the common names cotton aphid and melon aphid.

Jakinala et al. (2021) developed an eco-friendly and cost-effective method for the biosynthesis of AgNPs using wings extract of *M. mao* (winged termite) and evaluated their antioxidant and antimicrobial activity. The insect wings were hydrolyzed in 100 mM sodium hydroxide at 90 °C for an hour. It was later cooled and centrifuged at 8000 rpm for about 10 m. After adjusting the pH of the supernatant to 7, the wing extract was mixed with 1 mM $AgNO_3$ solution and incubated at 28 °C for 30 m. SEM images revealed that the nanoparticles were spherical, having dimensions of 40–60 nm, while EDX and XRD studies confirmed the crystalline structure of the AgNPs. The average particle size was observed to be about 43.9 nm. It was speculated that the various biomolecules existing in the wings extract played a major part in the synthesis and stability of AgNPs. The AgNPs also exhibited potent antimicrobial property against *Staphylococcus aureus* and *Fusarium oxysporum*, along with strong antioxidant activity.

13.3 Chitosan

Chitosan, a naturally occurring polycationic linear polysaccharide, is obtained by deacetylation of chitin (Fig. 13.1). It is a highly ordered biopolymer composed of repeating β(1,4)-*N*-acetylglucosamine units and is considered as the second most abundant natural polysaccharide on Earth. Chitin is abundantly found in the exoskeleton of members of phylum Arthropoda, apart from being produced by fungi and bacteria. Chitosan is biodegradable and exhibits unique properties like biocompatibility and nontoxicity that makes it suitable for various clinical applications notably as antimicrobial agent, antioxidant, carrier for drug delivery, and sensing material for biomolecule monitoring and in tissue engineering. Chitosan nanoparticles are developed from chitosan to enhance the performance of the chitosan like reducing

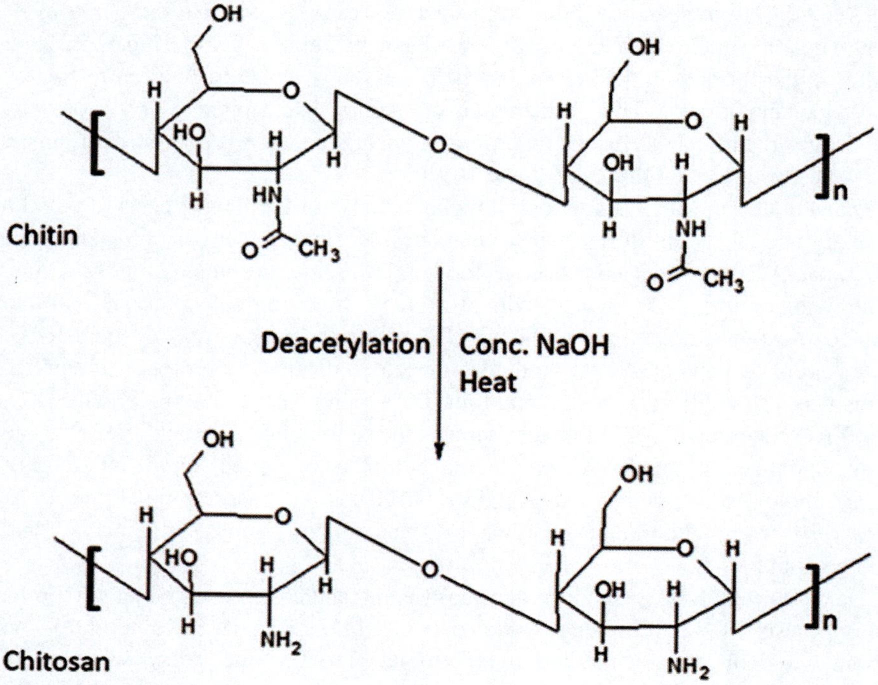

Fig. 13.1 Deacetylation of chitin to chitosan (reprinted from Mohammed et al. 2017)

agglomeration and improving overall stability. This makes the chitosan nanoparticles useful for several environmental applications like elimination of organic and inorganic contaminants from the environment.

13.4 Honeybee

13.4.1 Honey

Two substances associated with honeybee, viz., honey and propolis, have been used in the bio-nanofabrication process. Honey is a sweet viscous liquid that consists of 80–85% carbohydrate, vitamins (especially vitamin C), and minerals (especially potassium and magnesium), along with several active compounds like flavonoids, phenolic acids, tocopherols, catalase, superoxide dismutase, reduced glutathione, and a variety of enzymes. The rich antioxidant profile of honey makes it suited for the preservation of cancer and supporting longevity. The use of honeybee for the biosynthesis of nanoparticles is a recent phenomenon, especially utilized for biofabrication of metal nanoparticles (Table 13.1). Wu et al. (2013) carried out the first study on eco-friendly synthesis of carbon nanoparticles by suspending food-grade honey with an organic macromolecular passivating agent, purged with argon

Table 13.1 Use of bee honey in nanoparticle synthesis

Nanoparticle type	Metal salt used	Size (nm)	References
Calcium oxide	10 g $CaCO_3$ dissolved in 100 mL distilled water	<50	Arul et al. (2018)
Cerium oxide	12.6 g of $Ce(NO_3)_3.6H_2O$ dissolved in 30 mL of distilled water	23	Darroudi et al. (2014)
Copper oxide	1.705 g of $CuCl_2$ dissolved in 50 mL distilled water	Nr	Ali et al. (2014)
Gold	50 mg of $HAuCl_4$ dissolved in 120 mL distilled water	~15 9.9	Philip (2009) Sreelakshmi et al. (2011)
Gold-coated iron oxide	1:1, 2:1, and 3:1 M of Fe_3O_4/au-NPs	3.49–4.11	Rasouli et al. (2019)
Palladium	5×10^{-3} g of $PdCl_2$ and 0.5 g of honey dissolved in distilled water	5–40	Reddy et al. (2012)
Platinum	0.01 M H_2PtCl_6	2.2	Venu et al. (2011)
Silver	0.001 M $AgNO_3$ 0.1 M $AgNO_3$ 0.1 M $AgNO_3$ 1 mM $AgNO_3$ 1 mM $AgNO_3$ 1 mM $AgNO_3$ 1 mM $AgNO_3$	4–6 15.63–26.05 Nr 30.2–75.4 Nr 12–18 50.5–78.2	Philip (2010) Haiza et al. (2013) Obot et al. (2013) Kothai and Jayanthi (2014) Singh et al. (2014a, b) El-Deeb et al. (2015) Al-Brahim and Mohammed (2020)

Nr not reported.

and further heating. The polysorbate-coated carbon nanoparticles and polyethylene glycol$_{400}$-coated carbon nanoparticles had a diameter of 7 ± 3 nm and 8 ± 2 nm, respectively, with the polydispersity index and zeta potential values as 0.08 ± 0.03 and 8 ± 3 mV (Fig. 13.2).

13.4.2 Propolis

Propolis also known as "bee glue" is a sticky substance produced by honeybees from substances accumulated from diverse plant parts and mixed with salivary secretions. Propolis exhibits numerous medicinal properties like anti-inflammatory, antioxidant, antimicrobial, and cytostatic effects mainly attributed to a large number of active compounds.

The first report of using propolis for biogenic synthesis of nanomaterials was put forth by Roy et al. (2010) who confirmed the suitability of both ethanolic and water extracts of Indian propolis and its flavonoid components (Fig. 13.3) for the biosynthesis of stable Ag and Au nanoparticles under alkaline conditions. Silver nitrate and chloroauric acid were the silver and gold salts used for the study. It was observed that

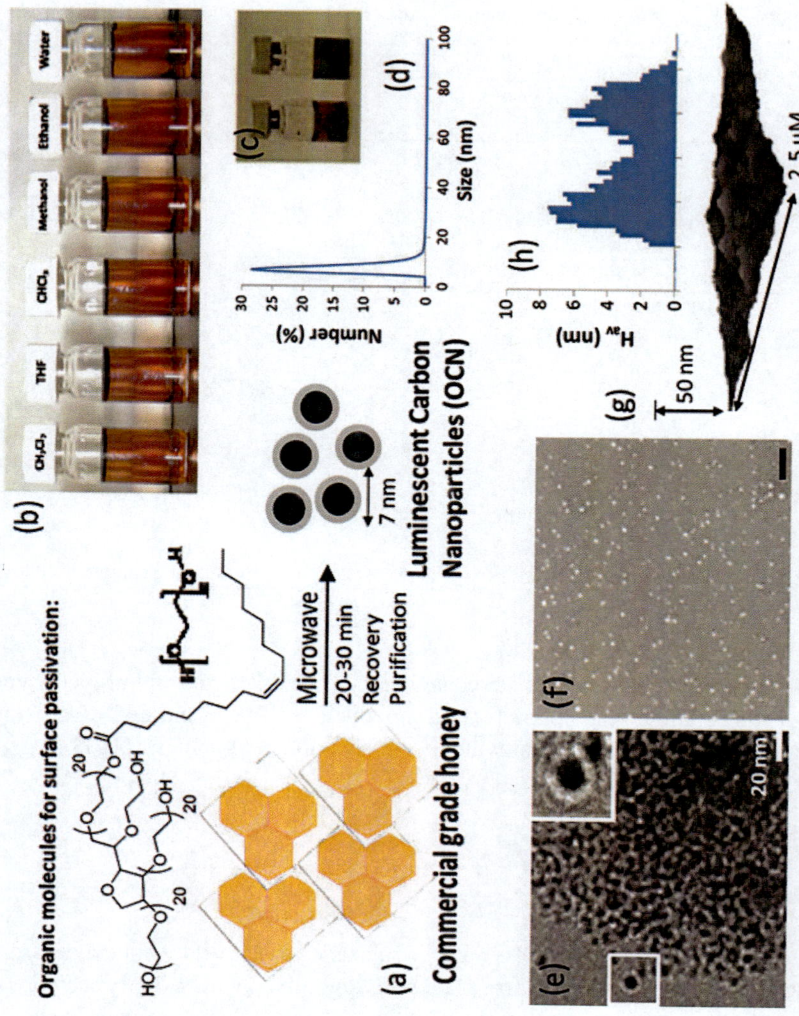

Fig. 13.2 Preparation of luminescent carbon nanoparticles from bee honey (reprinted with permission from Wu et al. 2013)

Fig. 13.3 Structures of pinocembrin and galangin (reprinted with permission from Roy et al. 2010)

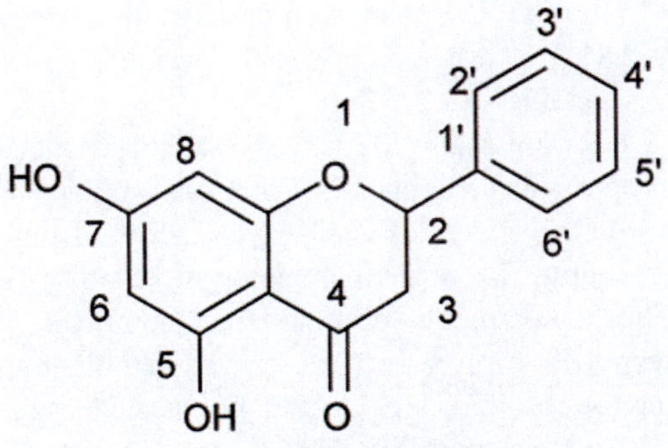

Pinocembrin (1).

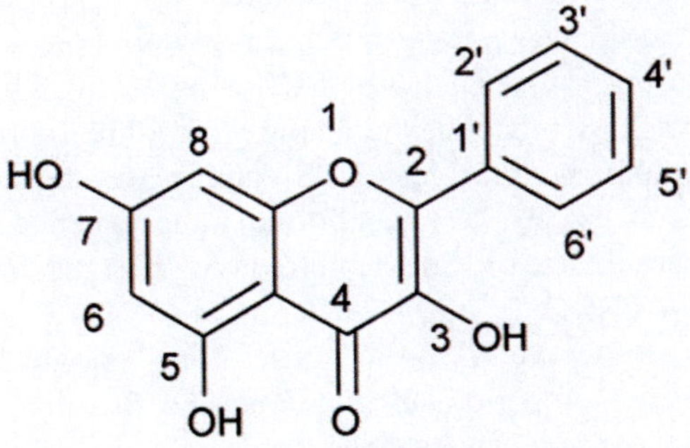

Galangin (2).

both extracts exhibited high efficiency for the biological synthesis of silver and gold nanoparticles. Similarly both pinocembrin and galangin were highly efficient in the nanofabrication process under similar chemical conditions. TEM analysis exhibited that most of the AgNPs obtained by propolis extract as well as pinocembrin and galangin were spheroidal with some having anisotropic morphology. Gold nanoparticles had a diameter of ~20 nm and were spherical, nano-rods, and nano-prisms with different cross sections. The presence of flavonoids had a dominating effect on the formation of gold nanoparticles by influencing the shape of the nanoentities.

Silver nanoparticles were biosynthesized using an ethanolic extract of stingless bee propolis (*Tetragonula iridipennis*) for the first time by Kothai and Jayanthi (2014). The size of AgNPs was calculated as 66–84 nm by SEM images and 91 nm using DLS.

Barbosa et al. (2019) prepared AgNps using Brazilian red propolis by mixing its ethanolic extract with $AgNO_3$ and NaOH solution. The Pareto plot pointed out that the agitation variable as well as $AgNO_3$ concentration had a bearing in reducing the size of the AgNPs. The nanoparticles were heterogeneous and tended to agglomerate, but propolis acted as the stabilizer and prevented them from aggregation. The TEM images showed two main particle groups: first comprising of well-dispersed particles (~5 nm) and the second one comprising of particles having size of about

11 nm. The PXRD showed peaks at 2θ angles at 38.04°, 44.80°, 64.54°, and 77.26° and 81.32° corresponding to the principal planes of cubic crystalline face structure of the metallic silver.

Bayrami et al. (2020) synthesized four metallic nanocomposites (zinc, silver, and copper) using ethanolic propolis extract (Ext), viz., ZnO/Ext, ZnO/Ag/Ext, ZnO/CuO/Ext, and ZnO/Ag/CuO/Ext, using the microwave method. The mean crystallite size of the nanocomposites was estimated in the range of 11.4–22.3 nm. The assessment of the nanocomposites for their wound healing and reepithelialization activity showed significant performance of ZnO/Ag/Ext as dressing for a wound care performance as well as in epidermis regeneration in comparison with other nanomaterials.

Ilk et al. (2020) developed an eco-friendly method for the production of silver nanoparticles by treating different volumes of ethanolic extract of propolis (PE) with 1 mM silver nitrate solution in sodium borohydride ($NaBH_4$) known as P-AgNPs1 (using 1 mL PE), P-AgNPs2 (using 2.5 mL PE), and P-AgNPs3 (using 5 mL PE). It was observed that the nanoparticle dimensions were greatly dependent on the PE volume with increasing PE volumes leading to decreased particle size and polydispersity index. The three formulations generated low polydispersity index with mean particle size of 150.90 nm for P-AgNPs1, 82.26 nm for P-AgNPs2, and 47.72 nm for P-AgNPs3.

Botteon et al. (2021) carried out biological synthesis of AuNPs using 0.5 mM tetrachloroauric acid trihydrate and Brazilian red propolis extract using different organic solvents, namely, hexane, dichloromethane, and ethyl acetate, using a green approach. The production of gold nanoparticles was confirmed by the appearance of an SPR band showing prominent peak in the range of 523–541 nm. TEM studies showed the presence of AuNPs ranging from 8 to 15 nm with narrow particle size distribution. The smaller nanoparticles were spherical, whereas the larger particles showed a variety of shapes like triangle, pentagon, hexagon, and rods. The gold nanoparticles exhibited an absorption peak of about 2.2 keV in the EDX spectroscopy which is characteristic of gold absorption. The crystallinity of the biologically synthesized AuNPs was also confirmed by the SAED pattern.

13.5 Spider Silk

Spider silk is a proteinaceous semicrystalline biopolymer spun by spiders comprising of protein polymers that possess the following properties:

- High tensile strength
- High extensibility similar to rubber
- High modulus
- High biocompatibility

Spider silk has a diameter of 0.2–1.0 mm and consists of highly ordered proteins. The proteins possess high amount of nonpolar and hydrophobic amino acids, and

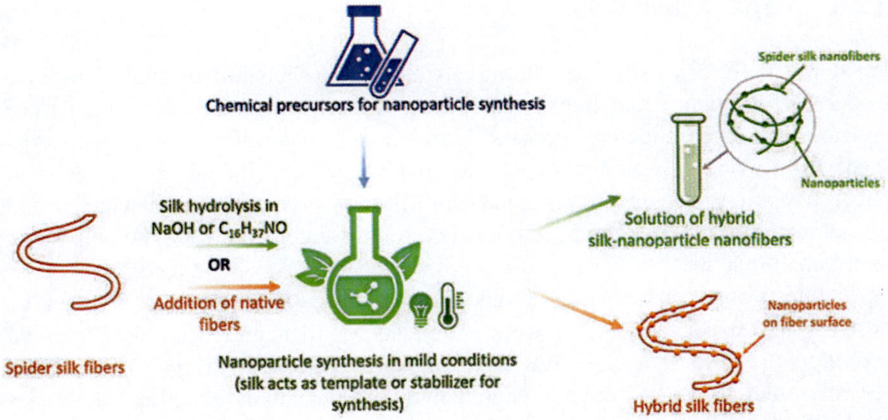

Fig. 13.4 Diagrammatic representation of nanofabrication utilizing spider silk (reprinted from Kiseleva et al. 2020)

about 90% of the proteins comprises of highly repetitive amino acid sequences. Such is the strength and resilience of spider silk that it is considered superior to silkworm silk. Spider silk has been the material of choice for diverse industrial applications like optics, photonics, and biomedical science.

Spider silk is increasingly being explored for the eco-friendly synthesis of nanoparticles (Fig. 13.4). Singh et al. (2007) used spider silk fibers as template for the bioproduction of gold nanoparticle bioconjugates. It was observed that the gold nanoparticles bonded strongly with the spider silk possibly due to the strong affinity of amine groups present in the silk proteins with the gold surface.

Hydrolyzed spider silk was combined with 1 mM $AgNO_3$ solution by Lateef et al. (2016a) for the synthesis of nanoparticles where the silk proteins served as capping agents. The AgNPs were crystalline, spherical, and polydispersed with size dimensions of 3–50 nm. These exhibited potent activity against multiple drug-resistant organisms which indicated their role in clinical applications as well as an additive in emulsion paints for reducing microbial attack. The carboxyl group present in spider silk was thought to have worked as an active reducing agent, while the carboxyl and amino groups provided stability to the metal nanoparticles. Biocompatible magnetite (Fe_3O_4) nanoparticles having antibacterial properties were synthesized by Singh et al. (2015) using silk fibers obtained from *Crossopriza lyoni* (family: Pholcidae), a cosmotropical spider commonly known as "tailed daddy longlegs." Spider silk was treated with a number of neoteric solvents of which tetrabutylammonium hydroxide ($C_{16}H_{37}NO$) gave the best results. Zhou et al. (2016) developed a cheap biogenic method for the synthesis of self-doped carbon nanofibers via pyrolysis of spider silk. The heteroatom-doped porous carbon nanofibers exhibited good oxygen reduction reaction (ORR) activity in both neutral and alkaline conditions.

13.6 Paper Wasp Nest

Paper wasps (*Polistes* spp.) are primitively eusocial insects in the family Vespidae (order Hymenoptera) that live in colonies comprising a single dominant female assisted by nonreproductive workers. Their nest is made of a paperlike material made from wood collected, chewed, and glued together with saliva. The nests of paper wasps are rich in proteins that can serve as efficient bioreductants in the biofabrication of nanoparticles. A typical colony is small with 100–200 cells, with less than 100 adults.

The nests of paper wasp were utilized for biogenic synthesis of AgNPs, and the newly synthesized nanoentities were examined for their antibacterial (*Klebsiella granulomatis* and *Pseudomonas aeruginosa*), antifungal (*Aspergillus niger*, *A. flavus*, and *A. fumigatus*), dye degradation, and thrombolytic activity by Lateef et al. (2016b). Nests of paper wasp were collected and hydrolyzed with 100 mM NaOH at high temperature (90 °C), cooled. After centrifugation at 4000 rpm for half an hour, 1 mL of nest extract was mixed with 40 mL 0.001 M $AgNO_3$ solution at 30 °C for 30 min. Hexagonal, rhomboid, 3D triangular, rod, and sphere-shaped nanoentities were evident in the TEM images having size ranging from 12.5 to 95.55 nm. The crystallinity of the AgNPs was confirmed by SAED as shown by the ringlike pattern, while XRD also indicated toward the prominent presence of silver. The formation of AgNPs was attributed to the presence of proteins and phenolics in the extract which functioned as efficient capping and stabilization agents.

13.7 Wasp Venom

The first such study utilizing venom of wasps in nanofabrication was carried out in 2018 by Jalaei et al. (2018) who successfully accomplished synthesis of peptide conjugated gold nanoparticles using Oriental hornet (*Vespa orientalis*) venom. The venom of *V. orientalis* contains high molecular weight, low molecular weight peptides, bioactive molecules, and allergens. For collection of venom, the venom glands were dissected, crushed in liquid nitrogen, and mixed with 100 mM slightly basic phosphate buffer. The suspension was then homogenized and centrifuged at 8000 × g and the supernatant lyophilized and stored at low temperature for peptide extraction. AuNPs were synthesized by citrate reduction oftetrachloroauric acid and using sodium dodecyl sulfate (SDS) as a surfactant to prevent agglomeration. Peptide conjugated gold nanoparticles (peptide AuNPs) were fabricated by adding SDS-NP suspension to the purified peptides. A mean size of 23.2 ± 2.7 nm was obtained through TEM, while it was calculated as 35 nm using XRD analysis. The FTIR analysis pointed toward the role of amides, amines, peptides, and protein compounds in nanoparticle biosynthesis.

13.8 Silk Moth and Its Derivatives

Bombyx mori (family: Bombycidae), the domestic silk moth, is the principal producer of commercial silk produced in the world. Silk, commonly known as the "queen of all fabrics," was first discovered in China which also is the largest silk producer followed by India. Silk is a continuous filament in the cocoon of the silkworm caterpillar. The caterpillar builds the cocoon by producing a continuous fiber as liquid secretions from specialized glands that harden on exposure to air forming twin filaments. These filaments are composed of fibroin, a proteinaceous material. Another pair of glands secretes sericin (10–300 kDa), a natural water-soluble macromolecular glue-like protein that cements the fibroin filaments to make silk fibers. Sericin contains about 18 amino acids, of which polar amino acids form the main components. The presence of polar amino acids imparts exceptional properties like hydrophilicity and reactivity to the silk fiber. A macromolecular polymer sericin, that is considered a waste product during silk processing, shows excellent biocompatibility and exhibits numerous medicinal properties such as an antioxidant and antibacterial, formulation of creams, and vehicle for drug delivery and also has wide applications in regenerative medicine.

13.8.1 Silk Fibroin

Su et al. (2007) carried out biosynthesis of nanocomposites comprising of evenly distributed cadmium sulfide quantum dots and silk fibroin obtained from cocoons of *B. mori*. The well-dispersed hybrid nanoparticles in the TEM images were spherical having a mean diameter of about 5 nm.

Fei et al. (2013) developed a green technique for the in situ preparation of AgNPs utilizing silk fibroin as a biotemplate. Silver nitrate ($AgNO_3$) was mixed with regenerated silk fibroin (RSF) in powder form leading to the formation of a transparent solution having final $AgNO_3$ concentration of 1–16 mg/mL. Thereafter this solution was exposed to light using an incandescent bulb and kept at ambient temperature for 1 day to produce RSF-AgNP composite solution. TEM images showed the mean nanoparticle size as 12.0 ± 2.1 nm, while HRTEM indicated the d-spacing of the crystallographic plane as 0.23 nm that conformed to the lattice plane of Ag. The EDX studies confirmed the single crystalline structure of the newly synthesized nanoentities.

Apart from silver nanoparticles, silk fibroin has also been used for the biological synthesis of other metal nanoparticles. A biogenic procedure to synthesize AuNP reduced graphene oxide composites (AuNPs/RGO composite) was proposed by Xu et al. (2013). The method was carried out using regenerated silk fibroin (RSF) both as a reducing agent for graphene oxide (GO) and chloroauric acid ($HAuCl_{4.}4H_2O$) (25 mM) and as a cementing agent for tightly binding AuNPs and RGO sheets (ca. 1 g/L). The AuNP/RGO composites were flowerlike in shape of 200–300 nm size having uneven surfaces and could be stably re-dispersed in water.

Das and Dhar (2014) developed a green method for the biosynthesis of gold, silver, palladium, and platinum nanoparticles through redox approach using thiol-modified silk fibroin (SF-SH). The aqueous SF-SH solution having basic pH (pH: 10) was treated at ambient temperature with metal precursor solution having concentration of 2 mM ($HAuCl_4$, $AgNO_3$, and Na_2PdCl_4). The reduction of gold salt by SF-SH and formation of gold nanoparticles were reflected as color of the reaction mixture from yellow to wine red. Similar change in reaction mixture was observed as the color changed from colorless to blackish brown in case of AgNPs and from yellow to colorless during formation of PdNPs. The gold and silver nanoparticles synthesized exhibited a particle size of 13.0 ± 2.5 nm and 5.0 ± 1.0 nm, respectively, while the palladium and platinum nanoparticles showed a narrow distribution particle size of 3.0 ± 1.0 and 2.0 ± 0.2, respectively. Thus, cysteine-modified silk fibroin served as efficient reducing and stabilizing agent for the synthesis of noble metal nanoparticles.

Water-soluble silk fibroin nanoparticles (WSSF-NPs) and silk fibroin were used for biosynthesis of gold nanoparticles (AuNPs) using 0.1 mM chloroauric acid ($HAuCl_4$) by Wongkrongsak et al. (2016). The reaction solution changed its color from light yellow to dark violet on addition of WSSF-NPs. The number of gold nanocolloid increased as the concentration of WSSF-NPs was increased that was indicated by a blue shift in the SPR band. It was concluded that the WSSF-NPs having size of 40 nm exhibited greater antioxidant capacity and reducing power for biofabrication of Au nanocolloid.

13.8.2 Silk Sericin

Sericin was used by Aramwit et al. (2014) for the biosynthesis of silk sericin (SS)-capped AgNPs. The extraction of sericin was carried out using a degumming technique performed under high pressure and temperature conditions, and its solution added to 1, 5, and 10 mM $AgNO_3$ solution at alkaline pH. Successful formation of SS-capped AgNPs was indicated by appearance of yellow color in the reaction mixture. The sericin-capped AgNPs were round, uniform sized having diameter ranging from 48 to 117 nm. It was observed that elevated levels of silk sericin provided higher stability to the silver ions. The FTIR analysis indicated that the carboxylate group played a major role in the generation of AgNPs by acting as the reducing agent, while COO^- and $NH_2{}^+$ groups acted as stabilizing agents.

He et al. (2017) followed a simple, one-step method for the biofabrication of sericin-silver nanoparticle composite where sericin served as the reductant, dispersant, and stabilizer. Sericin was diluted in hot water (0.1%), added to 20–160 mg $AgNO_3$ powder, and incubated in natural light for 36 hours for catalyzing the synthesis of AgNPs. The absorption peak at about 405 nm in UV-vis spectra along with electron microscopy techniques pointed toward the formation of AgNPs. TEM images confirmed the presence of multishaped Ag nanocrystals having a size range of 4–20 nm. The sericin-AgNP composite were stable and exhibited potent antibacterial activity and good biocompatibility which showed its potential for diverse biomedical applications.

A pioneer approach utilizing Ag^+-BH^{4-}-sericin for production of sericin attached silver nanoparticles (AgNPs-sericin) was developed by Lv et al. (2018). Different solutions were prepared by adding 0.008 M $AgNO_3$ solution with different concentrations of sericin followed by titration with 0.008 M $NaBH_4$ solution. The addition of sericin in the reaction mixture changed its color to yellow and finally to brown. Most of the nanoparticles were spherical and have size of 30–50 nm. It was observed that 0.3 mg/mL sericin was ideal for the bioproduction of AgNPs. Increased concentration of sericin led to corresponding increase in the size of the nanoparticles. The XRD analysis confirmed the presence of face-centered cubic silver crystals. It was concluded that sericin can serve as a potent dispersant and stabilizing agent for the biosynthesis of AgNPs.

Sericin-based AgNPs were synthesized by Muhammad Tahir et al. (2020) from sericin solution prepared by dissolving lyophilized sericin powder obtained from *B. mori* in distilled water. Thereafter $AgNO_3$ (5, 10, 20, and 30 mg) was mixed with 10 mL of 0.2% sericin solution and the mixture kept in light at room temperature. Bulk of the nanostructures had polydispersity index below 0.5 and mean size below 53.6 nm, which indicated toward their homogenous distribution in the medium. Spherical, hexagonal, and triangular-shaped nanoparticles with uniform distribution were observed in the SEM images.

Apart from *B. mori*, sericin obtained from non-mulberry silkworms like Eri silkworm (*Samia cynthia ricini*) (family: Saturniidae) has also been utilized for nanofabrication of AgNPs by Chaisabai et al. (2017). Sericin is extracted from the silk cocoons and treated with 1 M $AgNO_3$. TEM images clearly showed that the nanostructures were spherical-shaped particles having size ranging from 5.5 to 30.5 nm and mean diameter of around 14 nm. The AgNPs also exhibited strong antibacterial activity against *E. coli* and *S. aureus*.

Akturk et al. (2019) developed a green synthesis technique for the biological production of AuNPs using auric chloride (gold (III) chloride trihydrate, $HAuCl_4.3H_2O$) and sericin obtained after extracting cocoons in hot water containing 20 mM sodium carbonate. Sericin solution with concentrations of 0.25, 0.5, and 1% (wt/vol.) were added to 0.001 M of gold salt and exposed to ultraviolet C (UVC) radiation for 24 hr. at room temperature. A color change from yellow to bluish-red and reddish-pink, along with SPR peaks at 520–540 nm, strongly indicated toward the formation of AuNPs. The average size of gold nanoparticles calculated using TEM was <10 nm. An interesting observation was a phenomenal decrease in the nanoparticle size on increase in sericin concentrations.

Antheraea mylitta (family: Saturniidae), also known as the India tasar silkworm and Vanya silkworm, is an economically important wild silk moth species distributed across the Indian subcontinent (Dutta et al. 2020). The silk sericin protein of tasar silkworm's cocoon was fractionated into three fractions, viz., high molecular mass fraction-1 (50–300 kDa), intermediate molecular mass fraction-2 (30–50 kDa), and low molecular mass fraction-3 (10–30 kDa). The three fractions were used to synthesize crude sericin nanoparticles (CRSNPs) as well as cationic nanoparticles using poly-L-lysine solution. The fraction-specific cationic nanoparticles were termed as n-SNP1, n-SNP2, and n-SNP3, respectively. The particle sizes decreased gradually from n-CRSNPs (149 ± 21 nm) to n-SNP3 (16 ± 5 nm) and increased in all p-SNPs compared to n-SNPs after coating with PLL (156 ± 22 nm for p-CRSNPs and 21 ± 6 nm for p-SNP3). The polydispersity index showed that all the n-SNPs and p-SNPs were polydispersed in the solution. The zeta potential measurement showed surface potential of all n-SNPs and p-SNPS in the range of −24 mV to −34 mV and + 27 to +36 mV, respectively, indicating the stable nature of the SNPs.

Silver-coated tasar fibroin nanofibrous mats have also been produced using the leaf extract of tridax daisy (botanical name: *Tridax procumbens*; family: Asteraceae) and explored for various biomedical applications like tissue engineering and wound dressing.

13.9 Eggs

Recent approach in nanobiotechnology using animal systems also envisages the use of chicken eggs in nanoparticle production. However, the eggs of invertebrates like snails have also been utilized for fabrication of nanostructures. In a recent study Janthima et al. (2018) developed an eco-friendly process for synthesis of silver nanoparticles (AgNPs) using the egg extract of *Pomacea canaliculata*. This snail species, having the common name "golden apple snail" or the "channeled apple snail," is a native of South America and is listed as one of the world's highly invasive snail species. The extract of the snail eggs was prepared in three different concentrations and mixed with 1000 mM $AgNO_3$ separately both in the presence and absence of 2000 mM glucose, and the solutions kept in agitation in the dark for 2 days. The formation of silver nanoparticles was confirmed by the change in the color of the medium from yellow to brown. TEM images showed good dispersion of spherical nanoparticles with size ranging from 1 to 30 nm with mean diameter of 9.0 ± 5.9 nm. The crystalline AgNPs exhibited strong antibacterial activity against different bacterial species.

Chicken egg is a low cost protein-based nutritional component of millions of people around the world. The structure of chicken egg is divided into the following components (Fig. 13.5):

- Shell
- Shell membrane
- Yolk
- Egg white

Egg white, comprising about two-thirds of the component of the egg, is an abundant source of about 40 different proteins, the most important being ovalbumin and lysozyme. The yolk contains lipids, vitamins, minerals, and carotenoid pigments along with being the only source of bioavailable lutein and zeaxanthin.

Both egg white and yolk have been utilized in the production of nanoparticles.

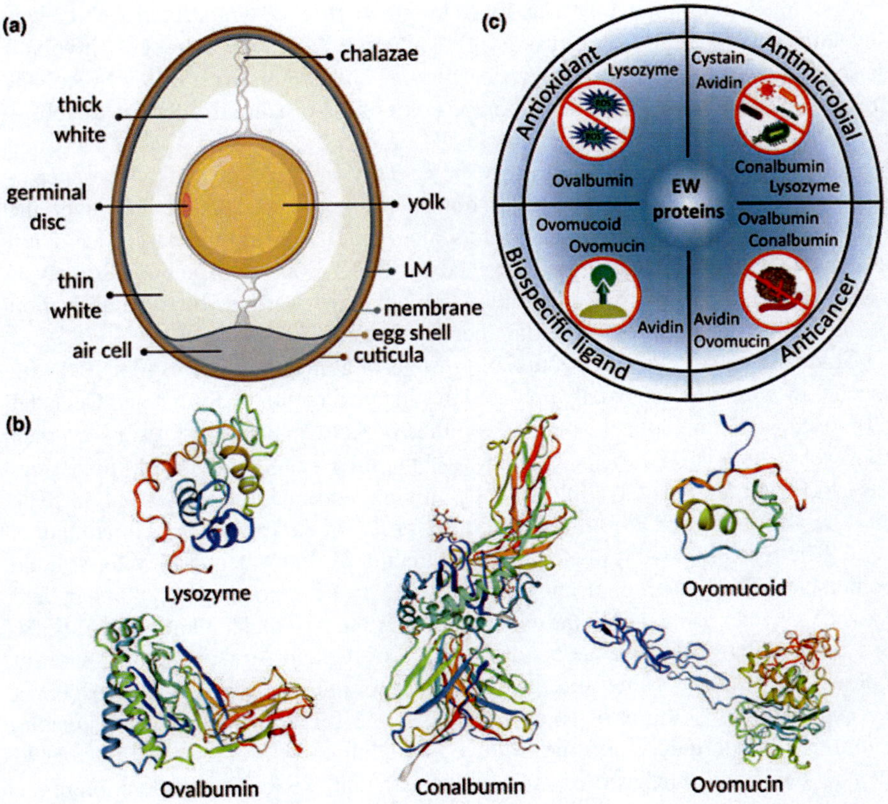

Fig. 13.5 (**a**) Different components of chicken egg. (**b**) Three-dimensional structures of five main egg white proteins. (c) Biological properties of egg white (reprinted with permission from Jalili-Firoozinezhad et al. 2020)

13.9.1 Egg White

Egg white has been successfully utilized for synthesis of a wide range of metal nanoparticles as well as carbon-based nanomaterials like carbon dots.

Yan et al. (2010) carried out synthesis of titanium dioxide (TiO_2) nanoparticles using denatured egg albumen gelatin network as a biotemplate. Egg albumin was added to deionized water, filtered, and added to tetrabutyl orthotitanate (TBOT) with vigorous stirring. The mixture was then crystallized in an autoclave at 180 °C for 48 h and cooled, and the solid product was isolated by centrifugation. The TiO_2 nanoparticles formed were uniform with mean diameter of about 9 nm. HRTEM images of TiO_2 nanoparticles showed the diameter of the particles as 9 nm, and the interplanar distance of 3.51 Å was observed that was the characteristic (101) plane of anatase. As per the Debye-Scherrer equation on (101) plane, the particle size of nanoparticles was calculated as 10.95 nm that was in consistence with the TEM observations.

A simple, environmentally friendly technique for the biosynthesis of AgNPs was developed by Lu et al. (2012) using egg white and $AgNO_3$. The process involved dissolution of egg white extract in water, filtering and treating with 10.0 mM $AgNO_3$ at ambient temperature. The change in the color of the mixture from white to yellow took place within 72 h. The SPR peak in UV-vis spectra present at 425 nm indicated the formation of spherical silver nanoparticles. TEM images depicted the presence of spherical silver nanoparticles of 12 nm diameter in homogenous state. However, the particle size calculated by dynamic light scattering (DLS) was quite large (154.2 nm in diameter) as compared to that obtained by TEM. Both TEM and DLS analysis revealed narrow particle size distribution that pointed toward the possible utilization of these silver nanoparticles in biomedical applications.

Gold nanoclusters are a collection of few tens of gold nanoparticles (AuNPs) with size below 1 nm that are usually prepared by chemical methods. Joseph and Geckeler (2014) carried out pioneer study to synthesize gold nanoclusters using a green approach. Aqueous solution of egg white (25 mg/mL) was added to 0.1 M potassium tetrachloroaurate ($KAuCl_4$) followed by mixing with 1 M NaOH solution. The mixture was incubated at room temperature for 6 h in dark with the pH maintained at 10. The reaction conditions led to the formation of fluorescent and water-soluble gold nanoclusters having mean size of 2 nm. The amino acids tryptophan and tyrosine played crucial part in the conversion of Au to Au^0 under alkaline conditions.

Li et al. (2016) reported the nanofabrication of Ag and Au@Ag microspheres at ambient temperatures using the aqueous extracts of the egg white. Ag microspheres were prepared by addition of 10 mM $AgNO_3$ to 15-fold diluted egg white aqueous solution with continuous stirring, followed by addition of ascorbic acid powder to the above mixture which served as the reducing agent. A significant alteration in the color of the mixture to black pointed toward the synthesis of Ag microspheres. For preparation of Au@Ag microspheres, bovine serum albumin was added to Ag microsphere aqueous solution, and thereafter addition of 1 mM $HAuCl_4$ was carried out with vigorous stirring for 30–40 min. Thereafter Au@Ag microspheres were obtained by centrifugation. The urchin like Ag microspheres having mean diameter of 600–800 nm were made up of nanopricks of 10–40 nm long as the building blocks. The Au@Ag nanostructures prepared by galvanic replacement exhibited monodispersity and were composed of nanostructures having diameter ranging from 10 to 40 nm.

Biosynthesis of highly fluorescent carbon dots was reported by Fu et al. (2019) using hydrothermal carbonization of ovalbumin, the main protein in hen eggs. A temperature of 230 °C and pH 9.0 for 12 h with 0.03 mg/mL ovalbumin concentration were optimal for the synthesis of carbon dots. Uniformly spherical particles with high-quality dispersion having size of 2.51 ± 0.45 nm were observed in the TEM images. The HRTEM images revealed the presence of carbon dots having lattice structure and a spacing of 0.22 nm. The atomic force microscopy showed that the particle height was about $3 \sim 4$ nm that conformed to the size distribution obtained from the TEM image. The synthesized carbon dots had high solubility, salt stability, low toxicity, good optical property, and biocompatibility that indicated their utilization potential as a superior fluorescent bioimaging agent.

Egg albumen was utilized as a biotemplate organic matrix for carrying out biosynthesis of iron oxide quantum dots by Moshafi et al. (2019). Iron(III) nitrate nonahydrate was used as the metal salt. EDX studies confirmed that the quantum dots were primarily composed of iron but had minor amounts of other metals as well. Spherically shaped homogenous iron oxide quantum dots with a diameter of about <80 nm were obtained which were dispersed uniformly. The results indicated that the use of albumen is quite beneficial for the production of nanostructures having little agglomeration and greater uniformity.

13.9.2 Yolk

Egg yolk has also been utilized for the production of nanostructures but to a lesser extent than egg white. Nadaroglu et al. (2017) successfully carried out one-step biofabrication of platinum nanoparticles (PtNPs) using quail egg yolk that is rich in vitamin and protein. A total of five different concentrations ranging from 0.5 to 7 mM of chloroplatinic acid (H_2PtCl_6) solution were used in the study. Optimization of reaction conditions showed that the synthesis of PtNPs was best accomplished at 20 °C, pH 6.0, and 0.5 mM salt concentration. The UV-vis spectra showed the peak at 329 nm which was characteristic of platinum. The XRD analysis of the dried and powdered platinum nanoparticles showed four distinct peaks corresponding to the crystal planes of cubic platinum. The SEM image showed the presence of spherical platinum nanoparticles having size range of 7–50 nm.

Egg yolk oil, obtained by heating cooked egg yolks of *Gallus gallus domesticus* under high temperatures (200 °C) and further extraction by water, was used for the production of quantum dots by Zhao et al. (2017). Morphology analyses performed using TEM indicated that the quantum dots were spherical having mean size of <10 nm, while HRTEM showed the lattice spacing as 0.267 nm. Elemental analysis and TEM established the presence of three elements, viz., C, O, and Fe, in the nanostructures. The quantum dots exhibited the potential to activate hemostasis since they stimulated intrinsic blood coagulation and activated the fibrinogen system which confirmed their role in the Chinese system of medicine.

13.10 Earthworms

Earthworm, also known as angleworm, are a group of detritivorous dwelling terrestrial worms of the class Oligochaeta (phylum Annelida) that are found around the world in soils having sufficient moisture and organic content. Earthworms participate in the circulation of organic matter in nature and alteration of soil structure and accelerate plant growth. In fact earthworms are brought into use for decomposition of organic matter by an important eco-friendly process known as vermicomposting. Recently some studies have been conducted that utilized different species of earthworms in the green synthesis of nanostructures.

The aqueous extract of *Eisenia andrei* earthworm (family: Lumbricidae) was used for the biofabrication of Ag and Au nanoparticles by Han et al. (2013) using $AgNO_3$ and $HAuCl_4.3H_2O$. As inferred from HR-TEM images, the mean diameter of the gold nanoparticles was 6.70 ± 2.69 nm with 4 ~ 6 nm being the most common diameters. For silver nanoparticles, the mean diameter was 12.19 ± 4.28 nm with maximum representation in 11 ~ 13 nm range. Both types of nanoparticles were mostly spherical and amorphous shaped with no agglomeration or aggregation.

Kim et al. (2013) successfully synthesized gold nanoparticles using a green approach with earthworm extracts and evaluated their anticoagulant activity. UV-vis spectrum showed the SPR peak at 533 nm. The HR-TEM images showed that gold nanoparticles were primarily spherical having diameters of around 6.13 ± 2.13 nm, with 4 ~ 5 and 7 ~ 8 nm being the two most abundant diameters. The AFM images revealed the presence of spherical, cubic, and block-shaped nanoparticles. The authors were of the view that proteins/peptides in the extract acted as stabilizers and reductants for reducing Au^{3+}.

Green synthesis of AgNPs was efficiently accomplished by Jaganathan et al. (2016) using *Eudrilus eugeniae* extract and 1 mM silver nitrate solution. *E. eugeniae* is native to tropical West Africa and is known as the "African nightcrawler" that is well known for its vermicomposting potential. The antimicrobial potential of AgNPs was evaluated against six pathogen bacteria and three fungi. The UV-vis images exhibited peak at 425 nm which was probably due to the existence of nanoparticles. Particle shape and size determination by FESEM showed mostly spherical nanoparticles of 4–10 nm dimensions in clusters. The AgNPs also exhibited significant potential against pathogenic microbes as well as mosquito vectors of deadly diseases.

Earthworms have also been used to produce semiconductor particles like quantum dots that have 2–10 nm diameters and exhibit size and composition-dependent optical and electronic (optoelectronic) properties. Stürzenbaum et al. (2013) synthesized cadmium telluride (CdTe) quantum dots using *Lumbricus rubellus* (family: Lumbricidae), a medium-sized epi-endogeic earthworm native to Western Europe. The organisms were exposed for more than 10 days to soil rich in cadmium chloride ($CdCl_2$) and sodium tellurite (Na_2TeO_3), and the quantum dots were isolated from chloragogenous tissues surrounding the gut of the earthworm. The UV-spectra showed an absorption peak of 450 nm which indicated the presence of CdTe quantum dots. HRTEM showed the existence of crystalline nanoentities having mean diameter of $2.33 + 0.59$ nm, while SAED analysis exhibited diffused rings consistent with CdTe nanoparticles. The quantum dots were luminescent and soluble in water and exhibited properties of semiconductors.

Talaeeshoar et al. (2018) described the biosynthesis and the impact of metal precursor concentrations on the synthesis of cadmium selenide quantum dots (CdSe QDs) using *Eisenia fetida* earthworm, also known by other names like "manure worm," "redworm," "trout worm," "tiger worm," and "red wiggler worm." Cadmium chloride and sodium selenite were the precursor salt sources for cadmium and selenium, respectively. The peak is obtained around 430 nm through UV spectrophotometry, and about 605 nm obtained through photoluminescence study indicated toward the successful formation of CdSe QDs within *E. fetida*. The CdSe QDs had mean diameter of 7 nm, a value confirmed by both effective mass model and TEM analysis.

13.11 Milk

Milk is the nutrient-rich primary source of nutrition for young ones which is produced by the mammary glands of mammals. Both cow and sheep milk has been exploited for biological synthesis of nanoparticles. Lee et al. (2013) accomplished biofabrication of silver nanoparticles using cow milk by adding it to 1 mM $AgNO_3$ solution and incubating the mixture in a rotary shaker at room temperature. An alteration in the color of reaction mixture from milky to brown indicated the reduction by cow milk and the resultant production of silver nanoparticles. Analysis of the samples using SEM-EDS showed strong prominent peaks around 3 keV indicating the presence of silver. TEM images showed 30–90 nm-sized nanoparticles which were circular and agglomerated. The authors were of the opinion that lipids prevalent in the cow milk had a crucial role in the aggregation of silver nanoparticles.

The pioneer report involving utilization of goat milk for nanofabrication was put forth by Gholami-Shabani et al. (2016). This study involved the biosynthesis of platinum nanostructures by treating goat milk with 0.001 M chloroplatinic acid (H_2PtCl_6) solution at ambient temperature. The platinum nanoparticles were formed rapidly within 3 hours which was evident by formation of brownish-red color reaction mixture. Spherical nanoparticles measuring 8.5 nm were observed in TEM images although some had isotropic nanostructures with irregular contours.

Ihum et al. (2019) carried out synthesis of AgNPs using goat milk as capping and stabilizing agent. The UV-vis spectra showed maximum absorbance at 417 nm, while spherical nanoparticles having size ranging from 10 to 100 nm were observed in TEM images. The XRD analysis confirmed the crystallinity of the AgNPs.

13.12 Future Prospects

The use of animal systems for nanofabrication is a recent advancement that has picked up speed only in the last decade. This area requires lots of consistent efforts since the majority of members of animal kingdom have not been explored till now. Urgent efforts are needed to develop this area and enrich the scant knowledge available till date for better utilization in green nanobiotechnology.

References

Akturk O, Gun Gok Z, Erdemli O, Yigitoglu M (2019) One-pot facile synthesis of silk sericin-capped gold nanoparticles by UVC radiation: investigation of stability, biocompatibility, and antibacterial activity. J Biomed Mater Res A 107(12):2667–2679

Al-Brahim JS, Mohammed AE (2020) Antioxidant, cytotoxic and antibacterial potential of biosynthesized nanoparticles using bee honey from two different floral sources in Saudi Arabia. Saudi J Biol Sci 27(1):363–373

Ali M, Rotte NK, Srikanth VSS (2014) Honey aided solution synthesis of polycrystalline Cu_2O particles. Mater Lett 128:253–255

Aramwit P, Bang N, Ratanavaraporn J, Ekgasit S (2014) Green synthesis of silk sericin-capped silver nanoparticles and their potent anti-bacterial activity. Nanoscale Res Lett 9(1):79

Arul E, Raja K, Krishnan S, Sivaji K, Das SJ (2018) Bio-directed synthesis of calcium oxide (CaO) nanoparticles extracted from limestone using honey. J Nanosci Nanotechnol 18(8):5790–5793

Babu KM (2019) Spider silks and their applications. In: Babu KM (ed) Silk: processing, properties and applications. Elsevier, Amsterdam, pp 235–253

Bandyopadhyay A, Chowdhury SK, Dey S, Moses JC, Mandal BB (2019) Silk: a promising biomaterial opening new vistas towards affordable healthcare solutions. J Indian Inst Sci 99: 445–487

Barbosa VT, JKC S, Alvino V, Meneghetti MR, Florez-Rodriguez PP, Moreira RE, GVB P, Landell MF, Basílio-Júnior ID, do Nascimento TG, LAM G, Dornelas CB (2019) Biogenic synthesis of silver nanoparticles using Brazilian propolis. Biotechnol Prog 35(6):e2888

Bayrami M, Bayrami A, Habibi-Yangjeh A, Shafeeyan MS, Feizpoor S, Arvanagh FM, Nourani MR, Taheri RA (2020) Biologically-synthesised ZnO/CuO/Ag nanocomposite using propolis extract and coated on the gauze for wound healing applications. IET Nanobiotechnol 14(7):548–554

Blouin M, Hodson ME, Delgado E, Baker G, Brussaard L, Butt KR, Dendoovan L, Peres G, Tondoh JE, Cluzeau D, Brun DJ-J (2013) A review of earthworm impact on soil function and ecosystem services. Eur J Soil Biol 64:161–182

Botteon CEA, Silva LB, Ccana-Ccapatinta GV, Silva TS, Ambrosio SR, Veneziani RCS, Bastos JK, Marcato PD (2021) Biosynthesis and characterization of gold nanoparticles using Brazilian red propolis and evaluation of its antimicrobial and anticancer activities. Sci Rep 11(1):1974

Buehler MJ (2013) Materials by design-a perspective from atoms to structures. MRS Bull 38:169–176

Campbell NA, Reece JB, Mitchell LG (1999) Biology. Benjamin/Cummings, Menlo Park, CA

Cao TT, Zhang YQ (2016) 2016. Processing and characterization of silk sericin from Bombyx mori and its application in biomaterials and biomedicines. Korean J Couns Psychother 61:940–952

Chaisabai W, Khamhaengpol A, Siri S (2017) Sericins of mulberry and non-mulberry silkworms for eco-friendly synthesis of silver nanoparticles. Artif Cells Nanomed Biotechnol 46(3):536–543

Chen J, Su H, Song F, Moon W-J, Kim Y-S, Zhang D (2012) Bioinspired Au/TiO2 photocatalyst derived from butterfly wing (Papilio paris). J Colloid Interface Sci 370(1):117–123

Darroudi M, Hoseini SJ, Oskuee RK, Hoseini HA, Gholami L, Gerayli S (2014) Food-directed synthesis of cerium oxide nanoparticles and their neurotoxicity effects. Ceram Int 40:7425–7430

Das S, Dhar BB (2014) Green synthesis of noble metal nanoparticles using cysteine-modified silk fibroin: catalysis and antibacterial activity. Royal Soc Chem Adv 4:46285–46292

Dongol Y, Dhananjaya BL, Shrestha RK, Aryal G (2014) Pharmacological and immunological properties of wasp venom. In: Gowder STJ (ed) Pharmacology and therapeutics. InTech, Rijeka, pp 47–81

Dutta S, Chowdhury T, Ghosh AK (2020) Green synthesis of poly-L-lysine-coated sericin nanoparticles and their molecular size-dependent antibacterial activity. Colloids Surf B Biointerfaces 188:110822

El-Deeb NM, El-Sherbiny IM, El-Aassara MR, Hafez EE (2015) Novel trend in colon cancer therapy using silver nanoparticles synthesized by honey bee. J Nanomed Nanotechnol 6:2

Fei X, Jia M, Du X, Yang Y, Zhang R, Shao Z, Zhao X, Chen X (2013) Green synthesis of silk fibroin-silver nanoparticle composites with effective antibacterial and biofilm-disrupting properties. Biomacromolecules 14(12):4483–4488

Fu X, Fu X, Li W, Chen Y, Cai Z (2019) Ovalbumin as a precursor for green synthesis of highly fluorescent carbon dots for cell imaging. J Biomed Nanotechnol 15(6):1232–1240

Gholami-Shabani M, Shams-Ghahfarokhi M, Gholami-Shabani Z, Akbarzadeh A, Riazi G, Abyaneh MR (2016) Biogenic approach using sheep milk for the synthesis of platinum nanoparticles: the role of milk protein in platinum reduction and stabilization. Int J Nanosci Nanotechnol 12:199–206

Guan Y, Su H, Yang C, Wu L, Chen S, Gu J, Zhang W, Zhang D (2018) Ordering of hollow ag-au nanospheres with butterfly wings as a bio-template. Sci Rep 8(1):9261

Haiza H, Azizan A, Mohidin AH, Halin DSC (2013) Green synthesis of silver nanoparticles using local honey. Nano Hybrids 4:87–98

Han J, Su H, Zhang D, Chen J, Chen Z (2009) Butterfly wings as natural photonic crystal scaffolds for controllable assembly of CdS nanoparticles. J Mater Chem 19(46):8741–8746

Han L, Kim YS, Cho S, Park Y (2013) Invertebrate water extracts as biocompatible reducing agents for the green synthesis of gold and silver nanoparticles. Nat Prod Commun 8(8):1149–1152

He H, Tao G, Wang Y, Cai R, Guo P, Chen L, Zuo H, Zhao P, Xia Q (2017) In situ green synthesis and characterization of sericin-silver nanoparticle composite with effective antibacterial activity and good biocompatibility. Korean J Couns Psychother 80:509–516

Hickman C, Roberts L, Keen S, Larson A, Eisenhour DJ (2000) Animal diversity. McGraw-Hill, New York

Ihum TA, Iheukwumere CC, Ogbonna IO, Gberikon GM (2019) Antimicrobial activity of silver nanoparticles synthesized using goat milk against pathogens of selected vegetables. Int J Biochem Res Rev 25(4):1–9

Ilk S, Tan G, Emül E, Sağlam N (2020) Investigation the potential use of silver nanoparticles synthesized by propolis extract as N-acyl-homoserine lactone-mediated quorum sensing systems inhibitor. Turkish J Med Sci 50(4):1147–1156

Jaganathan A, Murugan K, Panneerselvam C, Madhiyazhagan P, Dinesh D, Vadivalagan C, Aziz AT, Chandramohan B, Suresh U, Rajaganesh R, Subramaniam J, Nicoletti M, Higuchi A, Alarfaj AA, Munusamy MA, Kumar S, Benelli G (2016) Earthworm-mediated synthesis of silver nanoparticles: a potent tool against hepatocellular carcinoma, plasmodium falciparum parasites and malaria mosquitoes. Parasitol Int 65(3):276–284

Jakinala P, Lingampally N, Hameeda B, Sayyed RZ, Khan MY, Elsayed EA, El Enshasy H (2021) Silver nanoparticles from insect wing extract: biosynthesis and evaluation for antioxidant and antimicrobial potential. PLoS One 16(3):e0241729

Jalaei J, Layeghi-Ghalehsoukhteh S, Hosseini A, Fazeli M (2018) Antibacterial effects of gold nanoparticles functionalized with the extracted peptide from Vespa orientalis wasp venom. J Pept Sci 24:e3124–e3133

Jalili-Firoozinezhad S, Filippi M, Mohabatpour F, Letourneur D, Scherberich A (2020) Chicken egg white: hatching of a new old biomaterial. Mater Today 40:193–214

Janthima R, Khamhaengpol A, Siri S (2018) Egg extract of apple snail for eco-friendly synthesis of silver nanoparticles and their antibacterial activity. Artif Cells Nanomed Biotechnol 46(2):361–367

Joseph D, Geckeler KE (2014) Synthesis of highly fluorescent gold nanoclusters using egg white proteins. Colloids Surf B Biointerfaces 115:46–50

Kaewkorn W, Limpeanchob N, Tiyaboonchai W, Pongcharoen S, Sutheerawattananonda M (2012) Effects of silk sericin on the proliferation and apoptosis of colon cancer cells. Biol Res 45(1):45–50

Khatami M, Iravani S, Varma RS, Mosazade F, Darroudi M, Borhani F (2019) Cockroach wings-promoted safe and greener synthesis of silver nanoparticles and their insecticidal activity. Bioprocess Biosyst Eng 42(12):2007–2014

Kim HK, Choi MJ, Cha SH, Koo YK, Jun SH, Cho S, Park Y (2013) Earthworm extracts utilized in the green synthesis of gold nanoparticles capable of reinforcing the anticoagulant activities of heparin. Nanoscale Res Lett 8(1):542

Kiseleva AP, Krivoshapkin PV, Krivoshapkina EF (2020) Recent advances in development of functional spider silk-based hybrid materials. Front Chem 8:554

Kothai S, Jayanthi B (2014) Anticancer activity of silver nanoparticles bio-synthesized using stingless bee propolis (Tetragonula iridipennis) of Tamil Nadu. Asian J Biomed Pharm Sci 4: 30–37

Kunz RI, Brancalhão RMC, Ribeiro LFC, Natali MRM (2016) Silkworm sericin: properties and biomedical applications. Biomed Res Int 2016:8175701

Lateef A, Akande MA, Ojo SA, Folarin BI, Gueguim-Kana EB, Beukes LS (2016b) Paper wasp nest-mediated biosynthesis of silver nanoparticles for antimicrobial, catalytic, anticoagulant, and thrombolytic applications. 3 Biotechnol 6(2):140

Lateef A, Ojo SA, Azeez MA, Asafa TB, Yekeen TA, Akinboro A (2016a) Cobweb as novel biomaterial for the green and eco-friendly synthesis of silver nanoparticles. Appl Nanosci 6: 863–874

Lee K-J, Park S-H, Govarthanan M, Hwang P-H, Seo Y-S, Cho M, Lee W-H, Lee, J.-Y.: Kamala-Kannan S., Oh B.T. (2013) Synthesis of silver nanoparticles using cow milk and their antifungal activity against phytopathogens. Mater Lett 105:128–131

Li M, Zhang Y, Wang X, Cui D (2016) Egg white templated synthesis of ag and au@ag alloy microspheres for surface-enhanced raman spectroscopy research. J Nanosci Nanotechnol 16(1):939–943

Lu R, Yang D, Cui D, Wang Z, Guo L (2012) Egg white-mediated green synthesis of silver nanoparticles with excellent biocompatibility and enhanced radiation effects on cancer cells. Int J Nanomedicine 7:2101–2107

Lv X, Wang H, Su A, Chu Y (2018) A novel approach for sericin-conjugated silver nanoparticle synthesis and their potential as microbicide candidates. J Microbiol Biotechnol 28:1367–1375

Medina-Sauza RM, Álvarez-Jiménez M, Delhal A, Reverchon F, Blouin M, Guerrero-Analco JA, Cerdán CR, Guevara R, Villain L, Barois I (2019) Earthworms building up soil microbiota, a review. Front Environ Sci 7:81

Mohammed MA, Syeda JTM, Wasan KM, Wasan EK (2017) An overview of chitosan nanoparticles and its application in non-parenteral drug delivery. Pharmaceutics 9(4):53

Moshafi MH, Ranjbar M, Ilbeigi G (2019) Biotemplate of albumen for synthesized iron oxide quantum dots nanoparticles (QDNPs) and investigation of antibacterial effect against pathogenic microbial strains. Int J Nanomedicine 14:3273–3282

Mu Z, Zhao X, Xie Z, Zhao Y, Zhong Q, Bo L, Gu Z (2013) In situ synthesis of gold nanoparticles (AuNPs) in butterfly wings for surface enhanced Raman spectroscopy (SERS). J Mater Chem B 1(11):1607–1613

Muhammad Tahir H, Saleem F, Ali S, Ain QU, Fazal A, Summer M, Mushtaq R, Tariq Zahid M, Liaqat I, Murtaza G (2020) Synthesis of sericin-conjugated silver nanoparticles and their potential antimicrobial activity. J Basic Microbiol 60(5):458–467

Nadaroglu H, Gungor AA, Ince S, Babagil A (2017) Green synthesis and characterisation of platinum nanoparticles using quail egg yolk. Spectrochim Acta A Mol Biomol Spectrosc 172: 43–47

Obot IB, Umoren SA, Johnson AS (2013) Sunlight-mediated synthesis of silver nanoparticles using honey and its promising anticorrosion potentials for mild steel in acidic environments. J Mater Environ Sci 4:1013–1018

Philip D (2009) Honey mediated green synthesis of gold nanoparticles. Spectrochim Acta A Mol Biomol Spectrosc 73(4):650–653

Philip D (2010) Honey mediated green synthesis of silver nanoparticles. Spectrochim Acta A Mol Biomol Spectrosc 75(3):1078–1081

Porter D, Guan J, Vollrath F (2013) Spider silk: super material or thin fibre? Adv Mater 25:1275–1279

Rasouli E, Basirun WJ, Johan MR, Rezayi M, Darroudi M, Shameli K, Shanavaz Z, Akbarzadeh O, Izadiyan Z (2019) Facile and greener hydrothermal honey-based synthesis of Fe_3O_4/au core/shell nanoparticles for drug delivery applications. J Cell Biochem 120(4):6624–6631

Reddy SM, Datta KKR, Sreelakshmi C, Eswaramoorthy M, Reddy BV (2012) Honey mediated green synthesis of Pd nanoparticles for suzuki coupling and hydrogenation of conjugated olefins. Nanosci Nanotechnol Lett 4:420425

Reed HC, Landolt PJ (2019) Ants, wasps, and bees (hymenoptera). In: Mullen GR, Durden LA (eds) Medical and veterinary entomology. Elsevier, Amsterdam, pp 459–488

Rising A, Nimmervoll H, Grip S, Fernandez-Arias A, Storckenfeldt E, Knight DP, Vollrath F, Engström W (2005) Spider silk proteins-mechanical property and gene sequence. Zoolog Sci 22:273–281

Römer L, Scheibel T (2008) The elaborate structure of spider silk: structure and function of a natural high performance fiber. Prion 2(4):154–161

Roy N, Mondal S, Laskar RA, Basu S, Mandal D, Begum NA (2010) Biogenic synthesis of au and ag nanoparticles by Indian propolis and its constituents. Colloids Surf B Biointerfaces 76:317–325

Rozman NAS, Tong WY, Leong CR, Tan WN, Hasanolbasori MA, Abdullah SZ (2019) Potential antimicrobial applications of chitosan nanoparticles (ChNP). J Microbiol Biotechnol 29(7):1009–1013

Shitole M, Dugam S, Tade R, Nangare S (2020) Pharmaceutical applications of silk sericin. Ann Pharm Fr 78:469–486

Singh A, Hede S, Sastry M (2007) Spider silk as an active scaffold in the assembly of gold nanoparticles and application of the gold-silk bioconjugate in vapor sensing. Small 3:466–473

Singh N, Chaudhary A, Abraham J (2014a) Susceptibility testing of methicillin-resistant Staphylococcus aureus (MRSA) and biological role of silver nanoparticles of honey against MRSA. J Biol Active Prod Nat 4:332–342

Singh N, Mondal D, Sharma M, Devkar RV, Dubey S, Prasad K (2015) Sustainable processing and synthesis of nontoxic and antibacterial magnetic nanocomposite from spider silk in neoteric solvents. ACS Sustain Chem Eng 3:2575–2581

Singh R, Sahu SK, Thangaraj M (2014b) Biosynthesis of silver nanoparticles by marine invertebrate (polychaete) and assessment of its efficacy against human pathogens. J Nanoparticles 2014:718240

Sreelakshmi C, Datta KKR, Yadav JS, Subba Reddy BV (2011) Honey derivatized au and ag nanoparticles and evaluation of its antimicrobial activity. J Nanosci Nanotechnol 11:6995–7000

Stürzenbaum SR, Höckner M, Panneerselvam A, Levitt J, Bouillard JS, Taniguchi S, Dailey LA, Ahmad Khanbeigi R, Rosca EV, Thanou M, Suhling K, Zayats AV, Green M (2013) Biosynthesis of luminescent quantum dots in an earthworm. Nat Nanotechnol 8(1):57–60

Su H, Han J, Dong Q, Zhang D, Guo Q (2007) In situ synthesis and photoluminescence of QD-CdS on silk fibroin fibers at room temperature. Nanotechnology 19(2):025601

Talaeeshoar F, Delavari H, H., & Poursalehi, R. (2018) Can earthworms biosynthesize highly luminescent quantum dots? Luminescence 33(5):850–854

Tibbetts EA, Sheehan MJ (2013) Individual recognition and the evolution of learning and memory in Polistes paper wasps. In: Menzel R, Paul R (eds) Handbook of behavioral neuroscience. Elsevier, Amsterdam, pp 561–571

Trusheva B, Popova M, Bankova V, Simova S, Marcucci MC, Miorin PL, da Rocha PF, Tsvetkova I (2006 Jun) Bioactive constituents of brazilian red propolis. Evid Based Complement Alternat Med 3(2):249–254

Venu R, Ramulu TS, Anandakumar S, Rani VS, Kim CG (2011) Bio-directed synthesis of platinum nanoparticles using aqueous honey solutions and their catalytic applications. Colloid Surf A 384:733–738

Vollrath F (2000) Strength and structure of spiders' silks. J Biotechnol 74:67–83

Wang W, Meng Q, Li Q, Liu J, Zhou M, Jin Z, Zhao K (2020) Chitosan derivatives and their application in biomedicine. Int J Mol Sci 21(2):487

Wongkrongsak S, Tangthong T, Pasanphan W (2016) Electron beam induced water-soluble silk fibroin nanoparticles as a natural antioxidant and reducing agent for a green synthesis of gold nanocolloid. Radiat Phys Chem 118:27–34

Wu L, Cai X, Nelson K, Xing W, Xia J, Zhang R, Stacy AJ, Luderer M, Lanza GM, Wang LV, Shen B, Pan D (2013) A green synthesis of carbon nanoparticle from honey for real-time photoacoustic imaging. Nano Res 6(5):312–325

Xu S, Yong L, Wu P (2013) One-pot, green, rapid synthesis of flowerlike gold nanoparticles/reduced graphene oxide composite with regenerated silk fibroin as efficient oxygen reduction electrocatalysts. ACS Appl Mater Interfaces 5(3):654–662

Yan J, Wu G, Li L, Yu A, Sun X, Guan N (2010) Synthesis of uniform TiO_2 nanoparticles with egg albumen proteins as novel biotemplate. J Nanosci Nanotechnol 10(9):5767–5775

Zhao Y, Zhang Y, Liu X, Kong H, Wang Y, Qin G, Cao P, Song X, Yang X, Wang Q, Qu H (2017) Novel carbon quantum dots from egg yolk oil and their haemostatic effects. Sci Rep 7:4452

Zhou L, Fu P, Cai X, Zhou S, Yuan Y (2016) Naturally derived carbon nanofibers as sustainable electrocatalysts for microbial energy harvesting: a new application of spider silk. Appl Catal Environ 188:31–38

14 Applications of Biosynthesized Nanoparticles

Abstract

Green nanoparticles are the order of the day and find applications in translational research encompassing agriculture, environment, and biomedical science. Biologically synthesized nanoparticles have excellent insecticidal properties that have opened new avenues towards their utilization in efficient control of insect pests of pharmaceutical and veterinary importance. Nanobiotechnology has also opened new possibilities for improved environmental remediation, and nanoparticles are being used in the sensing, detection, and remediation of different environmental pollutants like heavy metals, dyes, pesticides, and even oil spills. Nanoparticles have been successfully explored in agriculture for improving germination, elemental uptake, use of nutrients, enhancement of crop yield, and stress tolerance in diverse plant species. Biogenic nanoparticles have been utilized for diverse biomedical uses comprising of diagnostics, drug delivery, biosensing, and bioimaging and as antimicrobial agents.

14.1 Introduction

Green nanobiotechnology has been in prominence as a cost-effective, dependable, and eco-friendly procedure for fabrication of a wide range of nanomaterials of various types, viz., metals, metal oxides, and nanocomposites. The biofabricated nanoparticles have diverse applications in translational research encompassing agriculture, environment, and biomedical science. Some of the applications of biogenically synthesized nanoparticles have been provided below:

S. Srivastava, A. Bhargava, *Green Nanoparticles: The Future of Nanobiotechnology*,
https://doi.org/10.1007/978-981-16-7106-7_14

14.2 Insecticidal Activities

14.2.1 Nanoparticles Against Storage Pests

A storage pest is a living entity (mostly insects) that has deleterious effects on stored food products. Nearly 1000 species are linked to damage to stored products in different parts of the world. These pests cause extensive damage to storage grains with a major chunk of the world's grain being lost during storage mostly due to insect pests. Worldwide losses in stored grains as a result of insect infestation is estimated to be around 5–10% in temperate and 20–30% in the tropical region. The insect pests of economically important food grains are categorized into primary and secondary pests. Primary pests have the ability to bite and pierce into whole as well as unbroken grains, while secondary pests do not attack whole, normal grains but induce considerable damage in damaged and/or processed products. The majority of insects and pests belong to the order of Coleoptera (beetles) and Lepidoptera (moths and butterflies), which account for about 60% and 8–9%, respectively, of the total number of species of stored product insects and pests. The major economic loss caused by grain infesting insects is not always the direct consumption of kernels but also the amount contaminated by them through accumulation of frass, exuviae, webbing, and insect cadavers, along with their excreta, which make food unfit for human consumption. Since insect pests induce extensive damage to stored grains, it becomes imperative to develop green and safer alternative methods of controlling them. Nanoparticles have shown great promise in this respect and have been explored as alternatives to synthetic pesticides, insecticides, and insect repellents. Different types of biologically produced nanopesticides like nanoemulsions, nanoencapsulation, nanosuspension, and nanoparticulates are being developed to mitigate the losses induced by pests in stored food. Table 14.1 shows the potential of nanoparticles in the control of stored grain pests.

14.2.2 Nanoparticles Against Mosquito Vectors

Vector-borne diseases are characterized by high morbidity and mortality around the world, accounting for about 17% of the global infectious diseases. These diseases account for infection in over one billion people and mortality of more than a million people each year. Among the vectors, mosquitoes are the most notorious accounting for the highest number of cases and mortality, especially in tropical regions of the world. Mosquitoes (Diptera: Culicidae) are a major threat to humankind since they transmit diverse infectious pathogens and parasites causing large-scale mortality worldwide due to spread of diseases like malaria, filariasis, dengue, Zika, eastern equine encephalitis, West Nile fever, chikungunya, and yellow fever. There are also some reports of possible association between mosquitoes, infectious diseases, and cancer. The family Culicidae encompasses 3556 valid species of mosquitoes distributed within the subfamilies Culicinae and Anophelinae. However, the mosquito vectors primarily belong to three genera, viz., *Anopheles*, *Aedes,* and *Culex*.

Table 14.1 Nanoparticles against different storage pests

Storage pest	Nanoparticle/nanostructure	Biological material used for nanostructure production	References
Callosobruchus analis	Zinc oxide nanoparticle	*Spinacia oleracea*	Wazid et al. (2018)
Callosobruchus maculatus	Chitosan nanoparticle Nanostructured silica Nickel oxide Silver nanoparticle	*Rauvolfia serpentine* *Nerium oleander*	Sahab et al. (2015) Arumugam et al. (2016) Rahman et al. (2020) Jafer and Annon (2018)
Corcyra cephalonica	Silica nanoparticle	Diatomaceous earth	Vani and Brindhaa (2013)
Ephestia kuehniella	Nanoemulsion	*Mentha longifolia*	Louni et al. (2018)
Rhyzopertha dominica	Nanostructured alumina Silver nanoparticle	*Myriostachya wightiana*	Teodoro et al. (2010) Vadlapudi and Amanchy (2017)
Sitophilus oryzae	Lead nanoparticle Silica nanoparticle Silver nanoparticle	*Avivennia marina* Diatomaceous earth *Euphorbia prostrata*	Vijaya Sankar and Abideen (2015) Debnath et al. (2011) Abduz Zahir et al. (2012)
Sitophilus granarius	Nanoemulsion	*Eucalyptus globulus*	Mossa et al. (2017)
Tribolium castaneum	Copper nanoparticles Nano-diatomaceous earth Nanoemulsion Silver nanoparticle	*Pseudomonas fluorescens* *Diatoms* *Pimpinella anisum*	El-Saadony et al. (2020) Sabbour and Shadia (2015) Hashem et al. (2018) Alif Alisha and Thangapandiyan (2019)
Tribolium confusum	Nano-diatomaceous earth Silver	Diatoms *Citrus sinensis*	Sabbour and Shadia (2015) Sedighi et al. (2019)
Trogoderma granarium	Silicon oxide nanoparticle Silver nanoparticle	*Peganum harmala*	Raduw and Mohammed (2020) Almadiy et al. (2018)

The overuse of currently marketed synthetic pesticides has led to serious environmental consequences including the development of multidrug resistance. Nanoparticles have come to the rescue in integrated pest management strategies in the development of nano-biopesticides, which have the capability to control mosquito populations and at the same time have negligible detrimental effect on other organisms and the environment. Biological synthesis of nanoparticles having excellent insecticidal properties have paved way toward efficient control of insect pests of pharmaceutical and veterinary importance. Nanobiotechnology-based approaches find application in vector control in the following ways:

1. Development of novel drugs having high activity, little toxicity, and sustained release
2. Production of novel insect repellent formulations using green approaches
3. Effective control of insect vectors using nanoparticles having insecticidal and/or larvicidal properties
4. Development of biosensors that can aid in efficient detection and diagnosis of the mosquito transmitted viral diseases

A number of nanoparticles have been obtained from microorganisms and higher taxa that have shown potent ovicidal, adulticidal, and oviposition deterrence effects. Table 14.2 lists the angiosperm species used for biofabrication of nanomaterials and have shown good activity against insect vectors. The biologically produced nanoparticles exhibit high toxicity against both mosquitoes and ticks (LC50: 1–35 μg/ml). The mode of action of nanoparticles against insect vectors has been provided in Fig. 14.1. Not just angiosperms but even some lower plants as ferns have been used for fabrication of nanoparticles which have shown significant activity against different insect vectors. Panneerselvam et al. (2016) assessed the mosquitocidal properties of the biologically synthesized AgNPs from *Pteridium aquilinum* against young instars of *Anopheles stephensi*, both in laboratory and actual field conditions. The AgNP were highly toxic against larvae and pupae, with LC50 of 7.48 ppm (larva I), 10.68 ppm (II), 13.77 ppm (III), 18.45 ppm (IV), and 31.51 ppm (pupae) (Table 14.3). The impact of *P. aquilinum* extract and AgNPs on longevity and fecundity of *An. stephensi* was also investigated. Ten-hour old male and female *An. stephensi* were exposed to different concentrations of AgNPs synthesized from *P. aquilinum*. The biologically synthesized AgNPs reduced the adult longevity and fecundity of the mosquito species. The female fecundity decreased sharply after application of *P. aquilinum*-synthesized AgNPs. As compared to 145 eggs in control, 130.00, 110.24, 98.06, 84.18, and 66.10 eggs were recorded at 3.125, 6.25, 12.5, 25, and 50 ppm treatments. In the field, the application of fern synthesized AgNP (10 × LC50) in water reservoirs cleared the larvae of *An. stephensi* after 3 days and displayed greater inhibition activity against *P. falciparum* in comparison with the antimalarial drug chloroquine. The results demonstrated that *P. aquilinum*-synthesized AgNP could be potential for reducing larval populations of mosquito vectors.

Table 14.2 Biologically produced nanoparticles from angiosperms used for mosquito vector control

Organism producing nanoparticle	Type of nanoparticle	Target organism (insect)	References
Annona muricata	Silver	*Aedes aegypti*, *Anopheles stephensi*, *Culex quinquefasciatus*	Santhosh et al. (2015)
Artemisia nilagirica	Silver	*Aedes aegypti*, *Anopheles stephensi*	Nalini et al. (2017)
Bauhinia acuminata	Silver	*Aedes aegypti*	Alharbi et al. (2018)
Belosynapsi kewensis	Silver	*Aedes aegypti*, *Anopheles stephensi*	Bhuvaneswari et al. (2016)
Cassia fistula	Silver	*Aedes albopictus*, *Culex pipiens pallens*	Fouad et al. (2018)
Catharanthus roseus	Silver	*Aedes aegypti*, *Culex quinquefasciatus*	Pavunraj et al. (2017)
Chenopodium ambrosioides	Silver	*Aedes albopictus*	Subramaniam et al. (2017)
Chomelia asiatica	Silver	*Aedes aegypti*, *Anopheles stephensi*, *Culex quinquefasciatus*	Muthukumaran et al. (2015)
Citrus limon	Gold, palladium	*Aedes aegypti*, *Anopheles stephensi*	Minal and Prakash (2020)
Cucurbita	Zinc oxide	*Culex tritaeniorhynchus*	Velsankar et al. (2019)
Eupatorium odoratum	Silver	*Culex quinquefasciatus*	Elemike et al. (2017)
Feronia elephantum	Silver	*Aedes aegypti*, *Anopheles stephensi*, *Culex quinquefasciatus*	Veerakumar et al. (2014)
Helitropium indicum	Silver	*Aedes aegypti*, *Anopheles stephensi*, *Culex quinquefasciatus*	Veerakumar et al. (2014)
Hyptis suaveolens	Silver	*Aedes aegypti*, *Anopheles stephensi*, *Culex quinquefasciatus*	Elumalai et al. (2017)
Jasminum nervosum	Gold	*Culex quinquefasciatus*	Lallawmawma et al. (2015)
Leucas aspera	Silver	*Aedes aegypti*, *Anopheles stephensi*, *Culex quinquefasciatus*	Elumalai et al. (2017)
Pedalium murex	Zinc oxide	*Aedes aegypti*	Ishwarya et al. (2017)
Punica granatum	Silver	*Spodoptera litura*	Bharani and Namasivayam (2017)
Rhazya stricta	Silver	*Aedes aegypti*, *Culex pipiens*	Aziz et al. (2020)
Ricinus communis	Silver	*Aedes albopictus*	Waris et al. (2020)
Rosa sp.	Cadmium	*Aedes albopictus*	Hajra et al. (2016)
Scadoxus multiflorus	Zinc oxide	*Aedes aegypti*	Al-Dhabi and Valan Arasu (2018)

(continued)

Table 14.2 (continued)

Organism producing nanoparticle	Type of nanoparticle	Target organism (insect)	References
Sida acuta	Silver	*Aedes aegypti*, *Anopheles stephensi*, *Culex quinquefasciatus*	Veerakumar et al. (2013)
Tagetus sp.	Cadmium	*Aedes albopictus*	Hajra et al. (2016)
Tridax procumbens	Copper oxide	*Aedes aegypti*	Muthamil Selvan et al. (2018)
Turbinaria ornata	Silver	*Aedes aegypti*, *Anopheles stephensi*, *Culex quinquefasciatus*	Deepak et al. (2017)
Ulva lactuca	Zinc oxide	*Aedes aegypti*	Ishwarya et al. (2018)
Zornia diphylla	Silver	*Anopheles subpictus*, *Aedes albopictus*, *Culex tritaeniorhynchus*	Govindarajan et al. (2016)

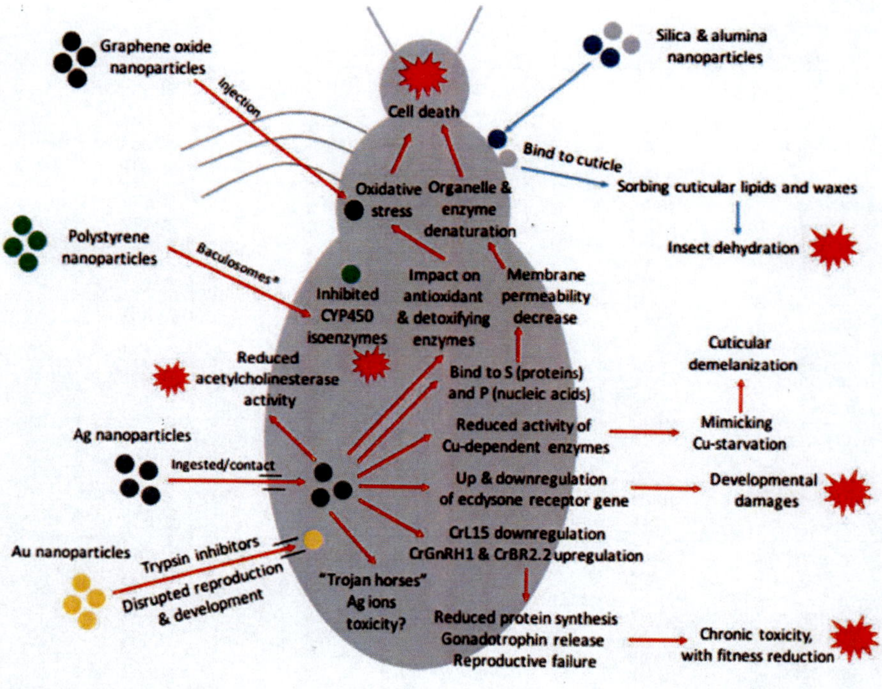

Fig. 14.1 Mode of action of nanoparticles against insect vectors (reprinted with permission from Benelli 2018)

Rajaganesh et al. (2016) evaluated the mosquitocidal properties of *Dicranopteris linearis* (family: Gleicheniaceae) extract and the biologically synthesized AgNPs against larvae of *A. aegypti*, along with assessment of the ovipositional deterrence of *A. aegypti*. The effect of smoke emanating from herbal coils prepared using different parts of *D. linearis* was also carried out. It was observed that the fern synthesized

Table 14.3 Larvicidal and pupicidal toxicity of the *Pteridium aquilinum*-synthesized silver nanoparticles against the malaria vector *Anopheles stephensi* (reprinted with permission from Panneerselvam et al. 2016)

Target	LC_{50} (LC_{90}) (ppm)	95% confidence limit LC_{50} (LCL-UCL)	LC_{90} (LCL-UCL)	Slope	χ2 (d.f. =4)
Larva I	7.48 (27.87)	(5.00–9.56)	(24.19–33.45)	1.27	1.41 n.s.
Larva II	10.68 (32.78)	(8.35–12.85)	(28.69–38.81)	1.39	3.34 n.s.
Larva III	13.77 (35.06)	(11.66–15.95)	(30.96–40.93)	1.54	4.43 n.s.
Larva IV	18.45 (45.05)	(15.91–21.14)	(40.05–52.02)	1.52	3.14 n.s.
Pupa	31.51 (73.72)	(21.26–54.31)	(52.10–156.1)	1.11	7.90 n.s.

AgNPs were highly toxic to both larvae and pupae of *A. aegypti* (Table 14.4). Toxicity studies against adult mosquitoes showed that the herbal coils prepared from different parts of fern caused high mortality among the mosquito species that was comparable to those obtained using permethrin. Experimental studies also showed that egg hatchability of the dengue vector reduced by 100% at doses of AgNPs as low as 30 ppm. Results obtained using oviposition deterrence assays indicated that the AgNPs reduced oviposition rates by >70% even at low concentrations (10 ppm). The study firmly indicated toward the possibility of using *D. linearis* for biofabrication of AgNPs and utilization of these biosynthesized nanoparticles for dengue control.

Although nanobiotechnology has been at the forefront in the development of biopesticides, biorepellents, and diagnostic devices for control of insect vectors, a lot needs to be done in reference to the following aspects:

1. Solving the bottlenecks related to the development of nano-based formulations which has been an impediment in the commercialization process
2. Better understanding of the behavior, fate, bioavailability, and toxicity of nanoparticles toward target as well as nontarget organisms
3. Availability of standard methods for the assessment of efficacy and toxicity of nanotechnology-based products
4. Reducing the cost associated with nano-based approaches
5. Development of a reliable arrangement for collection and unrestricted sharing of relevant information
6. Thorough understanding of the nature and interaction of nanomaterials in biological systems

Table 14.4 Acute toxicity of *Dicranopteris linearis*-synthesized silver nanoparticles against larvae and pupae of the dengue vector *Aedes aegypti* (reprinted with permission from Rajaganesh et al. 2016)

Target	LC50 (LC90) (ppm)	95% confidence limit LC50 (LC90)		Regression equation	$\chi 2$ (d.f. =4)
		LCL	UCL		
1st instar larva	18.905 (32.140)	17.622 (29.855)	20.143 (35.333)	y = 1.831 + 0.097x	4.961 n.s.
2nd instar larva	20.929 (35.489)	19.586 (32.666)	22.334 (39.574)	y = 1.842 + 0.088x	2.529 n.s.
3rd instar larva	23.187 (39.696)	21.662 (35.971)	24.978 (45.394)	y = 1.800 + 0.078x	1.493 n.s.
4th instar larva	26.312 (44.418)	24.444 (39.570)	28.855 (52.275)	y = 1.862 + 0.071x	0.793 n.s.
Pupa	29.328 (48.511)	26.979 (42.553)	32.905 (58.687)	y = 1.959 + 0.067x	0.119 n.s.

14.3 Environmental Applications

Anthropogenic activities and indiscriminate use of bioresources for domestic and industrial purpose have led to contamination of the environment. These harmful materials released into the environment have been a major source of pollution and are detrimental to microbial, plant, and animal systems. These have to be efficiently removed for a better and healthy environment. Nanobiotechnology has opened new possibilities for improved environmental remediation and is being used in the sensing, detection, and remediation of different environmental pollutants like heavy metals, dyes, pesticides, and even oil spills. Magnetite nanoparticles (average diameter: 9.6 $\pm$ 3 nm) biosynthesized using biocomponents extracted from *Matricaria aurea* (family: Asteraceae) and *Ochradenus baccatus* (family: Resedaceae) have successfully disperse crude oil and collect 95% and 91% of the oil spill, respectively. Similar encouraging results have been obtained during collection of oil spills by magnetite nanoparticles fabricated using green hydrophobic biocomponents of *Anthemis pseudocotula*, an evergreen perennial plant belonging to family Asteraceae.

14.3.1 Remediation of Pesticides

Pesticides have been extensively used throughout the world to improve production of plant-based products since they protect the plants against pests, fungi, and weeds. However, due to overuse, pesticide residues have contaminated drinking water, groundwater, and soils leading to serious consequences. The common classes of pesticides include insecticides, herbicides, fungicides, ascaricides, nematicides, molluscicides, and rodenticides. The pesticides show high persistence and bioaccumulative nature due to which they are found in diverse ecosystems and are a continuous health hazard to living organisms. Different methods involving adsorption, oxidation, catalytic degradation, membrane filtration, and biological treatments are in use for the removal or detoxification of harmful pesticides. The biological methods using bacteria and fungi are slow, expensive, and only partially successful since they have limited role in remediation of recalcitrant molecules. However, nanomaterials can be of use in this respect since they are semiconducting and can easily degrade pesticides photocatalytically, especially the organochlorines like DDT, methoxychlor, dieldrin, chlordane, lindane, and benzenehexachloride (Fig. 14.2) (Table 14.5). Several advanced nanomaterials especially of iron and titanium dioxide can serve as potential candidates for total degradation of pesticides within a few minutes. However, this research is in its nascent stage and requires more detailed efforts before it can be employed at the industrial scale. Also, a number of pesticides like quintozene, pentachlorophenol, and toxaphene have not been actively taken up using this approach.

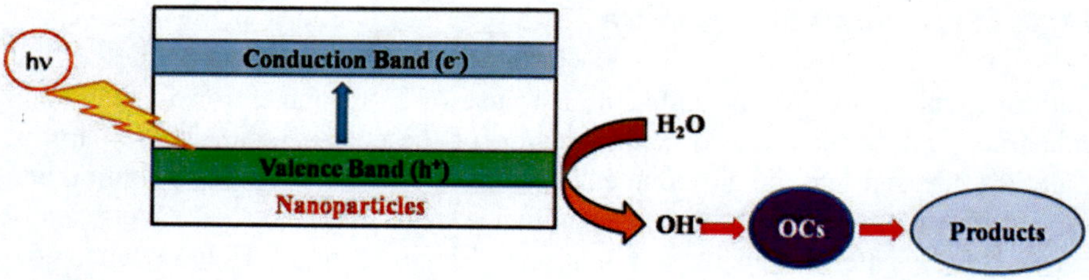

Fig. 14.2 Mechanism of action of organochlorine degradation using nanoparticles (reprinted with permission from Rani et al. 2017)

Table 14.5 Use of nanoparticles for pesticide removal/remediation (reprinted with permission from Rani et al. 2017)

S. No.	OC	Nanoparticles	Remarks	References
1.	DDT	Zerovalent Fe	Effective degradation of DDT was carried out in water (92%) and soil (22.4%)	El-Temsah et al. (2016)
2.	Lindane	Zerovalent Fe	Complete reductive degradation of lindane into benzene and chlorides was observed within 24 h	Elliott et al. (2009)
3.	DDT	Bimetallic Ni/Fe	Effective degradation of DDT was carried out using nanoparticles with higher efficiency in alkaline or weakly acidic medium	Tian et al. (2009)
4.	Dicofol	TiO_2	Completely degraded dicofol under UV light irradiation in 2 h	Bin et al. (2008)
5.	Lindane	FeS nanoparticles stabilized by biopolymers	Degradation efficiency of 94% was achieved in 8 h	Paknikar et al. (2005)
6.	Lindane	Fe-Pd bimetallic nanoparticles	Lindane was degraded effectively under both the conditions: aerobic (65%; Fenton-like process) and anaerobic (100%; reduction process)	Joo and Zhao (2008)

14.3.2 Remediation of Dyes

Dyes are stable natural or synthetic substances that linger on in the environment and have been categorized as potent xenobiotic. Organic dyes are of immense industrial importance since they are in high demand especially by paper mills and in the textile industry. Moderate estimates show that about 15% of total dyes used find their way into water bodies and have an adverse impact of aquatic flora and fauna. The toxic dyes are highly hazardous for the environment and are potentially toxic toward plants, animals, and human beings. Dyes significantly deteriorate water bodies, increase biochemical and chemical oxygen demand (BOD and COD), affect the photosynthetic activity, and inhibit plant growth. Dyes also enter the food chain, show recalcitrance and bioaccumulation, and may promote toxicity, mutagenicity, and carcinogenicity. Therefore, various methods for their detoxification and/or removal have been extensively explored, but most of them show several limitations. These range from being costly to having adverse impact on the environment, sludge production, complex pretreatment requirements, and release of potentially toxic by-products.

Removal of a number of dyes has been successfully accomplished using biologically synthesized nanoparticles (Table 14.6). The dyes remediated using green nanoparticles include both acidic and basic dyes and even the mutagenic and carcinogenic ones like methyl orange. A quick perusal of Table 14.6 reveals that primarily nanoparticles derived from metals and metal oxides are being explored for remediation of dyes from contaminated water bodies. Metal oxide nanoparticles have been the advantageous and the primary choice for photocatalytic detoxification of synthetic dyes due to their high surface area to mass ratio that increases the adsorption capacity with respect to pollutants (Fig. 14.3). The high surface energy of the nanoparticles due to several surface reactive sites on their surface leads to increased rate of dye removal even at low concentrations. As a result, a much smaller amount of nanocatalyst is required to treat large reservoirs of polluted water.

14.3.3 Heavy Metal Sensing and Remediation

A heavy metal is a member of an ill-defined subset of elements having a specific gravity of more than 5 g/cm^3 in their standard state and includes transition metals, some metalloids, lanthanides, and actinides. Heavy metals like lead, cadmium, and mercury are a serious health hazard and pose severe threat to human health since they accumulate biologically in the food chain. Released into the biosphere due to natural and anthropogenic activities, their removal has remained a challenging issue. Various efficient cleanup techniques are available, but most of them are costly, labor intensive, and commercially less feasible and have found limited acceptability among the communities. Nanoparticles have come to the rescue and have shown promise in removing heavy metals from contaminated sites in comparison with the conventional techniques. Another notable environmental application pertains to their use in the detection of heavy metal ions in contaminated water bodies which can be attributed to their adjustable size and size-dependent optical properties, especially of the metal nanoparticles. Karthiga and Anthony (2013) synthesized silver nanoparticles using fresh and dried extracts of neem (leaves and bark), mango (leaves), and pepper (seeds). The nanoparticles synthesized using fresh neem leaf extract detected mercury, and those synthesized using sun-dried neem leaf extract selectively sensed mercury and lead from aqueous solutions. Nanoparticles biosynthesized from neem bark demonstrated colorimetric sensing of zinc and mercury. Likewise nanoparticles synthesized from both dried and fresh mango leaves exhibited selective colorimetric sensing of mercury and lead. The silver nanoparticles obtained using pepper seed extracts exhibited high sensitivity to mercury, lead, and zinc detection. Similarly silver nanoparticles biosynthesized

Table 14.6 Use of nanoparticles for the removal of different contaminants

Source organism	Nanoparticle	Contaminant	References
Dyes			
Acorus calamus	Iron oxide, magnesium oxide	Methylene blue, Congo red	Arasu et al. (2019)
Amaranthus spinosus	Iron oxide	Methylene blue, methyl orange	Muthukumar and Matheswaran (2015)
Artemisia vulgaris	Iron oxide	Methyl orange	Kouhbanani et al. (2018)
Azadirachta indica	Zinc oxide	Methylene blue	Bhuyan et al. (2015)
Camillis sinensis	Iron	Malachite green, rhodamine B	Xiao et al. (2020)
Cupressus sempervirens	Zerovalent iron	Methyl orange	Ebrahinezhad et al. (2017)
Dolichos lablab	Iron oxide	Crystal violet	Basavaiah et al. (2018)
Eucalyptus tereticornis	Iron	Acid black 194	Wang et al. (2014)
Rubus glaucus	Magnetite	Congo red, methyl orange	Kumar et al. (2016)
Moringa oleifera	Nickel/iron oxide	Malachite green	Prasad et al. (2017a)
Parthenium hysterophorus	Titanium dioxide	Methyl orange, crystal violet	Thandapani et al. (2017)
Phaseolus vulgaris	Silver	4-Nitrophenol	Rani et al. (2020)
Pisum sativum	Iron oxide	Methyl orange	Prasad et al., (2017b)
Polyalthia longifolia	Silver	4-Nitrophenol	Muthu et al. (2019)
Punica granatum	Iron oxide	Congo red	Prasad et al., 2017c
Ruellia tuberosa	Copper oxide	Crystal violet	Vasantharaj et al. (2019)
Shorea robusta	Zerovalent iron	Congo red	Jha and Chakraborty (2020)
Sorghum sp.	Iron	Bromothymol blue	Njagi et al. (2011)
Syzygium aromaticum	Iron oxide	Methylene blue	Jain et al. (2021)
Syzygium cumini	Iron oxide	Blue 235	Natarajan and Ponnaiah (2017)
Teucrium polium	Iron oxide	Methyl orange	Kouhbanani et al. (2019)
Vitis sp.	Iron	Acid orange II	Luo et al. (2015)
Zanthoxylum armatum	Silver	Safranine O, methyl red	Jyoti and Singh (2016)
Heavy metals			
Arachis hypogaea	Iron	Chromium	Pan et al. (2020)
Aspergillus niger	Iron oxide	Chromium	Chatterjee et al. (2020)
Camellia sinensis	Iron	Chromium	Mystrioti et al. (2016)
Ilex paraguariensis	Iron	Chromium	Garcia et al. (2019)
Mentha spicata	Iron	Arsenic	Prasad et al. (2014)

(continued)

Table 14.6 (continued)

Source organism	Nanoparticle	Contaminant	References
Murraya koenigii	Iron oxide	Red	Prasad et al. (2017d)
Nephrolepis auriculata	Iron	Chromium	Yi et al. (2019)
Antibiotics			
Citrus limon	Magnetite	Ampicillin, erythromycin	Stan et al. (2017)
Desulfovibrio vulgaris	Palladium, platinum	Ciprofloxacin	Martins et al. (2017)
Excoecaria cochinchinensis	Iron oxide	Rifampicin	Cai et al. (2019)
Cucumis sativus	Magnetite	Ampicillin, erythromycin	Stan et al. (2017)
Punica granatum	Iron	Amoxicillin Cefixime	Ali et al. (2020) Esmaeili Bidhendi et al. (2020)
Urtica dioica	Zerovalent iron	Cephalexin	Leili et al. (2018)
Vitis vinifera	Magnetite	Ampicillin, erythromycin	Stan et al. (2017)

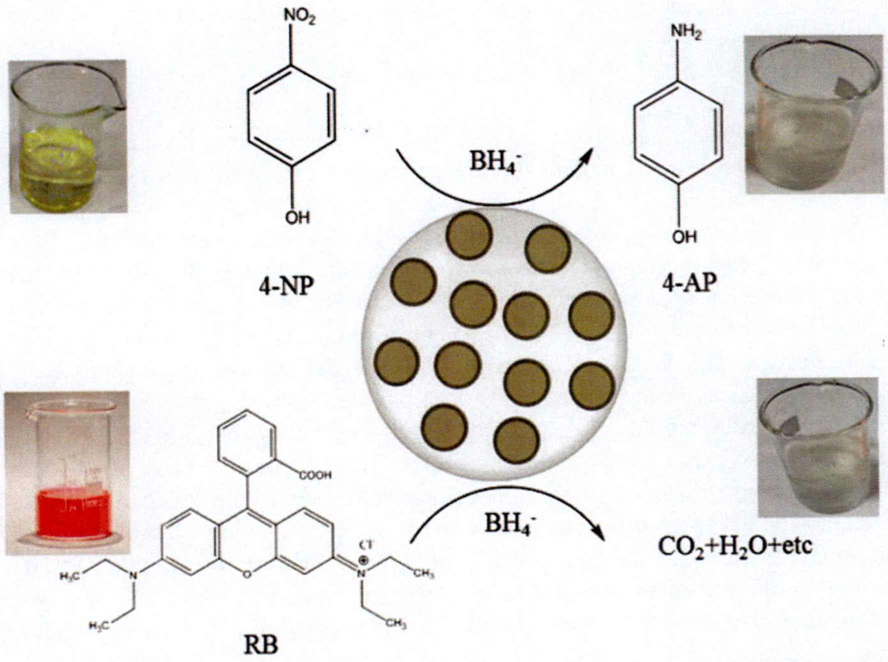

Fig. 14.3 Mechanism of action of the catalytic reduction and degradation of 4-NP and RhB dye with CuO/ZnO nanocomposite at room temperature (reprinted with permission from Bordbar et al. 2018)

from *Moringa oleifera* flower extract effectively detected the presence of copper ions from 1 to 12 mM concentrations.

The nanomaterials used for heavy metal remediation include carbon-based (carbon nanotubes, quantum dots, and graphene), silica-based, metal, and metal oxide nanoparticles. Due to their excellent adsorption effects, carbon nanotubes have been found useful in remediation of manganese, thallium, copper, lead, and chromium. The adsorption efficiency of these nanostructures toward heavy metals is further increased by introducing different functional groups (–COOH, $-NH_2$, and –OH) on their surface. For mercury sensing, quantum dots have been prepared from leaves of *Prosopis juliflora*, juice of *Citrus lemon*, peel of *Citrus maxima*, and fruits of *Gardenia* sp. Among the metal-based nanostructures, zerovalent iron, a composite consisting of Fe (0) and ferric oxide coating, has been extensively used as a novel adsorbent for remediation of metals like mercury, chromium, copper, nickel, and cadmium, all of which are potent water pollutants. The zerovalent iron provides advantages like high reducing capacity, controllable size, specific surface area, and presence of abundant reactive sites on the surface. However, zerovalent iron nanoparticles have the following shortcomings:

1. Readily oxidize with oxygen and water in an aqueous medium, which retards the reduction process.
2. They aggregate easily, which results in the decrease of the reaction surface area and mobility.
3. Difficulty in removal of zerovalent iron from wastewater.

These drawbacks have been reduced by implementing modification strategies such as surface chemical modification or doping the zerovalent iron nanoparticles with other metals like palladium, copper, nickel, and platinum.

Table 14.6 shows the potential of biologically synthesized nanoparticles in the removal of heavy metals from heavily polluted water bodies. It is evident that although there has been widespread utilization of nanoparticles synthesized through physical and chemical methods in heavy metal remediation, not much has been achieved regarding the use of nanoparticles fabricated through green approaches. Biologically synthesized nanomaterials have provided us with a promising alternative to the traditional adsorbents for remediation of heavy metals. A closer look reveals that the green approach has generally revolved around the use of iron and iron oxide nanoparticles. Also, the remediation capabilities of the nanoparticles have been mostly investigated in stimulated water in simple conditions. Detailed studies on actual contaminated sites are relatively rare. Thus, sincere efforts should be made to explore the remediation potential of green nanomaterials into practical heavy metal water treatment with respect to their removal capability, reusability, separation, and cost.

14.3.4 Removal of Antibiotics

Antibiotics are an important class of emerging pollutants that have the tendency to persist in the environment for a long period of time and alter natural food webs. Antibiotics are used in pharmaceutical industry, in animal farming, and for agricultural purposes. They have been found in several aquatic bodies and are a potential health hazard. After their entry in the human body, antibiotics are known to cause severe adverse impact in enzymatic, gastrointestinal, hormonal, and genetic systems of human beings. Thus, antibiotics need to be removed from the water bodies to avoid deleterious effects stemming out of term exposure. Nanoparticles have come quite handy in this respect. Studies have proven that nanoparticles have the potential to remove antibiotics with an acceptable efficiency of 61%. With respect to nanoparticles synthesized through green methods, less studies are available involving their use in removal of antibiotics from aquatic bodies (Table 14.6).

Thus, green nanotechnology has come handy for decontamination of environment. However, there is still a long way to go to put nanomaterials into on-site application.

14.4 Nanoparticles in Agriculture

Nanoparticles have been increasingly explored in agriculture for improving germination, elemental uptake, use of nutrients, enhancement of crop yield, and stress tolerance in diverse plant species (Fig. 14.4). Nanoparticles have been assessed as biofertilizers to enhance soil fertility and improve plant growth and development. Hydrogels, nanoclays, and nanozeolites considerably increase the water-holding capacity of soil leading to gradual release of water that lessens water shortage periods during dry spells.

Several promising results have been reported using metal oxide nanoparticles. Titanium dioxide nanoparticles have demonstrated to stimulate antioxidant system in soybean and increase dry weight of spinach. Iron oxide nanoparticles have been found to increase soybean pod yield and induce higher antioxidative enzyme activity in pumpkins. Copper oxide nanoparticles have been found to be useful in enhancing yield and controlling disease in tomato and brinjal. Similarly, application of zinc oxide and iron oxide nanoparticles enhanced anthocyanin, phenol, tannin, flavonoid, crude protein, and carbohydrate content in red radish plants. Carbon-based nanoparticles have also given some interesting results with respect to plant growth and metabolism. The role of carbon nanotubes in enhancing seed germination has been extensively researched in tomato, corn, garlic, and onion with results showing positive association between the two parameters at low concentrations of nanoparticles. Carbon nanotubes have also been reported to promote seedling growth in cabbage, tomato, radish, and red spinach. It has been speculated that nanoparticles break through the seed coat and increase water absorption which stimulates the enzymatic system and promotes the germination and growth of seedlings. Using "fullerol" (a fullerene derivative), an increase of about 20% in

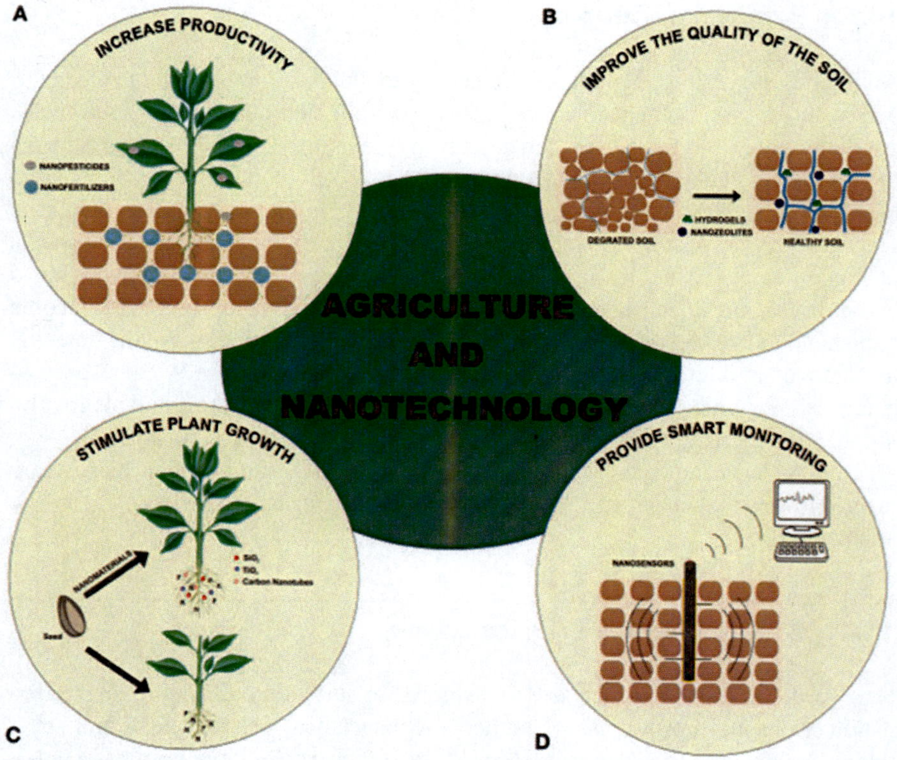

Fig. 14.4 Potential applications of nanotechnology in agriculture (reprinted from Fraceto et al. 2016)

fruit length, 59% in fruit number, and 70% in fruit weight has been reported in bitter melon (*Momordica charantia*) leading to an overall enhancement of about 128% in fruit yield.

A number of plant pathogens like fungi, viruses, bacteria, and nematodes are known to damage agricultural crops resulting in lessened agricultural production worldwide that impairs food quality and leads to economic losses worth millions of dollars. A number of biocides (pesticides, fungicides, and herbicides), mostly synthetic, have been used to control phytopathogens. However, most of these synthetic products represent a health hazard since they accumulate in the biosphere causing problems like decreased pollination due to insect deaths, development of antibiotic resistance, tolerance to several chemical products, and bioaccumulation-associated human disorders.

To mitigate the ill effects of use of chemical pesticides, nanoparticles are being explored for their use in control of plant pests and pathogens which is attributed to their strong inhibitory potential against a wide range of plant pathogens. Table 14.7 provides information on the use of green nanoparticles obtained from living organisms for the control of phytopathogens. The nanoparticles have large surface area to volume ratio, high surface reactivity, and capability of enhancing gene

Table 14.7 Use of biologically synthesized nanoparticles against phytopathogens

Source of nanoparticle	Type of nanoparticle	Plant pathogen	References
Microbes			
Bacillus sp.	Chitosan-Mg	*Acidovorax oryzae*, *Rhizoctonia solani*	Ahmed et al. (2021a)
Bacillus cereus	ZnO	*Burkholderia glumae*, *Burkholderia gladioli*	Ahmed et al. (2021b)
Nigrospora oryzae	Ag	*Fusarium moniliforme*, *Fusarium oxysporum*	Dawoud et al. (2021)
Paenibacillus polymyxa	MgO, MnO_2, ZnO	*Xanthomonas oryzae pv. oryzae*	Ogunyemi et al. (2020)
Pseudomonas poae	Ag	*Fusarium graminearum*	Ibrahim et al. (2020)
Pseudomonas rhodesiae	Ag	*Dickeya dadantii*	Hossain et al. (2019a)
Streptomyces sp.	Ag	*Alternaria alternata*, *Fusarium oxysporum*	Fouda et al. (2020)
Trichoderma harzianum	Ag, CuO, ZnO	*Alternaria alternata*, *Pyricularia oryzae*, *Sclerotinia sclerotiorum*	Consolo et al. (2020)
Plants			
Azadirachta indica	CuO Fe	*Alternaria mali*, *Diplodia seriata*, *Botryosphaeria dothidea* *Alternaria mali*, *Botryosphaeria dothidea*, *Diplodia seriata*	Ahmed et al. (2020) Ahmed et al. (2017)
Chenopodium quinoa	TiO_2	*Ustilago tritici*	Irshad et al. (2020)
Cinnamomum camphora	Ag	*Fusarium oxysporum*	Huang et al. (2019)
Citrus limon	ZnO or TiO_2	*Dickeya dadantii*	Hossain et al. (2019a, b)
Cymbopogon martini	Chitosan	*Fusarium graminearum*	Kalagatur et al. (2018)
Fragaria sp.	Ag	*Fusarium oxysporum*, *Ralstonia solanacearum*	Khan et al. (2021)
Matricaria chamomilla	MgO, MnO_2	*Acidovorax oryzae*	Ogunyemi et al. (2019a, b)
Melia azedarach	ZnO	*Cladosporium cladosporioides*, *Fusarium oxysporum*	Lakshmeesha et al. (2020)
Mentha piperita	Cu	*Colletotrichum capsici*	Iliger et al. (2021)
Olea europaea	ZnO	*Xanthomonas oryzae pv. oryzae*	Ogunyemi et al. (2019b)
Phyllanthus emblica	Ag	*Acidovorax oryzae*	Masum et al. (2019)
Pinus thunbergii	Ag	*Pseudomonas syringae*, *Burkholderia glumae*	Velmurugan et al. (2013)

(continued)

Table 14.7 (continued)

Source of nanoparticle	Type of nanoparticle	Plant pathogen	References
Piper nigrum	Ag	*Citrobacter freundii*, *Erwinia cacticida*	Paulkumar et al. (2014)
Rosmarinusofficinalis	MgO	*Xanthomonas oryzae pv. oryzae*	Abdallah et al. (2019)
Solanum torvum	Ag	*Xanthomonas axonopodis pv. punicae*, *Ralstonia solanacearum*	Vanti et al. (2020)
Syzygium aromaticum	ZnO	*Fusarium graminearum*	Lakshmeesha et al. (2019)

expression for redox processes, which facilitates their interaction with phytopathogen membranes resulting in their evolution as antimicrobial agents against major plant pathogens. Table 14.8 sheds light on the mechanism of action of nanoparticles against different plant pathogens. Thus, nanoparticles can play an important role in the control of agricultural diseases. However, although promising, the possible toxic effects of nanomaterials on plant systems need in-depth studies. Also, assessment of the interaction of nanoparticles with plant nutrients and other factors like plant species, age of plant, exposure time, prevalent environmental conditions, and nanomaterial attributes like shape, size, concentration, and surface charge needs to be carried out.

14.5 Green Nanoparticles in Medicine

Biogenic nanoparticles have been used for diverse biomedical uses comprising of diagnostics, drug delivery, biosensing, and bioimaging and as antimicrobial agents. Nanoparticles are also known to show potent antiviral properties. Biologically produced silver nanoparticles have been reported to exhibit significant activity against H7N3 influenza A virus, chikungunya virus, dengue virus, and Coxsackie B4 virus.

14.5.1 Antimicrobial Activity

Development of antimicrobial resistance (AMR) has become a major health issue with over 700,000 deaths reported worldwide due to multidrug resistant microorganisms. Pathogenic microbes have developed resistance against most of the currently used antibiotics. The situation is precarious since there has been a dearth of development of new antibiotics since it is a long and tedious process with less economic incentive. Pharmaceutical companies are less interested in research and development of new antibiotics which is likely to result in increased mortality from infections after complex surgeries. Nanobiotechnology has come to the rescue

Table 14.8 Mechanisms of action of ionic nanoparticles synthesized by green chemistry with antimicrobial activity against phytopathogens (reprinted with permission from Hernández-Díaz et al. 2021)

Nanoparticle type	Ionic form	Mechanism of action	Phytopathogen type
AgNPs	Ag+	Release of ions that are toxic toward pathogens Increased permeability of the bacterial membrane Disruption of the bacterial membrane Damage to the cell components (lipids, DNA, and proteins) Inhibition of enzyme activity Inhibition of DNA replication	Bacteria and fungi
Al_2O_3NPs	Al^{3+}	Increased reactive oxygen species (ROS) production Release of ions that are toxic toward pathogens Increased ROS production Depolarization of cell membranes	Bacteria
AuNPs	Au^{3+}	Disruption of bacterial membrane and alteration of metabolism Damage to cell organelles (cell wall and mitochondria) Inhibition of DNA uncoiling and transcription mediated by binding of AuNPs to bacterial DNA	Bacteria
CeO_2NPs	Ce^{4+}, Ce^{3+}	Inhibition of ion transport through pumps Induction of oxidative degradation of lipids and/or proteins of the pathogen's plasma membrane Impairment of electron flux and bacterial respiration Inhibition of fungal enzyme activity	Gram-positive bacteria and fungi
CdONPs	Cd^{2+}	Release of ions that are toxic against pathogens Induction of oxidative stress on bacterial cells Increased ROS production Interrupted transmembrane electron transport and mitochondria damage	Bacteria
CuNPs/ CuONPs	Cu^{2+}	Inhibition of enzyme activity essential to the microorganisms Increased ROS production	Bacteria and fungi
SeNPs	Se^{6+}, Se^{4+}	Intracellular ATP depletion Induction of oxidative stress through ROS production Alteration of bacterial membrane potential Disruption of bacterial membrane	Bacteria and fungi
SiNPs/ SiO_2NPs	Si^{4+}	Induction of mechanical damage to bacterial membrane Increased ROS production Induction of oxidative stress	Bacteria and fungi
TiO_2NPs	Ti^{4+}	Increase of ROS production Induction of photocatalytic damage	Bacteria and fungi
ZnONPs	Zn^{2+}	Release of ions that are toxic toward pathogens Increased ROS production Disruption of mitochondrial function	Bacteria

(continued)

Table 14.8 (continued)

Nanoparticle type	Ionic form	Mechanism of action	Phytopathogen type
		Induction of changes in cell morphology and release of cell components	
Fe_2O_3NPs	Fe^{3+}, Fe^{2+}	Induction of oxidative stress through ROS production Destruction of cell membranes, inducing changes in cell morphology and release of cell components Damage to essential molecules such as proteins and DNA Release of iron ions leading to oxidative damage by Fenton reaction	Bacteria and fungi

through metal and metal oxide nanoparticles that can serve as an alternative to the currently used antibiotic drugs (Table 14.9). The high surface to volume ratio of nanoparticles confers a selective advantage to them in interacting with the bacterial cell membrane with ease.

The first step in the antibacterial activity exhibited by nanoparticles involves their binding with the bacterial membrane through noncovalent interactions. This is followed by damage to the bacterial cell membrane and formation of pits or gaps that ultimately leads to its fragmentation. Thereafter, the nanoparticles interact with different functional groups present in the microbial proteins/DNA, which leads to interference in the metabolic processes and ultimately cell death. In fungi, the nanoparticles interact with the cell wall and cell membrane, diffuse inside, and inhibit production of enzymes that play a major role in synthesis of cell wall components. The nanoparticles are known to disrupt cellular integrity, for generation of superoxide radical and osmotic imbalance, and finally death of fungal cells.

14.5.2 Anticancer Activity

Cancer, also termed as a malignant tumor, or a neoplasm, arises from the uncontrolled division of abnormal cells through several steps, generally progressing from a precancerous lesion to a malignant tumor. These alterations involve interaction between the genetic profile of a person and three classes of external agents/substances termed as carcinogens:

1. Physical carcinogens: both naturally occurring and synthetic structures like fibers, particulate matter, ultraviolet, and ionizing radiation
2. Chemical carcinogens: including a wide range of structures like asbestos, benzene, tobacco smoke, aflatoxins, vinyl chloride, beryllium, and arsenic
3. Biological carcinogens: infections from certain oncogene viruses, fungi, bacteria, or parasites

Table 14.9 Activity of metal and metal oxide nanoparticles against different microorganisms (reprinted with permission from Patil and Chandrasekaran 2020)

S. No	Nanoparticles	Source	Microbes	Activity	References
1	TiO_2	*Hibiscus rosa-sinensis*	*Vibrio cholerae*, *Pseudomonas aeruginosa*, and *Staphylococcus aureus*	Disc diffusion method	Sahaya et al. (2014)
		M. citrifolia	*Staphylococcus aureus*, *Escherichia coli*, *Pseudomonas aeruginosa*, *Bacillus subtilis*	Agar well diffusion method	Sundrarajan et al. (2017)
			Candida albicans, *Aspergillus niger*		
2.	Pt	*Taraxacum laevigatum*	*Bacillus subtilis*, *Pseudomonas aeruginosa*	MHA well diffusion method	Tahir et al. (2017)
3.	Pd	*Garcinia Pedunculata*	*Cronobacter sakazakii*	Agar well diffusion method, MIC, and MBC	Hazarika et al. (2017)
4.	Se	*Emblica officinalis*	*Escherichia coli*, *Listeria monocytogenes*, *Staphylococcus aureus*, *Enterococcus faecalis*	Micro well dilution method	Gunti et al. (2019)
			Aspergillus brasiliensis, *A. flavus*, *A. oryzae*, *A. ochraceus*, *Fusarium anthophilum*, *Rhizopus stolonifer*		
5.	Ni	*Monsonia burkeana*	*Escherichia coli*, *Pseudomonas aeruginosa*	Broth dilution method	Kganyago et al. (2018)
6.	Iron oxide	*Acacia nilotica*	*Escherichia coli*, *Marsa*, *Salmonella*, *Staphylococcus aureus*	Gel diffusion assay	Da et al. (2018)
			Candida		
7.	Zinc oxide	*Albizia lebbeck*	*Bacillus cereus*, *Staphylococcus aureus*, *Escherichia coli*, *Klebsiella pneumonia*, *Salmonella typhi*	Disc diffusion method	Umar et al. (2019)

(continued)

Table 14.9 (continued)

S. No	Nanoparticles	Source	Microbes	Activity	References
8.	CuONPs	*Syzygium alternifolium (Wt.) Walp*	*Alternaria solani*, *Aspergillus flavus*, *Aspergillus niger*, *Penicillium chrysogenum*, and *Trichoderma harzianum*	Disc diffusion assay	Yugandhar et al. (2017)
9	Au	*Ziziphus zizyphus*	*Escherichia coli*	Radial diffusion assay	Aljabali et al. (2018)
			C. albicans	Micro dilution plate assay	
10.	Ag	*Erythrina suberosa (Roxb.)*	*Bacillus subtilis*, *Staphylococcus aureus*, *Pseudomonas aeruginosa*, *Escherichia coli*	Agar cup and micro broth dilution method	Mohanta et al. (2017)
			C. albicans, *C. kruseii*, *T. mentagrophytes*, *C. viswanathii*	Broth dilution method	

Besides the most frequently used metallic nanoparticles (Table 14.10), a number of nanomaterials like polymeric nanoparticles, polymeric conjugates, liposomes, polymeric micelles, microemulsion, and polymer-drug conjugates have also been explored for inducing autophagy and promoting cell death.

Three possible mechanisms have been put forward for explaining the anticancerous activity of biological nanoparticles:

1. Increased levels of ROS that generates oxidative stress and induces DNA disintegration in the tumor cell
2. Binding and interaction of metal nanoparticles to proteins or DNA resulting in distorted cell functions
3. Deviations in the cell permeability and mitochondrial dysfunction

Recently, biologically produced bimetallic nanoparticles have also gained attention for their anticancerous activity. For example, significant anticancer activity was exhibited by platinum-palladium (Pt-Pd) bimetallic nanoparticles synthesized using *Dioscorea bulbifera* tuber extract against HeLa cells and by gold-platinum (Au–Pt) bimetallic nanoparticles using tea phenols against human cervical cancer cells.

Table 14.10 Anticancer activity of metal and metal oxide nanoparticles (reprinted with permission from Patil and Chandrasekaran 2020)

Biological agent used for synthesis	Nanoparticle	Cell line	IC_{50} value	References
Marsilea quadrifolia	Gold	Human ovarian teratocarcinoma (PA-1) and lung adenocarcinoma (A549)	45.88 μg/mL and 52.015 μg/mL	Balashanmugam et al. (2018)
Morinda pubescens	Silver	HEP G2 (human epithelium cells of liver cancer)	937 μg/mL	Inbathamizh et al. (2012)
Nepeta deflersiana	Silver	Human cervical cancer cells (HeLa)	5 μg/mL	Al-Sheddi et al. (2018)
Murraya koenigii	Silver	HT-29 colon cancer	26.05 μg/mL	Cancer et al. (2018)
Cyanobacterium Nostoc sp. strain HKAR-2.	Silver	MCF-7 cells	27.5 μg/mL.	Sonker et al. (2017)
Piper nigrum	Silver	MCF-7 and Hep2 cells	52 mL, 54 μg/mL	Krishnan et al. (2016)
Ficus religiosa	Silver	A549 and Hep2 cells	1.9 μg/mL and 1.6 μg/mL	Ju et al. (2016)
Solanum xanthocarpum	Gold	HCT 15 human colon cancer	–	Zhang (2018)
Streptomyces sp.	Gold	HeLa	350 μg/mL	Manivasagan and Oh (2015)
Evolvulus alsinoides	Palladium	Human ovarian cancer A2780 cells	–	Gurunathan et al. (2015)
Gloriosa superb	Platinum and palladium	MCF-7 (human breast adenocarcinoma)	49.65 ± 1.99% and 36.26 ± 0.91%	Rokade et al. (2018)
Barleria prionitis	Platinum and palladium	Human breast adenocarcinoma (MCF-7)	–	Rokade et al. (2017)
Sargassum muticum	Iron oxide nanoparticles	HeLa cells, MCF-7 cells, HepG2 cells, and human cell lines for leukemia Jurkat	1.1 μg/mL (HepG2), 18.75 ± 2.1 μg/mL (MCF-7), 12.5 ± 1.7 μg/mL (HeLa), and 6.4 ± 2.3 μg/mL (Jurkat)	Namvar et al. (2014)
Albizia lebbeck	Zinc oxide	MDA-MB 231 and MCF-7	48.5, 48.7, and 60.2 μg/mL	Umar et al. (2019)
Costus pictus D. Don	Zinc oxide	DLA	–	Suresh et al. (2018)
Ficus religiosa	Copper oxide	A549 cells	200 μg/mL	Sankar et al. (2015)

14.5.3 Anti-inflammatory Activity

Inflammation is a pathological condition in which the inflamed part of the body develops swelling, redness, pain, etc. in response to an infection or injury and thereafter restores cellular homeostasis and tissue microenvironment. Several chronic inflammatory diseases like atherosclerosis, rheumatoid arthritis, cancer, diabetes mellitus, chronic kidney disease, nonalcoholic fatty liver disease (NAFLD), and autoimmune and neurodegenerative conditions have been implicated in increased mortality over the years. Excessive inflammation is a common thread during disease development and progression in all the abovementioned disorders. The conventional method of treatment of inflammation involves the use of both steroidal and nonsteroidal anti-inflammatory drugs which have multiple side effects. Recently several studies have shown that metal-based nanoparticles are quite efficient in controlling inflammation and can serve as potential agents in the development of anti-inflammatory drugs (Table 14.11).

14.5.4 Drug Delivery

Drug delivery systems increase the administration and efficacy of different pharmaceutically important compounds like drugs, vaccines, antibodies, enzymes, peptides, and proteins. However, conventional delivery systems are marred by limitations/disadvantages like poor bioavailability, poor solubility, poor absorption, lack of selectivity, first pass effect, intolerance, in vivo instability, adverse effects, fluctuations in drug levels, and limited effectiveness. Nanotechnology can play a major role in advanced drug formulations, drug targeting, and controlled drug release to the specific site of the body system without affecting healthy cells. Diverse nanostructures have been exhaustively researched for design of target specific drug delivery systems, especially for drugs that display poor solubility and less absorption

Table 14.11 Use of nanoparticles in inflammatory disorders

Nanoparticle	Source organism	References
Copper	*Millettia pinnata*	Thiruvengadam et al. (2019)
Copper oxide	*Abies spectabilis*	Liu et al. (2020)
Gold	*Curcuma pseudomontana* *Euphrasia officinalis* *Prunus serrulata*	Muniyappan et al. (2021) Liu et al. (2019) Singh et al. (2018)
Selenium	*Kluyveromyces lactis* *Ulva lactuca*	Song et al. (2021) Zhu et al. (2017)
Silver	*Cornus sanguinea* *Phoenix dactylifera* *Selaginella myosurus*	David et al. (2020) Alajmi et al. (2019) Kedi et al. (2018)
Zinc oxide	*Andrographis paniculata* *Hyssops officinalis* *Polygala tenuifolia*	Rajakumar et al.(2018) Mohammad et al. (2019) Nagajyothi et al. (2015)

ability. The green approach to production of drug loaded nanoparticles is highly advantageous since it minimizes the use of unsafe substances in the biosynthetic process. A number of nanostructures like liposomes, micelles, polymeric nanoparticles, solid lipid nanoparticles, inorganic nanoparticles (of various shapes like nanospheres, nanorods, nanostars, nanoshells, and nanocages), and dendrimers have been used as nanocarriers in drug delivery. FDA has approved a number of lipid-based (Doxil, Visudyne, Onivyde, Vyxeos, Onpattro), polymer-based (Oncaspar, Copaxone, Eligard, Abraxane, Cimiza, Plegridy), and inorganic (DexFerrum, Ferrlecit, Venofer, Feraheme, Injectafer) nanomedicines for drug delivery.

Biocompatible gold nanoparticles (AuNPs) prepared by utilizing the natural polysaccharide gum karaya were conjugated with an FDA-approved anticancer drug gemcitabine on the surface to develop an efficient drug delivery system (DDS) by Pooja et al. (2015). The administration of the DDS to human lung cancerous cells A549 exhibited better potential for inhibition of cell proliferation as compared to the pristine gemcitabine due to increased production of intracellular ROS. Devendiran et al. (2016) undertook biofabrication of gold nanoparticles using pectin polysaccharide followed by the encapsulation of anticancer drug doxorubicin and targeting ligand folic acid (FA) on the surface. This delivery system significantly inhibited the proliferation of folate receptor over expressed HT-29 colon cancer cells through induction of apoptosis as compared to free doxorubicin. A number of green nanoparticle-based drug delivery systems have shown great promise in control of a number of diseases. Some of these have been provided below:

1. Gold nanoparticles (fucoidan) + doxorubicin: efficient against human breast cancer cells MDA-MB-231
2. Gold nanoparticles (obtained from *Eclipta alba* leaf extract) + doxorubicin: efficient against human breast cancer cells MCF-7 and MDA-MB-231
3. Gold and silver nanoparticles (obtained from *Butea monosperma* leaf extract) + doxorubicin: efficient against B16F10 and MCF-7 cells
4. Platinum and silver nanoparticles (obtained from *Azadirachta indica* and *Murraya koenigii* leaf extract) + doxorubicin: efficient against MCF-7 cell lines

Similarly quantum dots have been researched for use in biomedical imaging, as biosensors and as carrier molecules for drug delivery. Some of the successful application of biogenically synthesized quantum dots and their role in drug delivery is provided below:

1. Luminescent carbon dots (using pasteurized milk) + lisinopril: efficient against HeLa cells
2. Bismuth subcarbonate quantum dots (using walnut skin extract) + sorafenib: efficient in liver targeting
3. Carbon dots (using cyanobacteria) + doxorubicin: efficient against MCF-7 and HepG2 cancer cells

Table 14.12 Use of green nanoparticles in bioimaging

Nanoparticle	Source organism	References
Cadmium sulfide quantum dots	*Ocimum tenuiflorum*	Naranthatta et al. (2021)
Carbon dots	*Ocimum sanctum*	Kumar et al. (2017)
Graphene quantum dots	*Mangifera indica*	Kumawat et al. (2017)
Graphene oxide quantum dots	*Calotropis gigantea*	Balajia et al. (2018)
Nitrogen-doped carbon dots	*Prunus persica*	Atchudan et al. (2016)
Zinc sulfide-manganese quantum dots	*Halomonas maura*	Raveendran et al. (2014)

Green polymeric nanoparticles synthesized from natural polymers like gelatin, chitosan, sodium alginate, and albumin have also gained significant attention in drug delivery owing to their unique optical, electrical, and optoelectrical qualities.

14.5.5 Bioimaging

Green nanoparticles on account of their high fluorescence, noninvasiveness, high sensitivity, long half-life, and target-specific binding have been used in bioimaging as contrast agents, for cell labeling and tracking; labeling implants, transplants, grafts, and organs; and image-guided interventions. Table 14.12 provides some examples of the use of green nanoparticles in bioimaging. Silica nanoparticles are quite important since they can be used in imaging techniques, such as magnetic resonance imaging (MRI), positron emission tomography (PET), single-photon emission computed tomography (SPECT), computed tomography (CT), ultrasound, photoacoustic imaging, and Raman imaging. Silica nanoparticles (SiNPs) have also been explored as contrast agents in medical imaging to encapsulate contrast agent particles like iron oxide, gold, silver, organic dyes, and quantum dots. Apart from conventional methods of preparation of silica nanoparticles, several plant-based waste materials have been utilized for their biosynthesis. Rice (*Oryza sativa*) husk has been the biomaterial of choice for the synthesis of silica nanoparticles. Pandey et al. (2014) successfully utilized rice husk for the fabrication of mesoporous silica oxide microparticles (*meso*-SiO_2)/C-dot complex for the theranostic application of cellular imaging. Rare-earth ion-integrated SiNPs using rice husk were prepared by Araichimani et al. (2020) using a rapid and simple microwave-assisted technique for making low cost bioimaging contrast agents.

References

Abdallah Y, Ogunyemi SO, Abdelazez A, Zhang M, Hong X, Ibrahim E, Hossain A, Fouad H, Li B, Chen J (2019) The green synthesis of MgO nano-flowers using *Rosmarinus officinalis* L. (Rosemary) and the antibacterial activities against *Xanthomonas oryzae* pv. *oryzae*. Biomed Res Int 2019:5620989

Abduz Zahir A, Bagavan A, Kamaraj C, Elango G, Abdul Rahuman A (2012) Efficacy of plant-mediated synthesized silver nanoparticles against *Sitophilus oryzae*. J Biopest 5:95–102

Ahmad H, Rajagopal K, Shah AH, Bhat AH, Venugopal K (2017) Study of bio-fabrication of iron nanoparticles and their fungicidal property against phytopathogens of apple orchards. IET Nanobiotechnol 11(3):230–235

Ahmad H, Venugopal K, Bhat AH, Kavitha K, Ramanan A, Rajagopal K, Srinivasan R, Manikandan E (2020) Enhanced biosynthesis synthesis of copper oxide nanoparticles (CuO-NPs) for their antifungal activity toxicity against major phyto-pathogens of apple orchards. Pharm Res 37(12):246

Ahmed T, Noman M, Luo J, Muhammad S, Shahid M, Ali MA, Zhang M, Li B (2021a) Bioengineered chitosan-magnesium nanocomposite: A novel agricultural antimicrobial agent against Acidovorax oryzae and Rhizoctonia solani for sustainable rice production. Int J Biol Macromol 168:834–845

Ahmed T, Wu Z, Jiang H, Luo J, Noman M, Shahid M, Manzoor I, Allemailem KS, Alrumaihi F, Li B (2021b) Bioinspired green synthesis of zinc oxide nanoparticles from a native *Bacillus cereus* strain RNT6: characterization and antibacterial activity against rice panicle blight pathogens *Burkholderia glumae* and *B. gladioli*. Nanomaterials 11(4):884

Alajmi RA, Al-Megrin WA, Metwally D, Al-Subaie H, Altamrah N, Barakat AM, Abdel Moneim AE, Al-Otaibi TT, El-Khadragy M (2019) Anti-*Toxoplasma* activity of silver nanoparticles green synthesized with *Phoenix dactylifera* and *Ziziphus spina-christi* extracts which inhibits inflammation through liver regulation of cytokines in Balb/c mice. Biosci Rep 39(5): BSR20190379

Al-Dhabi NA, Valan Arasu M (2018) Environmentally-friendly green approach for the production of zinc oxide nanoparticles and their anti-fungal, ovicidal, and larvicidal properties. Nanomedicine 8:500

Alharbi NS, Govindarajan M, Kadaikunnan S, Khaled JM, Almanaa TN, Alyahya SA et al (2018) Nanosilver crystals capped with *Bauhinia* phytochemicals as new antimicrobials and mosquito larvicides. J Trace Elem Med Biol 50:146–153

Ali I, Afshinb S, Poureshgh Y, Azari A, Rashtbari Y, Feizizadeh A, Hamzezadeh A, Fazlzadeh M (2020) Green preparation of activated carbon from pomegranate peel coated with zero-valent iron nanoparticles (nZVI) and isotherm and kinetic studies of amoxicillin removal in water. Environ Sci Pollut Res Int 27(29):36732–36743

Alif Alisha AS, Thangapandiyan S (2019) Comparative bioassay of silver nanoparticles and malathion on infestation of red flour beetle, *Tribolium castaneum*. J Basic Appl Zool 80:55

Aljabali AAA, Akkam Y, Salim M, Zoubi A, Al-batayneh KM, Id BA, Alrob OA, Alkilany AM, Benamara M, Id DJE (2018) Synthesis of gold nanoparticles using leaf extract of Ziziphus zizyphus and their antimicrobial activity. Nanomedicine 8(2018):1–15

Almadiy AA, Nenaah GE, Shawer DM (2018) Facile synthesis of silver nanoparticles using harmala alkaloids and their insecticidal and growth inhibitory activities against the khapra beetle. J Pestic Sci 91:727–737

Al-Sheddi ES, Farshori NN, Al-Oqail MM, Al-Massarani SM, Saquib Q, Wahab R, Musarrat J, Al-Khedhairy AA, Siddiqui MA (2018) Anticancer potential of green synthesized silver nanoparticles using extract of Nepeta deflersiana against human cervical cancer cells. Bioinorg Chem Appl 5:1–12

Araichimani P, Prabu K, Kumar GS, Karunakaran G, Van Minh N, Karthi S, Girija E, Kolesnikov E (2020) Rare-earth ions integrated silica nanoparticles derived from rice husk via microwave-assisted combustion method for bioimaging applications. Ceram Int 46(11):18366–18372

Arasu MV, Arokiyaraj S, Viayaraghavan P, Kumar TSJ, Duraipandiyan V, Al-Dhabi NA, Kaviyarasu K (2019) One step green synthesis of larvicidal, and azo dye degrading antibacterial nanoparticles by response surface methodology. J Photochem Photobiol B 190:154–162

Arumugam G, Velayutham V, Shanmugavel S, Sundaram J (2016) Efficacy of nanostructured silica as a stored pulse protector against the infestation of bruchid beetle, *Callosobruchus maculatus* (Coleoptera: Bruchidae). Appl Nanosci 6:445–450

Atchudan R, Edison TNJI, Lee YR (2016) Nitrogen-doped carbon dots originating from unripe peach for fluorescent bioimaging and electrocatalytic oxygen reduction reaction. J Colloid Interface Sci 482:8–18

Athinarayanan J, Periasamy VS, Alhazmi M, Alshatwi AA (2017) Synthesis and biocompatibility assessment of sugarcane bagasse-derived biogenic silica nanoparticles for biomedical applications. J Biomed Mater Res B Appl Biomater 105(2):340–349

Aziz AT, Alshehri MA, Alanazi NA, Panneerselvam C, Trivedi S, Maggi F, Sut S, Dall'Acqua S (2020) Phytochemical analysis of Rhazya stricta extract and its use in fabrication of silver nanoparticles effective against mosquito vectors and microbial pathogens. Sci Total Environ 700:134443

Balajia M, Jegatheeswarana S, Nithyaa P, Boomib P, Selvamc S, Sundrarajana M (2018) Photoluminescent reduced graphene oxide quantum dots from latex of Calotropis gigantea for metal sensing, radical scavenging, cytotoxicity, and bioimaging in Artemia salina: a greener route. J Photochem Photobiol 178:371–379

Balashanmugam P, Mosachristas K, Kowsalya E (2018) In vitro cytotoxicity and antioxidant evaluation of biogenic synthesized gold nanoparticles from *Marsilea quadrifolia* on lung and ovarian cancer. Int J App Pharm 10:153–158

Barui AK, Das S, Patra CR (2019) Biomedical applications of green-synthesized metal nanoparticles using polysaccharides. In: Functional polysaccharides for biomedical applications. Elsevier, Cambridge, pp 329–355

Basavaiah K, Kahsay MH, RamaDevi D (2018) Green synthesis of magnetite nanoparticles using aqueous pod extract of *Dolichos lablab* L for an efficient adsorption of crystal violet. Emergent Mater 1:121–132

Benelli G (2015) Research in mosquito control: current challenges for a brighter future. Parasitol Res 114:2801–2805

Benelli G (2016) Plant-mediated biosynthesis of nanoparticles as an emerging tool against mosquitoes of medical and veterinary importance: a review. Parasitol Res 115:23–34

Benelli G (2018) Mode of action of nanoparticles against insects. Environ Sci Pollut Res 25(13):12329–12341

Benelli G, Lo Iacono A, Canale A, Mehlhorn H (2016) Mosquito vectors and the spread of cancer: an overlooked connection? Parasitol Res 115(6):2131–2137

Benelli G, Maggi F, Pavela R, Murugan K, Govindarajan M, Vaseeharan B, Petrelli R, Cappellacci L, Kumar S, Hofer A, Youssefi MR, Alarfaj AA, Hwang JS, Higuchi A (2018) Mosquito control with green nanopesticides: towards the One Health approach? A review of non-target effects. Environ Sci Pollut Res Int 25(11):10184–10206

Berhanu H, Emana (2018) Host type and textures on the survival of *Tribolium castaneum* (Coleoptera: Tenebrionidae) parental and filial generations. J Entomol Zool Stud 6(1):622–626

Bharani RA, Namasivayam SKR (2017) Biogenic silver nanoparticles mediated stress on developmental period and gut physiology of major lepidopteran pest Spodoptera litura (Fab.)(Lepidoptera: Noctuidae)—An eco-friendly approach of insect pest control. J Environ Chem Eng 5(1):453–467

Bhuvaneswari R, Xavier RJ, Arumugam M (2016) Larvicidal property of green synthesized silver nanoparticles against vector mosquitoes (*Anopheles stephensi* and *Aedes aegypti*). J King Saud Univers Sci 28:318–323

Bhuyan T, Mishra K, Khanuja M et al (2015) Biosynthesis of zinc oxide nanoparticles from *Azadirachta indica* for antibacterial and photocatalytic applications. Mater Sci Semicond Process 32:55–61

Bin YB, Bin ZJ, Fen GL, Qing YX, Mei ZL, Xi C (2008) Photocatalytic degradation investigation of dicofol. Chin Sci Bull 53:27–32

Bordbar M, Negahdar N, Nasrollahzadeh M (2018) Melissa officinalis L. Leaf extract assisted green synthesis of CuO/ZnO nanocomposite for the reduction of 4-nitrophenol and Rhodamine B. Sep Purif Technol 191:295–300

Brisola Marcondes C, Benelli G (2019) Mosquitoes, infectious diseases, and cancer: a connection to study? Int J Environ Res Public Health 16(23):4859

Cai W, Weng X, Chen Z (2019) Highly efficient removal of antibiotic rifampicin from aqueous solution using green synthesis of recyclable nano-Fe_3O_4. Env Pollution 247:839–846

Campos EVR, de Oliveira JL, Abrantes DC, Rogério CB, Bueno C, Miranda VR, Monteiro RA, Fraceto LF (2020) Recent developments in nanotechnology for detection and control of *Aedes aegypti*-Borne diseases. Front Bioeng Biotechnol 8:102

Cancer HC, Line C, Roshni K, Younis M, Ilakkiyapavai D, Basavaraju P, Puthamohan VM (2018) Cancer science & therapy anticancer activity of biosynthesized silver nanoparticles using *Murraya koenigii* leaf extract against HT-29 colon cancer cell line. J Cancer Sci Ther 10:72–75

Castillo-Henríquez L, Alfaro-Aguilar K, Ugalde-Álvarez J, Vega-Fernández L, Montes de Oca-Vásquez G, Vega-Baudrit JR (2020) Green synthesis of gold and silver nanoparticles from plant extracts and their possible applications as antimicrobial agents in the agricultural area. Nanomaterials 10(9):1763

Chatterjee S, Mahanty S, Das P, Chaudhuri P, Das S (2020) Biofabrication of iron oxide nanoparticles using manglicolous fungus Aspergillus niger BSC-1 and removal of Cr (VI) from aqueous solution. Chem Eng J 2020(385):123790

Consolo VF, Torres-Nicolini A, Alvarez VA (2020) Mycosinthetized Ag, CuO and ZnO nanoparticles from a promising *Trichoderma harzianum* strain and their antifungal potential against important phytopathogens. Sci Rep 10:20499

Da E, Taha A, Afkar E (2018) Green synthesis of iron nanoparticles by Acacia nilotica pods extract and its catalytic, adsorption, and antibacterial activities. Appl Sci 2018:8

David L, Moldovan B, Baldea I, Olteanu D, Bolfa P, Clichici S, Filip GA (2020) Modulatory effects of Cornus sanguinea L. mediated green synthesized silver nanoparticles on oxidative stress, COX-2/NOS2 and NFkB/pNFkB expressions in experimental inflammation in Wistar rats. Korean J Couns Psychother 110:110709

Dawoud TM, Yassin MA, El-Samawaty ARM, Elgorban AM (2021) Silver nanoparticles synthesized by *Nigrospora oryzae* showed antifungal activity. Saudi J Biol Sci 28(3):1847–1852

De Jong WH, Borm PJ (2008) Drug delivery and nanoparticles:applications and hazards. Int J Nanomedicine 3(2):133–149

Debnath N, Das S, Seth D, Chandra R, Bhattacharya SC, Goswami A (2011) Entomologic effect of silica nanoparticles against Sitophilus oryzae (L.). J Pestic Sci 81(1):99–105

Deepak P, Sowmiya R, Ramkumar R, Balasubramani G, Aiswarya D, Perumal P (2017) Structural characterization and evaluation of mosquito-larvicidal property of silver nanoparticles synthesized from the seaweed. Artif Cells Nanomed Biotechnol 45(5):990–998

Devendiran RM, Chinnaiyan SK, Yadav NK, Moorthy GK, Ramanathan G, Singaravelu S et al (2016) Green synthesis of folic acid-conjugated gold nanoparticles with pectin as reducing/stabilizing agent for cancer theranostics. RSC Adv 6:29757–29768

Dinesh D, Murugan K, Madhiyazhagan P, Panneerselvam C, Nicoletti M, Jiang W, Benelli G, Chandramohan B, Suresh U (2015) Mosquitocidal and antibacterial activity of green-synthesized silver nanoparticles from Aloe vera extracts: towards an effective tool against the malaria vector Anopheles stephensi? Parasitol Res 114:1519–1529

Durán N, Islan GA, Durán M, Castro GR (2016) Nanobiotechnology solutions against *Aedes aegypti*. J Braz Chem Soc 27:1139–1149

Ebrahiminezhad A, Zare-Hoseinabadi A, Sarmah AK, Taghizadeh S, Ghasemi Y, Berenjian A (2017) Plant-mediated synthesis and applications of iron nanoparticles. Mol Biotechnol 60(2):154–168

El-Bahnasawy MM, Fadil EE, Morsy TA (2013) Mosquito vectors of infectious diseases: are they neglected health disaster in Egypt? J Egypt Soc Parasitol 43(2):373–386

Elemike EE, Onwudiwe DC, Ekennia AC, Sonde CU, Ehiri RC (2017) Green synthesis of Ag/Ag_2O nanoparticles using aqueous leaf extract of Eupatorium odoratum and its antimicrobial and mosquito larvicidal activities. Molecules 22(5):674

Elliott DW, Lien HL, Zhang WX (2009) Degradation of lindane by zero-valent iron nanoparticles. J Environ Eng 135:317–324

El-Saadony MT, Abd El-Hack ME, Taha AE, Fouda M, Ajarem JS, Maodaa S, Allam AA, Elshaer N (2020) Ecofriendly synthesis and insecticidal application of copper nanoparticles against the storage pest *Tribolium castaneum*. Nanomaterials 10(3):587

El-Temsah YS, Sevcu A, Bobcikova K, Cernik M, Joner EJ (2016) DDT degradation efficiency and ecotoxicological effects of two types of nano-sized zero-valent iron (nZVI) in water and soil. Chemosphere 144:2221–2228

Elumalai D, Hemavathi M, Deepaa CV, Kaleena PK (2017) Evaluation of phytosynthesised silver nanoparticles from leaf extracts of Leucas aspera and *Hyptis suaveolens* and their larvicidal activity against malaria, dengue and filariasis vectors. Parasite Epidemiol Control 2:15–26

Esmaeili Bidhendi M, Poursorkh Z, Sereshti H, Rashidi Nodeh H, Rezania S, Afzal KM (2020) Nano-size biomass derived from pomegranate peel for enhanced removal of cefixime antibiotic from aqueous media: kinetic, equilibrium and thermodynamic study. Int J Environ Res Public Health 17(12):4223

Fouad H, Hongjie L, Hosni D, Wei J, Abbas G, Ga'al H, Jianchu M (2018) Controlling Aedes albopictus and Culex pipiens pallens using silver nanoparticles synthesized from aqueous extract of Cassia fistula fruit pulp and its mode of action. Artif Cells Nanomed Biotechnol 46: 558–567

Fouda A, Hassan SE, Abdo AM, El-Gamal MS (2020) antimicrobial, antioxidant and larvicidal activities of spherical silver nanoparticles synthesized by endophytic Streptomyces spp. Biol Trace Elem Res 195(2):707–724

Fraceto LF, Grillo R, de Medeiros GA, Scognamiglio V, Rea G, Bartolucci C (2016) Nanotechnology in agriculture: which innovation potential does it have? Front Environ Sci 4:20

García FE, Senn AM, Meichtry JM, Scott TB, Pullin H, Leyva AG, Halac EB, Ramos CP, Sacanell J, Mizrahi M, Requejo FG, Litter MI (2019) Iron-based nanoparticles prepared from yerba mate extract. Synthesis, characterization and use on chromium removal. J Environ Manage 235:1–8

Govindarajan M, Rajeswary M, Muthukumaran U, Hoti SL, Khater HF, Benelli G (2016) Single-step biosynthesis and characterization of silver nanoparticles using Zornia diphylla leaves: a potent eco-friendly tool against malaria and arbovirus vectors. J Photochem Photobiol B 161: 482–489

Gunti L, Dass RS, Kalagatur NK (2019) Phytofabrication of selenium nanoparticles from Emblica officinalis fruit extract and exploring its biopotential applications: antioxidant, antimicrobial, and biocompatibility. Front Microbiol 10:1–17

Gurunathan S, Kim E, Han JW, Park JH, Kim J (2015) Green chemistry approach for synthesis of effective anticancer palladium nanoparticles. Molecules 20:22476–22498

Hajra A, Dutta S, Mondal NK (2016) Mosquito larvicidal activity of cadmium nanoparticles synthesized from petal extracts of marigold (*Tagetes* sp.) and rose (*Rosa* sp.) flower. J Parasit Dis 40(4):1519–1527

Hashem AS, Awadalla SS, Zayed GM, Maggi F, Benelli G (2018) Pimpinella anisum essential oil nanoemulsions against *Tribolium castaneum*—insecticidal activity and mode of action. Environ Sci Pollut Res 25(9):18802–18812

Hazarika M, Borah D, Bora P, Silva AR, Das P (2017) Biogenic synthesis of palladium nanoparticles and their applications as catalyst and antimicrobial agent. PLoS One 2017:1–19

Hernández-Díaz JA, Garza-García JJ, Zamudio-Ojeda A, León-Morales JM, López-Velázquez JC, García-Morales S (2021) Plant-mediated synthesis of nanoparticles and their antimicrobial activity against phytopathogens. J Sci Food Agric 101(4):1270–1287

Hossain A, Abdallah Y, Ali MA, Masum MMI, Li B, Sun G, Meng Y, Wang Y (2019b) An Q. Lemon-fruit-based green synthesis of zinc oxide nanoparticles and titanium dioxide nanoparticles against soft rot bacterial pathogen *Dickeya dadantii*. Biomolecules 9(12):863

Hossain A, Hong X, Ibrahim E, Li B, Sun G, Meng Y, Wang Y, An Q (2019a) Green synthesis of silver nanoparticles with culture supernatant of a bacterium *Pseudomonas rhodesiae* and their antibacterial activity against soft rot pathogen *Dickeya dadantii*. Molecules 24(12):2303

Hua S, de Matos MBC, Metselaar JM, Storm G (2018) Current trends and challenges in the clinical translation of nanoparticulate nanomedicines: pathways for translational development and commercialization. Front Pharmacol 9:14

Huang W, Xu M, Duan H, Bi Y, Yu H (2019) Inhibition of *Fusarium oxysporum* by AgNPs biosynthesised using *Cinnamomum camphora* fruit extract. IET Nanobiotechnol 13(1):42–45

Ibrahim E, Zhang M, Zhang Y, Hossain A, Qiu W, Chen Y, Wang Y, Wu W, Sun G, Li B (2020) Green-synthesization of silver nanoparticles using endophytic bacteria isolated from garlic and its antifungal activity against wheat *Fusarium* head blight pathogen *Fusarium Graminearum*. Nanomaterials 10(2):219

Iliger KS, Sofi TA, Bhat NA, Ahanger FA, Sekhar JC, Elhendi AZ, Al-Huqail AA, Khan F (2021) Copper nanoparticles: Green synthesis and managing fruit rot disease of chilli caused by *Colletotrichum capsici*. Saudi J Biol Sci 28(2):1477–1486

Inbathamizh L, Ponnu TM, Mary EJ (2012) In vitro evaluation of antioxidant and anticancer potential of Morinda pubescens synthesized silver nanoparticles free radicals. JOPR J Pharm Res 6:32–38

Irshad MA, Nawaz R, Zia Ur Rehman M, Imran M, Ahmad J, Ahmad S, Inam A, Razzaq A, Rizwan M, Ali S (2020) Synthesis and characterization of titanium dioxide nanoparticles by chemical and green methods and their antifungal activities against wheat rust. Chemosphere 258:127352

Ishwarya R, Vaseeharan B, Anuradha R, Rekha R, Govindarajan M, Alharbi NS et al (2017) Eco-friendly fabrication of Ag nanostructures using the seed extract of Pedalium murex, an ancient Indian medicinal plant: histopathological effects on the Zika virus vector Aedes aegypti and inhibition of biofilm-forming pathogenic bacteria. J Photochem Photobiol B Biol 174:133–143

Ishwarya R, Vaseeharan B, Kalyani S, Banumathi B, Govindarajan M, Alharbi NS et al (2018) Facile green synthesis of zinc oxide nanoparticles using *Ulva lactuca* seaweed extract and evaluation of their photocatalytic, antibiofilm and insecticidal activity. J Photochem Photobiol Biol 178:249–258

Jafer FS, Annon MR (2018) Larvicidal effect of pure and green-synthesized silver nanoparticles against *Tribolium castaneum* (herb.) and *Callosobruchus maculatus* (fab.). J Global Pharma Technol 10(3):448–454

Jahangirian H, Lemraski EG, Webster TJ, Rafiee-Moghaddam R, Abdollahi Y (2017) A review of drug delivery systems based on nanotechnology and green chemistry: green nanomedicine. Int J Nanomedicine 12:2957–2978

Jain A, Wadhawan S, Mehta SK (2021) Biogenic synthesis of non-toxic iron oxide NPs via Syzygium aromaticum for the removal of methylene blue. Environmental Nanotechnology, Monitoring & Management 16:100464

Jha AK, Chakraborty S (2020) Photocatalytic degradation of Congo Red under UV irradiation by zero valent iron nano particles (nZVI) synthesized using *Shorea robusta* (Sal) leaf extract. Water Sci Technol 82(11):2491–2502

Joo SH, Zhao D (2008) Destruction of lindane and atrazine using stabilized iron nanoparticles under aerobic and anaerobic conditions: effects of catalyst and stabilizer. Chemosphere 70:418–425

Josephrajkumar A, Subrahmanyam B, Devakumar C (2000) Growth regulatory activity of silver fern extract on the cotton bollworm, Helicoverpa armigera (Hübner). Int J Tropic Insect Sci 20(04):295–302

Ju S, Jun I, Tettey CO, Mo K, Mook H (2016) In-vitro anticancer activity of green synthesized silver nanoparticles on MCF-7 human breast cancer cells. Mater Sci Eng C 68:430–435

Jyoti K, Singh A (2016) Green synthesis of nanostructured silver particles and their catalytic application in dye degradation. J Genet Eng Biotechnol 14:311–317

Kalagatur NK, Nirmal Ghosh OS, Sundararaj N, Mudili V (2018) Antifungal Activity of Chitosan Nanoparticles Encapsulated With *Cymbopogon martinii* Essential Oil on Plant Pathogenic Fungi *Fusarium graminearum*. Front Pharmacol 9:610

Kanwar R, Rathee J, Salunke DB, Mehta SK (2019) Green nanotechnology-driven drug delivery assemblies. ACS Omega 4(5):8804–8815

Karthiga D, Anthony SP (2013) Selective colorimetric sensing of toxic metal cations by green synthesized silver nanoparticles over a wide pH range. RSC Adv 37:16765–16774

Kedi PBE, Meva FE, Kotsedi L, Nguemfo EL, Zangueu CB, Ntoumba AA, Mohamed HEA, Dongmo AB, Maaza M (2018) Eco-friendly synthesis, characterization, in vitro and in vivo anti-inflammatory activity of silver nanoparticle-mediated Selaginella myosurus aqueous extract. Int J Nanomedicine 13:8537–8548

Kganyago P, Mahlaule-Glory LM, Mathipa MM, Ntsendwana B, Mketo N, Mbita Z, Hintsho-Mbita NC (2018) Synthesis of NiO nanoparticles via a green route using Monsonia burkeana: the physical and biological properties. J Photochem Photobiol B 182:18–26

Khan M, Khan AU, Bogdanchikova N, Garibo D (2021) Antibacterial and antifungal studies of biosynthesized silver nanoparticles against plant parasitic nematode *Meloidogyne incognita,* Plant pathogens *Ralstonia solanacearum* and *Fusarium oxysporum*. Molecules 26(9):2462

Kouhbanani MAJ, Beheshtkhoo N, Amani AM, Taghizadeh S, Beigi V, Bazmandeh AZ, Khalaf N (2018) Green synthesis of iron oxide nanoparticles using Artemisia vulgaris leaf extract and their application as a heterogeneous Fenton-like catalyst for the degradation of methyl orange. Mater Res Express 5(11):115013

Kouhbanani MAJ, Beheshtkhoo N, Taghizadeh S, Amani AM, Alimardani V (2019) One-step green synthesis and characterization of iron oxide nanoparticles using aqueous leaf extract of Teucrium polium and their catalytic application in dye degradation. Adv Nat Sci Nanosci Nanotechnol 10(1):015007

Krishnan V, Bupesh G, Manikandan EKTA, Magesh S, Kalyanaraman R, Maaza M (2016) Green synthesis of silver nanoparticles using Piper nigrum concoction and its anticancer activity against MCF-7 and Hep-2 cell lines. J Antimicrobial Agents 2(2016):8–12

Kumar A, Chowdhuri AR, Laha D, Mahto TK, Karmakar P, Sahu SK (2017) Green synthesis of carbon dots from Ocimum sanctum for effective fluorescent sensing of Pb^{2+} ions and live cell imaging. Sens Actuators B 242:679–686

Kumar B, Smita K, Cumbal L, Debut A, Galeas S, Guer-rero VH (2016) Phytosynthesis and photocatalytic activity of magnetite (Fe3O4) nanoparticles using the Andean blackberry leaf. Mater Chem Phys 179:310–315

Kumar D, Kumar P, Singh H, Agrawal V (2020) Biocontrol of mosquito vectors through herbal-derived silver nanoparticles: prospects and challenges. Environ Sci Pollut Res Int 27(21):25987–26024

Kumawat MK, Thakur M, Gurung RB, Srivastava R (2017) Graphene quantum dots from Mangifera indica: application in near-infrared bioimaging and intracellular nanothermometry. ACS Sustain Chem Eng 5:1382–1391

Lakshmeesha TR, Kalagatur NK, Mudili V, Mohan CD, Rangappa S, Prasad BD, Ashwini BS, Hashem A, Alqarawi AA, Malik JA, Abd Allah EF, Gupta VK, Siddaiah CN, Niranjana SR (2019) Biofabrication of zinc oxide nanoparticles with *Syzygium aromaticum* flower buds extract and finding its novel application in controlling the growth and mycotoxins of *Fusarium graminearum*. Front Microbiol 10:1244

Lakshmeesha TR, Murali M, Ansari MA, Udayashankar AC, Alzohairy MA, Almatroudi A, Alomary MN, Asiri SMM, Ashwini BS, Kalagatur NK, Nayak CS, Niranjana SR (2020) Biofabrication of zinc oxide nanoparticles from *Melia azedarach* and its potential in controlling soybean seed-borne phytopathogenic fungi. Saudi J Biol Sci 27(8):1923–1930

Lallawmawma H, Sathishkumar G, Sarathbabu S, Ghatak S, Sivaramakrishnan S, Gurusubramanian G, Kumar NS (2015) Synthesis of silver and gold nanoparticles using *Jasminum nervosum* leaf extract and its larvicidal activity against filarial and arboviral vector *Culex quinquefasciatus* Say (Diptera: Culicidae). Environ Sci Pollut Res 22:17753–17768

Leili M, Fazlzadeh M, Bhatnagar A (2018) Green synthesis of nano-zero-valent iron from Nettle and Thyme leaf extracts and their application for the removal of cephalexin antibiotic from aqueous solutions. Environ Technol 39(9):1158–1172

Liu H, Zheng S, Xiong H, Alwahibi MS, Niu X (2020) Biosynthesis of copperoxide nanoparticles using Abies spectabilis plant extract and analyzing its antinociceptive and anti-inflammatory potency in various mice models. Arab J Chem 13:6995–7006

Liu Y, Kim S, Kim YJ, Perumalsamy H, Lee S, Hwang E, Yi TH (2019) Green synthesis of gold nanoparticles using *Euphrasia officinalis* leaf extract to inhibit lipopolysaccharide-induced inflammation through NF-κB and JAK/STAT pathways in RAW 264.7 macrophages. Int J Nanomedicine 14:2945–2959

Louni M, Shakarami J, Negahban M (2018) Insecticidal efficacy of nanoemulsion containing *Mentha longifolia* essential oil against *Ephestia kuehniella* (Lepidoptera: Pyralidae). J Crop Prot 7(2):171–182

Luo F, Yang D, Chen Z, Megharaj M, Naidu R (2015) The mechanism for degrading Orange II based on adsorption and reduction by ion-based nanoparticles synthesized by grape leaf extract. J Hazard Mater 296:37–45

Magro M, Bramuzzo S, Baratella D, Ugolotti J, Zoppellaro G, Chemello G et al (2019) Self-assembly of chlorin-e6 on γ-Fe2O3 nanoparticles: application for larvicidal activity against *Aedes aegypti*. J Photochem Photobiol 194:21–31

Malaikozhundan B, Vinodhini J, Vaseeharan B (2019) Nanopesticides for the management of insect pests of stored grains. In: Panpatte D, Jhala Y (eds) Nanotechnology for agriculture: crop production & protection. Springer, Singapore

Malakootian M, Yaseri M, Faraji M (2019) Removal of antibiotics from aqueous solutions by nanoparticles: a systematic review and meta-analysis. Environ Sci Pollut Res Int 26(9):8444–8458

Manivasagan P, Oh J (2015) Production of a novel fucoidanase for the green synthesis of gold nanoparticles by Streptomyces sp. and its cytotoxic effect on HeLa cells. Mar Drugs 13(2015):6818–6837

Martinez JL (2009) Environmental pollution by antibiotics and by antibiotic resistance determinants. Environ Pollut 157(11):2893–2902

Martins M, Mourato C, Sanches S, Noronha JP, Crespo MTB, Pereira IAC (2017) Biogenic platinum and palladium nanoparticles as new catalysts for the removal of pharmaceutical compounds. Water Res 108:160–168

Masum MMI, Siddiqa MM, Ali KA, Zhang Y, Abdallah Y, Ibrahim E, Qiu W, Yan C, Li B (2019) Biogenic synthesis of silver nanoparticles using *Phyllanthus emblica* fruit extract and its inhibitory action against the pathogen *Acidovorax oryzae* strain RS-2 of rice bacterial brown stripe. Front Microbiol 10:820

Matea CT, Mocan T, Tabaran F, Pop T, Mosteanu O, Puia C, Iancu C, Mocan L (2017) Quantum dots in imaging, drug delivery and sensor applications. Int J Nanomedicine 12:5421–5431

Mehlhorn H, Al-Rasheid KA, Al-Quraishy S, Abdel-Ghaffar F (2012) Research and increase of expertise in arachno-entomology are urgently needed. Parasitol Res 110:259–265

Mesterházy Á, Oláh J, Popp J (2020) Losses in the grain supply chain: causes and solutions. Sustainability 12(6):2342

Minal SP, Prakash S (2020) Laboratory analysis of Au-Pd bimetallic nanoparticles synthesized with Citrus limon leaf extract and its efficacy on mosquito larvae and non-target organisms. Sci Rep 10(1):21610

Mitchell MJ, Billingsley MM, Haley RM et al (2021) Engineering precision nanoparticles for drug delivery. Nat Rev Drug Discov 20:101–124

Mohammad GRKS, Tabrizi MH, Ardalan T, Yadamani S, Safavi E (2019) Green synthesis of zinc oxide nanoparticles and evaluation of anti-angiogenesis, anti-inflammatory and cytotoxicity properties. J Biosci 44(2):30

Mohanta YK, Panda SK, Jayabalan R, Sharma N (2017) Antioxidant and cytotoxic activity of silver nanoparticles synthesized by leaf extract of Erythrina suberosa (Roxb). Front Mol Biosci 4:1–9

Mossa ATH, Abdelfattah NAH, Mohafrash SMM (2017) Nanoemulsion of camphor (*Eucalyptus globulus*) essential oil, formulation, characterization and insecticidal activity against wheat weevil, *Sitophilus*. Asian J Crop Sci 9(3):50–62

Muniyappan N, Pandeeswaran M, Amalraj A (2021) Green synthesis of gold nanoparticles using Curcuma pseudomontana isolated curcumin: Its characterization, antimicrobial, antioxidant and anti-inflammatory activities. Environ Chem Ecotoxicol 3:117–124

Murugan K, Roni M, Panneerselvam C, Aziz AT, Suresh U, Rajaganesh R, Aruliah R, Mahyoub JA, Trivedi S, Rehman H, Naji Al-Aoh HA, Kumar S, Higuchi A, Vaseeharan B, Wei H, Senthil-Nathan S, Canale A, Benelli G (2018) Sargassum wightii-synthesized ZnO nanoparticles reduce the fitness and reproduction of the malaria vector Anopheles stephensi and cotton bollworm Helicoverpa armigera. Physiol Mol Plant Pathol 101:202–213

Muthamil Selvan S, Vijai Anand K, Govindaraju K, Tamilselvan S, Kumar VG, Subramanian KS, Kannan M, Raja K (2018) Green synthesis of copper oxide nanoparticles and mosquito larvicidal activity against dengue, zika and chikungunya causing vector *Aedes aegypti*. IET Nanobiotechnol 12(8):1042–1046

Muthu K, Akilandaeaswari B, Nagasundar SM (2019) Green synthesis of silver nanoparticles using polyalthia longifoliastem bark extract and its catalytic reduction of 4-nitrophenol. Asian J Chem 31:2439–2442

Muthukumar H, Matheswaran M (2015) Amaranthus spinosus leaf extract mediated FeO nanoparticles: physicochemical traits, photocatalytic and antioxidant activity. ACS Sustain Chem Eng 3(12):3149–3156

Muthukumaran U, Govindarajan M, Rajeswary M (2015) Mosquito larvicidal potential of silver nanoparticles synthesized using Chomelia asiatica (Rubiaceae) against Anopheles stephensi, Aedes aegypti, and Culex quinquefasciatus (Diptera: Culicidae). Parasitol Res 114(3):989–999

Mystrioti C, Xanthopoulou TD, Tsakiridis P, Papassiopi N, Xenidis A (2016) Comparative evaluation of five plant extracts and juices for nanoiron synthesis and application for hexavalent chromium reduction. Sci Total Environ 539:105–113

Nagajyothi PC, Cha SJ, Yang IJ, Sreekanth TVM, Kim KJ, Shin HM (2015) Antioxidant and anti-inflammatory activities of zinc oxide nanoparticles synthesized using Polygala tenuifolia root extract. J Photochem Photobiol B Biol 146:10–17

Nalini M, Lena M, Sumathi P, Sundaravadivelan C (2017) Effect of phyto-synthesized silver nanoparticles on developmental stages of malaria vector, *Anopheles stephensi* and dengue vector, Aedes aegypti. Egyptian J Basic Appl Sci 4:212–218

Namvar F, Rahman HS, Mohamad R, Baharara J, Mahdavi M, Amini E, Chartrand MS, Yeap SK (2014) Cytotoxic effect of magnetic iron oxide nanoparticles synthesized via seaweed aqueous extract. Int J Nanomedicine 2014:2479–2488

Naranthatta S, Janardhanan P, Pilankatta R, Nair SS (2021) Green synthesis of engineered CdS nanoparticles with reduced cytotoxicity for enhanced bioimaging application. ACS Omega 6(12):8646–8655

Natarajan E, Ponnaiah GP (2017) Optimization of process parameters for the decolorization of Reactive Blue 235 dye by barium alginate immobilized iron nanoparticles synthesized from aluminum industry waste. Environ Nanotechnol Monit Manage 7:73–88

Njagi EC, Huang H, Stafford L, Genuino H, Galindo HM, Collins JB, Hoag GE, Suib SL (2011) Biosynthesis of iron and silver nanoparticles at room temperature using aqueous Sorghum bran extracts. Langmuir 27:264–271

Nune SK, Gunda P, Thallapally PK, Lin YY, Forrest ML, Berkland CJ (2009) Nanoparticles for biomedical imaging. Expert Opin Drug Deliv 6(11):1175–1194

Ogunyemi SO, Abdallah Y, Zhang M, Fouad H, Hong X, Ibrahim E, Masum MMI, Hossain A, Mo J, Li B (2019b) Green synthesis of zinc oxide nanoparticles using different plant extracts and their antibacterial activity against Xanthomonas oryzae pv. oryzae. Artif Cells Nanomed Biotechnol 47(1):341–352

Ogunyemi SO, Zhang F, Abdallah Y, Zhang M, Wang Y, Sun G, Qiu W, Li B (2019a) Biosynthesis and characterization of magnesium oxide and manganese dioxide nanoparticles using

Matricaria chamomilla L. extract and its inhibitory effect on *Acidovorax oryzae* strain RS-2. Artif Cells Nanomed Biotechnol 47:2230–2239

Ogunyemi SO, Zhang M, Abdallah Y, Ahmed T, Qiu W, Ali MA, Yan C, Yang Y, Chen J, Li B (2020) The bio-synthesis of three metal oxide nanoparticles (ZnO, MnO_2, and MgO) and their antibacterial activity against the bacterial leaf blight pathogen. Front Microbiol 11:588326

Omodior O, Luetke MC, Nelson EJ (2018) Mosquito-borne infectious disease, risk-perceptions, and personal protective behavior among U.S. international travelers. Prev Med Rep 12:336–342

Paknikar KM, Nagpal V, Pethkar AV, Rajwade JM (2005) Degradation of lindane from aqueous solutions using iron sulfide nanoparticles stabilized by bio-polymers. Sci Technol Adv Mater 6: 370–374

Pan Z, Lin Y, Sarkar B, Owens G, Chen Z (2020) Green synthesis of iron nanoparticles using red peanut skin extract: synthesis mechanism, characterization and effect of conditions on chromium removal. J Colloid Interface Sci 558:106–114

Pandey S, Mewada A, Thakur M, Pillai S, Dharmatti R, Phadke C, Sharon M (2014) Synthesis of mesoporous silica oxide/C-dot complex (meso-SiO 2/C-dots) using pyrolysed rice husk and its application in bioimaging. RSC Adv 4(3):1174–1179

Panneerselvam C, Murugan K, Roni M, Aziz AT, Suresh U, Rajaganesh R, Madhiyazhagan P, Subramaniam J, Dinesh D, Nicoletti M, Higuchi A, Alarfaj AA, Munusamy MA, Kumar S, Desneux N, Benelli G (2016) Fern-synthesized nanoparticles in the fight against malaria: LC/MS analysis of *Pteridium aquilinum* leaf extract and biosynthesis of silver nanoparticles with high mosquitocidal and antiplasmodial activity. Parasitol Res 115(3):997–1013

Patil S, Chandrasekaran R (2020) Biogenic nanoparticles: a comprehensive perspective in synthesis, characterization, application and its challenges. J Genet Eng Biotechnol 18(1):67

Patra JK, Das G, Fraceto LF, Campos EVR, Rodriguez-Torres MDP, Acosta-Torres LS, Diaz-Torres LA, Grillo R, Swamy MK, Sharma S, Habtemariam S, Shin HS (2018) Nano based drug delivery systems: recent developments and future prospects. J Nanobiotechnol 16(1):71

Paulkumar K, Gnanajobitha G, Vanaja M, Rajeshkumar S, Malarkodi C, Pandian K, Annadurai G (2014) Piper nigrum leaf and stem assisted green synthesis of silver nanoparticles and evaluation of its antibacterial activity against agricultural plant pathogens. Sci World J 2014:829894

Pavoni L, Pavela R, Cespi M, Bonacucina G, Maggi F, Zeni V, Canale A, Lucchi A, Bruschi F, Benelli G (2019) Green micro- and nanoemulsions for managing parasites, vectors and pests. Nanomaterials 9(9):1285

Pavunraj M, Baskar K, Duraipandiyan V, Al-Dhabi NA, Rajendran V, Benelli G (2017) Toxicity of Ag nanoparticles synthesized using stearic acid from Catharanthus roseus leaf extract against Earias vittella and mosquito vectors (Culex quinquefasciatus and Aedes aegypti). J Clust Sci 28: 2477–2492

Pereira L, Alves M (2012) Dyes- environmental impact and remediation. In: Malik A, Grohmann E (eds) Environmental protection strategies for sustainable development, strategies for sustainability. Springer, New York, pp 111–162

Pooja D, Panyaram S, Kulhari H, Reddy B, Rachamalla SS, Sistla R (2015) Natural polysac-charide functionalized gold nanoparticles as biocompatible drug delivery carrier. Int J Biol Macromol 80:48–56

Prabha S, Durgalakshmi D, Rajendran S, Lichtfouse E (2021) Plant-derived silica nanoparticles and composites for biosensors, bioimaging, drug delivery and supercapacitors: a review. Environ Chem Lett 19:1667–1691

Prasad C, Karlapudi S, Venkateswarlu P, Bahadur I, Kumar S (2017c) Green arbitrated synthesis of Fe3O4 magnetic nanoparticles with nanorod structure from pomegranate leaves and Congo red dye degradation studies for water treatment. Mol Liq 240:322–328

Prasad C, Krishna Murthy P, Hari Krishna RH, Sreenivasa Rao R, Suneetha V, Venkateswarlu P (2017d) Bio-inspired green synthesis of RGO/Fe3O4 magnetic nanoparticles using Murraya koenigii leaves extract and its application for removal of Pb (II) from aqueous solution. J Environ Chem Eng 5(5):4374–4380

Prasad C, Sreenivasulu K, Gangadhara S, Venkateswarlu P (2017a) Bio inspired green synthesis of Ni/Fe3O4 magnetic nanoparticles using Moringa oleifera leaves extract: a magnetically recoverable catalyst for organic dye degradation in aqueous solution. J Alloys Compd 700:252–258

Prasad C, Yuvaraja G, Venkateswarlu P (2017b) Biogenic synthesis of Fe3O4 magnetic nanoparticles using Pisum sativum peels extract and its effect on magnetic and Methyl orange dye degradation studies. J Magn Magn Mater 424:376–381

Prasad KS, Gandhi P, Selvaraj K (2014) Synthesis of green nano iron particles (GnIP) and their application in adsorptive removal of As(III) and As(V) from aqueous solution. Appl Surf Sci 317:1052–1059

Prasad R, Bhattacharyya A, Nguyen QD (2017) Nanotechnology in sustainable agriculture: recent developments, challenges, and perspectives. Front Microbiol 8:1014

Raduw GG, Mohammed AA (2020) Insecticidal efficacy of three nanoparticles for the control of khapra beetle (*Trogoderma granarium*) on different grains. J Agric Urban Entomol 36(1):90–100

Rahman MA, Parvin A, Khan MSH, Lingaraju K, Prasad R, Das S, Hussain B, War AR, Bhattacharyya A (2020) Efficacy of the green synthesized nickel-oxide nanoparticles against pulse beetle, *Callosobruchus maculatus* (F.) in black gram (*Vigna mungo* L.). Int J Pest Manage. https://doi.org/10.1080/09670874.2020.1773572

Rajaganesh R, Murugan K, Panneerselvam C, Jayashanthini S, Aziz AT, Roni M, Suresh U, Trivedi S, Rehman H, Higuchi A, Nicoletti M, Benelli G (2016) Fern-synthesized silver nanocrystals: towards a new class of mosquito oviposition deterrents? Res Vet Sci 109:40–51

Rajakumar G, Thiruvengadam M, Mydhili G, Gomathi T, Chung IM (2018) Green approach for synthesis of zinc oxide nanoparticles from Andrographis paniculata leaf extract and evaluation of their antioxidant, anti-diabetic, and anti-inflammatory activities. Bioprocess Biosyst Eng 41(2018):21–30

Rani M, Shanker U, Jassal V (2017) Recent strategies for removal and degradation of persistent & toxic organochlorine pesticides using nanoparticles: A review. J Environ Manage 190:208–222

Rani P, Kumar V, Singh PP, Matharu AS, Zhang W, Kim KH, Singh J, Rawat M (2020) Highly stable AgNPs prepared via a novel green approach for catalytic and photocatalytic removal of biological and non-biological pollutants. Environ Int 143:105924

Raveendran S, Girija AR, Balasubramanian S, Ukai T, Yoshida Y, Maekawa T et al (2014) Green approach for augmenting biocompatibility to quantum dots by extremo-philic polysaccharide conjugation and nontoxic bioimaging. ACS Sustain Chem Eng 2:1551–1558

Rokade SS, Joshi KA, Mahajan K, Patil S, Tomar G, Dubal DS, Parihar VS, Kitture R, Bellare JR (2018) Gloriosa superba mediated synthesis of platinum and palladium nanoparticles for induction of apoptosis in breast cancer citation Gloriosa superba mediated synthesis of platinum and palladium nanoparticles for induction of apoptosis in breast cancer. Bioinorg Chem Appl 2018:4924186

Rokade SS, Joshi KA, Mahajan K, Tomar G, Dubal DS, Singh V, Kitture R, Bellare J, Ghosh S (2017) Novel anticancer platinum and palladium nanoparticles from Barleria prionitis. Glob J Nano 2(5):555600

Roni M, Murugan K, Panneerselvam C, Subramaniam J, Nicoletti M, Madhiyazhagan P, Dinesh D, Suresh U, Khater HF, Wei H, Canale A, Alarfaj AA, Munusamy MA, Higuchi A, Benelli G (2015) Characterization and biotoxicity of Hypnea musciformis-synthesized silver nanoparticles as potential eco-friendly control tool against Aedes aegypti and Plutella xylostella. Ecotoxicol Environ Saf 121:31–38

Rueda LM (2008) Global diversity of mosquitoes (Insecta: Diptera: Culicidae) in freshwater. Dev Hydrobiol 595:477–487

Sabbour M, Shadia E (2015) Efficacy of Nano-diatomaceous earth against red flour beetle, *Tribolium castaneum* and confused flour beetle, *Tribolium confusum* (Coleoptera, Tenebrionidae) under laboratory and storage conditions. Bull Environ Pharmacol Life Sci 4(7):54–59

Sahab AF, Waly AI, Sabbour MM, Nawar LS (2015) Synthesis, antifungal and insecticidal potential of Chitosan (CS)-g-poly (acrylic acid) (PAA) nanoparticles against some seed borne fungi and insects of soybean. Int J Chem Tech Res 8:589–598

Sahaya P, Kumar M, Francis AP, Devasena T (2014) Biosynthesized and chemically synthesized titania nanoparticles: comparative analysis of antibacterial activity. J Environ Nanotechnol 3: 73–81

Salama DM, Abd El-Aziz ME, Rizk FA, Abd Elwahed MSA (2021) Applications of nanotechnology on vegetable crops. Chemosphere 266:129026. https://doi.org/10.1016/j.chemosphere.2020.129026

Sankar R, Baskaran A, Shivashangari KS, Ravikumar V (2015) Inhibition of pathogenic bacterial growth on excision wound by green synthesized copper oxide nanoparticles leads to accelerated wound healing activity in Wistar Albino rats. J Mater Sci Mater Med 26(7):214

Santhosh SB, Yuvarajan R, Natarajan D (2015) Annona muricata leaf extract-mediated silver nanoparticles synthesis and its larvicidal potential against dengue, malaria and filariasis vector. Parasitol Res 114(8):3087–3096

Sedighi A, Imani S, Moshtaghi Kashanian GR, Najafi H, Fathipour Y (2019) Efficiency of green synthesized silver nanoparticles with sweet orange, Citrus sinensis (L.) (Rutaceae, Sapindales) against Tribolium confusum Duval. (Coleoptera, Tenebrionidae). JAST 21(6):1485–1494

Singh P, Ahn S, Kang JP, Veronika S, Huo Y, Singh H, Chokkaligam M, El-Agamy Farh M, Aceituno VC, Kim YJ, Yang DC (2018) In vitro anti-inflammatory activity of spherical silver nanoparticles and monodisperse hexagonal gold nanoparticles by fruit extract of Prunus serrulata: a green synthetic approach. Artif Cells, Nanomed Biotechnol 46(2018):2022–2032

Song X, Qiao L, Yan S, Chen Y, Dou X, Xu C (2021) Preparation, characterization, and *in vivo* evaluation of anti-inflammatory activities of selenium nanoparticles synthesized by *Kluyveromyces lactis* GG799. Food Funct. https://doi.org/10.1039/d1fo01019k

Sonker AS, Pathak J, Vinod K, Sinha RP (2017) Characterization and in vitro antitumor, antibacterial and antifungal activities of green synthesized silver nanoparticles using cell extract of Nostoc sp. strain. Can J Biotech 1:26–37

Stan M, Lung I, Soran M, Leostean C, Popa A, Stefan M, Lazar MD, Opris O, Silipas T-D, Porav AS (2017) Removal of antibiotics from aqueous solutions by green synthesized magnetite nanoparticles with selected agro-wasteextracts. Process Saf Environ Prot 107:357–372

Subramaniam J, Murugan K, Jebanesan A, Pontheckan P, Dinesh D, Nicoletti M, Wei H, Higuchi A, Kumar S, Canale A, Benelli G (2017) Do *Chenopodium ambrosioides*-synthesized silver nanoparticles impact Oryzias melastigma predation against Aedes albopictus larvae? J Clust Sci 28:413–436

Sukumar K, Perich MJ, Booba LR (1991) Botanical derivatives in mosquito control: a review. J Am Mosq Control Assoc 7:210–237

Sundrarajan M, Bama K, Bhavani M, Jegatheeswaran S, Ambika S, Sangili A, Nithya P, Sumathi R (2017) Obtaining titanium dioxide nanoparticles with spherical shape and antimicrobial properties using M. citrifolia leaves extract by hydrothermal method. J Photochem Photobiol B Biol 171(2017):117–124

Suresh J, Pradheesh G, Alexramani V (2018) Green synthesis and characterization of zinc oxide nanoparticle using insulin plant (Costus pictus D. Don) and investigation of its antimicrobial as well as anticancer activities.Advances in Natural Sciences: Nanoscience and. Nanotechnology 2018:9

Suresh U, Murugan K, Benelli G, Nicoletti M, Barnard DR, Panneerselvam C, Mahesh Kumar P, Subramaniam J, Dinesh D, Chandramohan B (2015) Tackling the growing threat of dengue: Phyllanthus niruri-mediated synthesis of silver nanoparticles and their mosquitocidal properties against the dengue vector Aedes aegypti (Diptera: Culicidae). Parasitol Res 114:1551–1562

Tahir K, Nazir S, Ahmad A, Li B, Ullah A, Ul Z, Khan H, Ullah F, Ullah Q, Khan A (2017) Ur A (2017) Facile and green synthesis of phytochemicals capped platinum nanoparticles and in vitro their superior antibacterial activity. JPB 166:246–251

Teodoro S, Micaela B, David KW (2010) Novel use of nano-structured alumina as an insecticide. Pest Manag Sci 66(6):577–579

Thandapani K, Kathiravan M, Namasivayam E et al (2017) Enhanced larvicidal, antibacterial, and photocatalytic efficacy of TiO_2 nanohybrids green synthesized using the aqueous leaf extract of *Parthenium hysterophorus*. Environ Sci Pollut Res 25:1–12

Thiruvengadam M, Chung I-M, Gomathi T, Ansari MA, Gopiesh Khanna V, Babu V, Rajakumar G (2019) Synthesis, characterization and pharmacological potential of green synthesized copper nanoparticles. Bioprocess Biosyst Eng 42:1769–1777

Tian H, Li J, Mu Z, Li L, Hao Z (2009) Effect of pH on DDT degradation in aqueous solution using bimetallic Ni/Fe nanoparticles. Sep Purif Technol 66:84–89

Umar H, Kavaz D, Rizaner N (2019) Biosynthesis of zinc oxide nanoparticles using albizia lebbeck stem bark, and evaluation of its antimicrobial, antioxidant, and cytotoxic activities on human breast cancer cell lines. Int J Nanomedicine 14:87–100

Vadlapudi V, Amanchy R (2017) Phytofabrication of silver nanoparticles using Myriostachya wightiana as a novel bioresource, and evaluation of their biological activities. Braz Arch Technol 60:e17160329

Vani C, Brindhaa U (2013) Silica nanoparticles as nanocides against *Corcyra cephalonica* (S.), the stored grain pest. Int J Pharm Bio Sci 4:1108–1118

Vanti GL, Kurjogi M, Basavesha KN, Teradal NL, Masaphy S, Nargund VB (2020) Synthesis and antibacterial activity of solanum torvum mediated silver nanoparticle against Xanthomonas axonopodis pv.punicae and Ralstonia solanacearum. J Biotechnol 309:20–28

Vasantharaj S, Sathiyavimal S, Saravanan M, Senthilkumar P, Gnanasekaran K, Shanmugavel M, Manikandan E, Pugazhendhi A (2019) Synthesis of ecofriendly copper oxide nanoparticles for fabrication over textile fabrics: Characterization of antibacterial activity and dye degradation potential. J Photochem Photobiol B 191:143–149

Veerakumar K, Govindarajan M (2014) Adulticidal properties of synthesized silver nanoparticles using leaf extracts of Feronia elephantum (Rutaceae) against filariasis, malaria, and dengue vector mosquitoes. Parasitol Res 113(11):4085–4096

Veerakumar K, Govindarajan M, Rajeswary M (2013) Green synthesis of silver nanoparticles using Sida acuta (Malvaceae) leaf extract against Culex quinquefasciatus, Anopheles stephensi, and Aedes aegypti (Diptera: Culicidae). Parasitol Res 112(12):4073–4085

Veerakumar K, Govindarajan M, Rajeswary M, Muthukumaran U (2014) Mosquito larvicidal properties of silver nanoparticles synthesized using Heliotropium indicum (Boraginaceae) against Aedes aegypti, Anopheles stephensi, and Culex quinquefasciatus (Diptera: Culicidae). Parasitol Res 113(6):2363–2373

Velmurugan P, Lee SM, Iydroose M, Lee KJ, Oh BT (2013) Pine cone-mediated green synthesis of silver nanoparticles and their antibacterial activity against agricultural pathogens. Appl Microbiol Biotechnol 97(1):361–368

Velsankar K, Sudhakar S, Maheshwaran G, Krishna KM (2019) Effect of biosynthesis of ZnO nanoparticles via Cucurbita seed extract on Culex tritaeniorhynchus mosquito larvae with its biological applications. J Photochem Photobiol B 200:111650

Vijaya Sankar M, Abideen S (2015) Pesticidal effect of Green synthesized silver and lead nanoparticles using *Avicennia marina* against grain storage pest *Sitophilus oryzae*. Int J Nanomater Bios 5(3):32–39

W.H.O (2017) Vector-borne diseases factsheet. World Health Organization, Geneva

Wang Z, Fang C, Megharaj M (2014) Characterization of iron–polyphenol nanoparticles synthesized by three plant extracts and their fenton oxidation of azo dye. ACS Sustain Chem Eng 2(4):1022–1025

Waris M, Nasir S, Rasule A, Yousaf I (2020) Evaluation of larvicidal efficacy of Ricinus communis (Castor) plant extract and synthesized green silver nanoparticles against Aedes albopictus. J Arthropod Borne Dis 14(2):162–172

Wazid SN, Prabhuraj A, Harishchanra NR, Shakuntala NM, Sharanagouda H (2018) Effect of biosynthesized zinc oxide green nanoparticles on pulse beetle, Callosobruchus analis (Coleoptera: Chrysomelidae). Int J Curr Microbiol App Sci 7(09):503–512

WHO (2014) A global brief on vector-borne diseases. World Health Organization, Geneva

Xiao C, Li H, Zhao Y, Zhang X, Wang X (2020) Green synthesis of iron nanoparticle by tea extract (polyphenols) and its selective removal of cationic dyes. J Environ Manage 275:111262

Yang J, Hou B, Wang J, Tian B, Bi J, Wang N, Li X, Huang X (2019) Nanomaterials for the removal of heavy metals from wastewater. Nanomaterials 9(3):424

Yi Y, Tu G, Tsang PE, Xiao S, Fang Z (2019) Green synthesis of iron-based nanoparticles from extracts of Nephrolepis auriculata and applications for Cr(VI) removal. Mater Lett 2019(234):388–391

Yugandhar P, Vasavi T, Uma P, Devi M (2017) Bioinspired green synthesis of copper oxide nanoparticles from Syzygium alternifolium (Wt.) Walp: characterization and evaluation of its synergistic antimicrobial and anticancer activity. Appl Nanosci 7(2017):417–427

Zhang P (2018) Synthesis of gold nanoparticles with Solanum xanthocarpum extract and their in vitro anticancer potential on nasopharyngeal carcinoma cells. Int J Nanomedicine 13(2018):7047–7059

Zhu C, Zhang S, Song C, Zhang Y, Ling Q, Hoffmann PR, Li J, Chen T, Zheng W, Huang Z (2017) Selenium nanoparticles decorated with Ulva lactuca polysaccharide potentially attenuate colitis by inhibiting F-B mediated hyper inflammation. J Nanobiotechnol 15(2017):1–15

Toxicity Aspects of Biologically Synthesized Nanoparticles

15

Abstract

The surge in nanobiotechnology has given rise to successful utilization of nanoparticles in several novel applications in agriculture, industry, and the biomedical field. However, nanoparticles have a more pronounced effect in systems with respect to inflammation and cell damage as compared to an equal mass of particles having greater size. They are known to induce pulmonary, cutaneous, cardiovascular, and xenotoxicity. The engineered nanoparticles, owing to their high reactivity, have the capacity to initiate inflammatory responses that act as a trigger for metastasis. Huge knowledge gaps exist with respect to the nature of nanoparticles and their interaction with different environmental components. Apart from beneficial reports, there have been numerous studies implicating nanoparticles in plant toxicity through chromosomal abnormalities, oxidative damages to biological membranes, decreased photosynthetic rate, growth retardation, reduction in seed germination, biomass, leaf number, root elongation, and even plant death.

15.1 Introduction

The distribution of nanoparticles in the environment depends on their ability to make metastable aqueous suspensions or aerosols in the environmental fluids. Thus, the stability of nanomaterials can be assessed on their ability to aggregate and/or interact with the surrounding environments. Aggregation is considered as a time-dependent phenomenon that has significant association with the rate of particle collision. Similarly, the stability of the suspension depends on the nanoparticle dimensions and their affinity toward the environmental components. One important attribute to consider while assessing the toxicity of nanostructures is their transformation. One glaring example is the sulfurization of silver nanoparticles occurring in both urban

S. Srivastava, A. Bhargava, *Green Nanoparticles: The Future of Nanobiotechnology*,
https://doi.org/10.1007/978-981-16-7106-7_15

sewage systems and natural surroundings rich in sulfur. Sulfurization leads to the creation of silver sulfide nanoparticles which greatly reduce their toxicity primarily due to the low solubility of silver sulfide (Ag_2S). Apart from this, the toxicity of nanoparticles also depends on their size, shape, conformation, and composition as well as on the intrinsic toxic properties of compounds contained in the nanoparticle.

15.2 Health Risks

Recent advancements in nanobiotechnology have given rise to the successful utilization of nanoparticles in several novel applications in agriculture, industry, and the biomedical field. However, detailed studies concentrated around the safety aspect of nanomaterials have not kept pace with their usage. It has been observed that nanoparticles have a more pronounced effect in animal systems with respect to inflammation and cell damage as compared to an equal mass of particles having greater size. Some of the human health risks associated with the use of nanoparticles are provided below:

15.2.1 Pulmonary and Respiratory Effects

The lungs and the respiratory system represent a unique target for study of the toxicological parameters of nanoparticles. This is due to the fact that apart from being the entry point for inhaled particles, it is also an important connecting link to the circulatory system. Therefore, there are greater chances of the lungs being exposed to countless number of nanomaterials by incidental/accidental or intentional means. Once inside the lungs, the nanoparticles become widely distributed with the lungs becoming an important site for their accumulation. After inhalation, the nanoparticles are carried by electrostatic force to the respiratory tract and exhibit different fractional depositions within the respiratory tract based on the particle's aerodynamic or thermodynamic diameter as shown below:

Ultrafine particles ($<$100 nm): deposition in almost all regions
Particles $<$10 nm in size: highest deposition in the tracheobronchial region
Particles approximately 10–20 nm in size: highest in alveolar region
Particles $<$20 nm in size: high deposition in the nasopharyngeal-laryngeal region

In spite of such rapid exposure and deposition, toxicological studies with respect to the adverse effects of different nanoparticles on the pulmonary system are rather limited. Sporadic studies have been carried out with respect to the deposition patterns of inhaled nanoparticles in healthy and diseased lungs. Lung disorders arising from nanoparticle exposure include pulmonary fibrosis, pneumoconiosis, and exacerbation of asthma. Pulmonary fibrosis is a condition that causes lung scarring and stiffness and occurs as a result of accumulation of fibrous connective tissue due to increased tissue reactivity. Among the nanoparticles, multiwalled carbon nanotubes (MWCNTs) have been implicated in fibrosis, scarring, inflammation, and mesothelioma formation. Pneumoconiosis refers to a class of interstitial lung diseases caused by the inhalation of organic or nonorganic airborne dust and fibers. Different nanoparticles varying in size and concentration influence the disease progression. The particulate clouds are mineral and combustion-derived, and nanoparticles found in such clouds have been implicated in fibrogenicity. Nanoparticles like carbon rods behave like asbestos in vivo due to the virtue of having similar shape, size, and properties. Studies have shown that the asbestos-like pathogenic behavior of MWNTs in mice leads to inflammation and formation of granulomas. Intratracheal instillation of SWNTs on rats and mice causes pulmonary inflammation, formation of collagen-rich granulomas, and fibrosis. Intrapharyngeal aspiration of purified single-walled carbon nanotubes on mice leads to acute inflammation, early onset of formation of granulomas, and progressive fibrosis. Individuals having asthma, an inflammatory disease of the airways to the lungs, are more prone to nanoparticle-induced lung toxicity. In the asthmatic patients, obstruction of the airway causes air trapping, increased alveolar volume and a resultant increase in the deposition of ultra-fine particles. About 74% more particles have been reported in the respiratory system of asthmatic patients in comparison to healthy individuals. Therefore, greater deposition of nanoparticles in the pulmonary system would worsen airway inflammation in asthmatic subjects. It has been observed that carbon nanotubes, nickel and titanium dioxide nanoparticles exacerbate asthma, while gold nanoparticles have the property to attenuate asthma.

15.2.2 Cutaneous Toxicity

Skin, the largest and the outermost organ of human body, protects the underlying internal organs against external interferences. Four pathways through which extraneous materials gain access in the human body by crossing the skin barrier are intercellular, transcellular, through hair follicles and sweat glands. Humans are exposed to nanoparticles through the dermal route since nanoparticles are a component of adhesives, polishes, varnishes, photocopier toner, food stabilizing agents and cosmetic formulations. For example titanium and zinc oxides form an important component of many sunscreen lotions and creams for protection against UV-induced skin damage. Damaged and weak skin is a good entry point for both finer and larger particles. It has been observed that small (<600 Da) lipophilic molecules can gain easy access to the skin through passive means. Penetration of nanomaterials into healthy skin entails the generation of free radical, oxidative stress and collagen depletion. Quantum dots are known to penetrate skin through abrasions which is of significance for people engaged in the synthesis of quantum dots and for researchers studying their biomedical applications. However, reliable information on passage of nanoparticles through the stratum corneum or through the circulatory system, and their accumulation in the dermal tissue is lacking.

Nanoparticles in cosmetics have caused significant health-related risks. Most of the nanoparticles stimulate the immune system, and result in inflammatory response via. Abnormal secretion of different cytokines and other chemicals. Inflammation leads to differential cutaneous physiology and impairment of the skin's barrier function. However, most of the studies carried out on skin absorption and toxicity of nanomaterials have provided conflicting results. In fact the European project NANODERM where several research labs were participants also provided inconclusive results with some studies suggesting no adverse health effects of sunscreen lotions containing titanium dioxide nanoparticles on healthy skin, while others confirming an association between dermal cells and nanoparticles leading to illeffects.

15.2.3 Cardiovascular Toxicity

The detrimental effect of nanostructures to the cardiovascular system have been increasingly researched in recent years (Table 15.1). This is due to two reasons:

1. The small size of nanoparticles enables them to cross several physiological barriers and enter the blood stream. This assumes more importance for people who are exposed to nanoparticles during production and use.
2. The circulatory system is required for delivering the nanoparticles to their targets, which makes it imperative to carefully assess their effects for ensuring their safe use in biomedicine.

The harmful effects of some commonly used nanoparticles on the cardiovascular system and the underlying mechanisms have been provided in Fig. 15.1. Some studies have also indicated towards cardiomyocyte injury and impairment of their physiological functions on nanoparticle exposure.

Several mechanisms have been proposed with respect to the toxic effects of nanoparticles towards the circulatory system. Some of these have been provided below:

1. Oxidative stress: Nanoparticles are highly reactive and have been reported to aid on the formation of reactive oxygen species that induces oxidative damage in various biomolecules associated with the circulatory system. Additionally, an oxidative milieu due to nanoparticle exposure could also lead to endothelial NO synthase (eNOS) uncoupling, which in turn generate oxidants rather than nitric oxide (NO).
2. Modulation of inflammation: It is believed that nanoparticles, especially metal ones, are crucial in modulating the immune function and altering inflammatory responses. Since strong correlation exists between cardiovascular diseases and inflammatory responses, it has been speculated that metal nanoparticles might induce adverse cardiovascular effects via modulation of inflammatory responses. This is supported by studies using animal models wherein exposure to metal-based nanoparticles bring about inflammatory damage to the cardiovascular system.

Table 15.1 Cardiovascular toxicity of ZnO, Ag, and TiO_2NPs as assessed by in vivo models (Reprinted with permission from Cao et al. 2018)

NPs	Physicochemical properties	Exposure	Models	Cardiovascular toxicity	References
TiO_2	Anatase, 5 nm	IP injection of 5, 10, 50,100, and 150 mg/kg daily for 14 days	Female CD-1 mice	Damage to myocardium, increased serum glucose and lipids	Liu et al. (2009)
TiO_2	Fine TiO_2,~1 μm; nano TiO_2, 21 nm	Inhalational exposure to 1.5–16 mg/m^3 particles for 240–720 min	Male SD rats	Microvascular dysfunction, ROS and decreased NO	Nurkiewicz et al. (2009)
Ag	5–35 nm, PVA coated	10, 25, 50, 75, 100 μg/mL	Zebrafish embryos	Decreased heart rate, abnormal cardiac morphology, circulatory defects	Asharani et al. (2011)
TiO_2	Rutile fine TiO_2, 288 nm; photocatalytic 92/8 anatase/ rutile TiO_2, 12 nm; rutile nano TiO_2, 21.6 nm	IT instillation of 0.5 mg/kg, once a week for 4 weeks	ApoE−/−mice	Plaque progression in aorta	Mikkelsen et al. (2011)
ZnO	50 nm	Oral exposure to 600 mg/kg, 1 g/ kg for 5 days	Wistar albino rats	Increased serum inflammatory cytokines; DNA damage, caspase-3 activation, increased calcium concentration in hearts	Baky et al. (2013)
TiO_2	Anatase, 5–10 nm	IT instillation of 10, 50,100 μg, once a week for 6 weeks	ApoE−/−mice	Systemic inflammation, endothelial dysfunction, lipid metabolism dysfunction, increased plaque area and lipid-rich core area	Chen et al. (2013)
Ag	33 or 39 nm mean aerodynamic diameter	Inhalational exposure to wet ag colloid for 5 h (estimated lung burdens 1.4, 14 μg ag/rat)	Rats	Increased blood monocytes, decreased dilation of tail artery, elevated heart rate	Roberts et al. (2013)
TiO_2	Nano TiO_2, rodlike, 15 x 60 nm; micro,1–5 μm	IP injection of 0.5, 5, 50 mg/kg	SD rats	Decreased heart function, pathological changes, and oxidative damage in the heart tissue	Sha et al. (2013)

(continued)

Table 15.1 (continued)

NPs	Physicochemical properties	Exposure	Models	Cardiovascular toxicity	References
Ag	20–30, 50–60 nm	50 mg/kg, IP injection once daily for 1 week	Male inbred SD rats (young and old)	Disruption of BBB and BCSFB, brain damage (old > young)	Sharma et al. (2013)
TiO_2	P25 (80% anatase, 20% rutile, 21 nm)	Inhalational exposure (daily deposition 43.8 ± 1.2 μg) for 5 h per day, average 6.8 ± 0.5 days	Pregnant SD rats	Endothelial dysfunction in coronary and uterine arterioles, reduced maximal mitochondrial respiration in the left ventricle and uterus	Stapleton et al. (2015)
ZnO	100 nm	Oropharyngeal aspiration, once a week for 5 weeks, cumulative dose 32, 64 μg	Bmal1−/−mice	Increased coagulation factor VII	Luyts et al. (2014)
Ag	Commercial ag product (10 ppm, 5–10 nm; 32 ppm, 25–40 nm)	Oral exposure, 10 ppm for 3, 7, 14 days; 32 ppm for 14 day	Human volunteers	No significant effect in serum inflammation, oxidative stress, heart beat, or morphological changes	Munger et al. (2014)
TiO_2	Rutile, 1–3 nm (ultrasmall NPs); 30 nm (NPs)	100–1000 mg/L in water or 1 ng/embryo	Zebrafish embryos	Mortality, inhibited hatching, developmental abnormalities, increased Myo1c expression (ultrasmall NPs only)	Bayat et al. (2015)
TiO_2	24 ± 5 nm	Oral exposure of 2, 10, 50 mg/kg for 30 and 90 days	SD rats	Reduced heart rate and systolic blood pressure, increased diastolic blood pressure, injury of cardiac function, serum inflammation	Chen et al. (2015)
TiO_2 ZnO Ag	ZnO 150 nm; ag 15 nm; TiO_2 anatase 38 nm, Rutile 67 nm	Injection of 1 mg/kg in left femoral artery	C57BL/6 mice	Thrombus formation (TiO_2 anatase only)	Haberl et al. (2015)
Ag	20 nm, citrate coated	IT instillation of 0.2 mg	Male SD rat	Increased serum cytokines, cardiac ischemia-reperfusion	Holland et al. (2015)

				injury, impaired vascular responsiveness	
TiO_2	Average size 20.6 ± 0.3 nm	IT instillation of 18 and 162 μg	C57BL/6 mice	Complement cascade, inflammation in the heart, complement factor 3 in the blood	Husain et al. (2015)
Ag	20–30, 50–60 nm	50 mg/kg, IP injection once daily for 1 week	Male inbred SD rats (sleep deprivation)	Exacerbate disruption of BBB and BCSFB, brain damage	Sharma et al. (2015a)
Ag	50–60 nm	50 mg/kg, IP injection once daily for 1 week	Male inbred SD rats (hot and cold environments, METH treatment)	Exacerbate disruption of BBB and BCSFB, brain damage	Sharma et al. (2015b)
Ag	50 nm, PVP coated	0.1, 1.0, 10 mg/L	Zebrafish embryos/larvae	Transient VEGF related gene expression, delayed vascular development, and smaller larvae size at hatching	Gao et al. (2016)
Ag	20 or 110 nm, citrate or PVP coated	IT instillation (200 μg for 1 day or 7 days)	Male SD rats	Increased serum cytokines, endothelial vasomotor dysfunction, cardiac ischemia reperfusion injury	Holland et al. (2016)
Ag	20 or 110 nm, PVP coated	IV injection, 1 mg/kg; gavage 10 mg/kg	Pregnant SD rat	No effect on cardiovascular injury	Fennell et al. (2017)
TiO_2	21 nm	Oral exposure to 150 mg/kg for 6 weeks	Male SD rats	Histological changes in the heart	Hassanein and El-Amir (2017)
Ag	75 nm, citrate coated	IV injection of 2, 3, 4, 6, and 8 mg/kg	Adult male C56BL6/J mice	Heart rate decrease, sinus bradycardia, complete atrioventricular conduction block, and cardiac asystole	Lin et al. (2017)

(continued)

Table 15.1 (continued)

NPs	Physicochemical properties	Exposure	Models	Cardiovascular toxicity	References
ZnO	30 nm	IT instillation (0, 1.25, 2.5, and 5.0 mg/kg) once a week, 12 weeks	Male Wistar rats	Dyslipidemia (increased TC, LDL; decreased HDL), systemic inflammation (increased serum IL-8, TNF-a), atherosclerotic changes (increased HO-1, PECAM-1; aortic EC damage)	Yan et al. (2017)
ZnO	Mobility diameter ~ 10 nm	0.5, 1.0, 2.0 mg/m^3 for 4 h	Healthy human volunteers	Increased serum acute phase proteins and neutrophils	Monse et al. (2018)

ApoE apolipoprotein E, *BBB* blood-brain barrier, *BCSFB* blood-cerebrospinal fluid barrier, *Bmal1* brain and muscle ARNT-likeprotein-1, *EC* endothelial cell, *HDL* high-density lipoprotein, *HO-1* heme oxygenase-1, *IL-8* interleukin-8, *IP* intraperitoneal, *IT* intratracheal, *IV* intravenous, *LDL* low-density lipoprotein, *METH* methamphetamine, *NO* nitric oxide, *NP* nanoparticle, *PECAM-1* platelet endothelial cell adhesion molecules-1, *PVA* polyvinyl alcohol, *PVP* polyvinylpyrrolidone, *SD* Sprague-Dawley, *TC* total cholesterol, *TNF-a* tumor necrosis factor-a, *VEGF* vascular endothelial growth factor

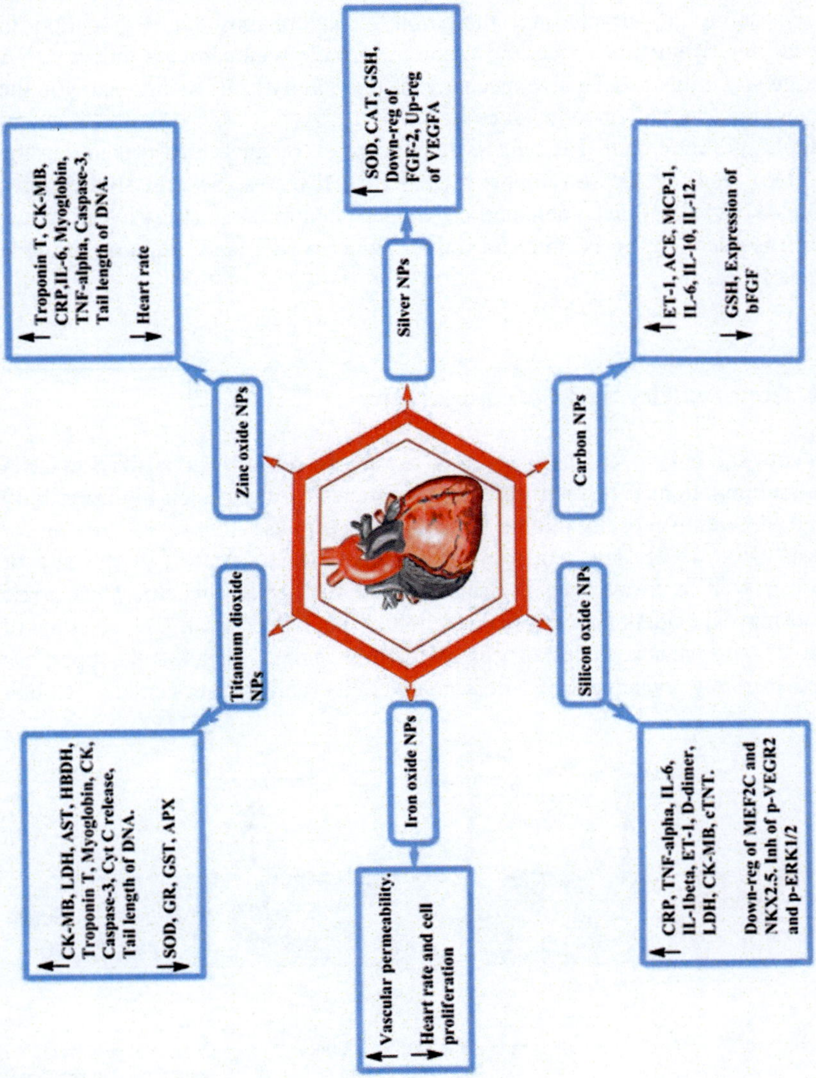

Fig. 15.1 Cardiovascular toxicity and the inherent mechanisms of six highly used nanoparticles (Reprinted with permission from Bostan et al. 2016)

3. Dysfunction of autophagy: The term 'autophagy', first coined by Christian de Duve, refers to a self-digesting mechanism in which selected endogenous and exogenous substances are sequestered through a lysosome-dependent regulated mechanism. Dysfunction of autophagy due to metal nanoparticles has been considered as one of the mechanisms for cardiovascular disorders. Nanoparticles either inhibit or overstimulate autophagy via alteration of lysosome functions i.e. lysosome impairment and inhibition of autophagosome degradation in hepatocytes. Both iron oxide and titanium dioxide nanoparticles induce DNA damage and initiates defensive mechanisms that finally leads to autophagy in the human brain-derived endothelial cells.
4. Endoplasmic reticulum (ER) stress: Accumulation of misfolded proteins in the ER often leads to the development of human diseases. Several studies have indicated the role of metal nanoparticle-induced endoplasmic stress in cardiovascular disorders. This is true for silver, zinc oxide, and titanium dioxide nanoparticles.

15.2.4 Genotoxicity and Carcinogenicity

Both in vitro and in vivo studies have hinted toward the role of nanoparticles in DNA damage and mutations. The engineered nanoparticles owing to their high reactivity and good adsorption capacity initiate inflammatory responses that act as a trigger for metastasis (Fig. 15.2). The tendency of nanoparticles to deposit at the site of injection as well as away from it is also a matter of great concern. Thus, toxic responses may be evident far away from the initial site of deposition. The potential of metal/metal oxide nanoparticles, especially titanium, cobalt, nickel, and copper, has been exhaustively evaluated for their genotoxicity and carcinogenicity. Cobalt

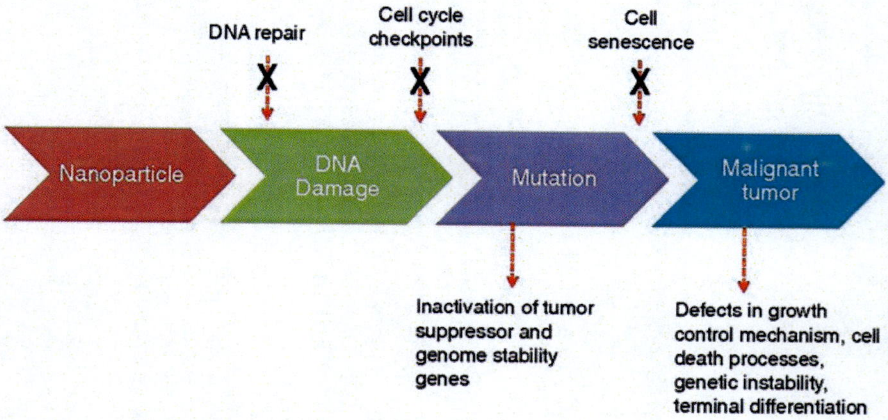

Fig. 15.2 Figure depicting the involvement of cellular processes in engineered nanoparticle-induced genotoxicity and carcinogenicity (Reprinted with permission from Kumar and Dhawan 2013)

nanoparticles lead to elevated cellular inflammatory response and induce breaks in nucleic acids, sister chromatid exchanges, aneuploidy, and interference with DNA repair processes. Likewise, cobalt-chrome nanoparticles are highly genotoxic and generate free radicals and induce DNA damage, deletions, tetraploidy, and DNA double-strand breaks. In this respect, tungsten carbide-cobalt (WC-Co) nanoparticles are known to induce more breaks in the nucleic acids than cobalt nanoparticles alone and are highly genotoxicity. Both nickel and nickel oxide nanoparticles possibly activate the transcription factor HIF-1α pathway that leads to cell alterations and carcinogenicity. Hodgkin's lymphoma is a typical condition associated with exposure to nickel nanoparticles. Copper nanoparticle-induced oxidative stress induces DNA damage and helps in tumor progression.

Titanium dioxide nanoparticles find extensive use in paint, paper, dye, pharmaceutical industries, and development of cosmetic products. Several researches have brought to light new evidence that titanium dioxide nanoparticles can cause inflammation, pulmonary damage, fibrosis, and lung cancer in rodents, besides showing carcinogenicity in human beings. Table 15.2 provides information on in vivo studies related to biological effects of TiO_2 NPs.

Some studies have suggested carbon nanoparticles result in lipid membrane peroxidation, gene downregulation, mutagenesis, DNA damage, and increased cell death.

15.3 Environmental Hazards

The recent surge in nanotechnology and its utilization in different sectors have introduced large quantities of nanomaterials into the environment that end up in the sediments, soil, and water. It has been estimated that landfills (~ 63–91%) and soils (~ 8–28%) receive a large chunk of the nanomaterials followed by aquatic bodies (7%) and the atmosphere (1.5%). Sometimes nanomaterials are also directly introduced in the environment when they are used as nanopesticides and for nanoremediation.

Three aspects are of prime importance while evaluating the impact of nanoparticles as pollutants:

1. Mobility: the movement of nanoparticles from one site to another.
2. Ecotoxicity: the possible harmful effects of nanoparticles toward organisms inhabiting different environments.
3. Modification: the extent to which nanoparticles are modified in the environment and how it affects their different parameters.

The interaction of nanomaterials with substances present in the environment may lead to reduction of enhancement of their toxicity and influences their fate and ecotoxicological potential. Therefore, some nanomaterials which are otherwise harmful can have less drastic effects on the environment. In fact sometimes they may even be beneficial for the environment. The organic and inorganic substances

Table 15.2 In vivo studies on genotoxicity of titanium dioxide nanoparticles (Reprinted with permission from Chen et al. 2014a, b)

Size and crystalline structure	Dose	Test system	Result	References
In vivo comet assay				
33 nm	40–1000 mg/kg	CBAB6F1 mice were gavaged daily for 7 d, and their bone marrow, liver, and brain were assayed	Positive in the bone marrow and liver; negative in the brain	Sycheva et al. (2011)
21 nm; 75% anatase and 25% rutile	500 mg/kg	Mice were exposed via drinking water for 5 d, and their blood cells were used for the assay	Positive	Trouiller et al. (2009)
80 nm; 74% anatase and 26% brookite	Up to 28.5 mg/m^3	Rats were exposed via inhalation for 5 d, 4 h/d, and their lung cells were assayed	Negative	Lindberg et al. (2012)
10 × 50 nm; rutile in T-lite	10 mg/kg	Rats were exposed by inhalation, and their lung cells were assayed	Negative	Landsiedel et al. (2010)
5 nm; anatase	Up to 5 mg/kg	Intratracheally exposure to male Sprague-Dawley rats, and the lung cells were assayed	Negative	Naya et al. (2012)
Micronucleus assay				
21 nm; 75% anatase and 25% rutile	500 mg/kg	Mice were exposed via drinking water for 5 d. the mouse blood cells were tested	Positive	Trouiller et al. (2009)
80 nm; 74% anatase and 26% brookite	Up to 28.5 mg/m^3	Rats were exposed via inhalation for 5 d, 4 h/d. the peripheral blood polychromatic erythrocytes were tested	Negative	Lindberg et al. (2012)
10 nm; anatase	Three daily dose of 50 mg/kg	Intravenously exposure to male B6C3F1 mice; the blood cells were assayed	Negative	Sadiq et al. (2012)
40 nm; anatase	Up to 1387 mg/kg	One dose intravenous injection to mice, and the blood cells were assayed	Negative	Xu et al. (2013)
Mutation assay				
2.3 nm; anatase	Fed, 0.1–10 mM	*Drosophila melanogaster*, the wing somatic mutation and recombination assay	Negative	Demir et al. (2013)
10 nm; anatase	Three daily dose of 50 mg/kg	Intravenously exposure to male B6C3F1 mice and *pig-a* mutation assay in blood cells were conducted	Negative	Sadiq et al. (2012)

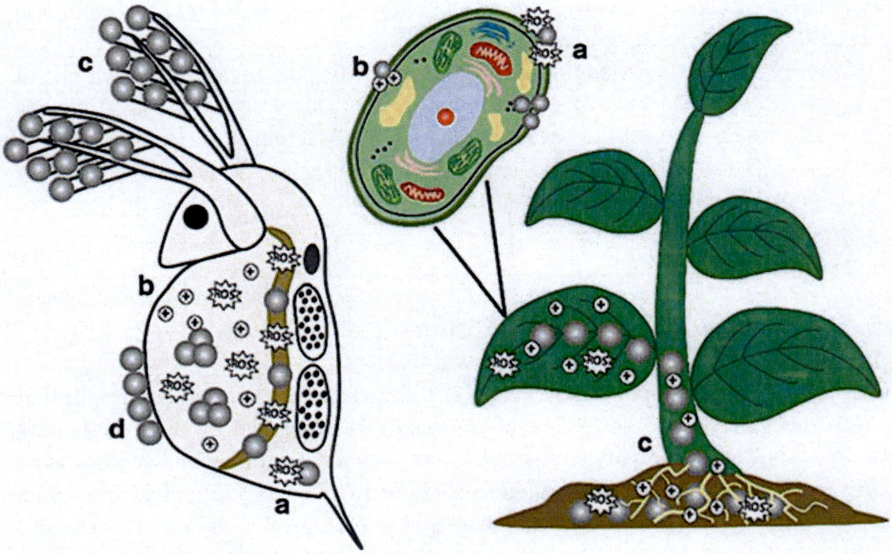

Fig. 15.3 Ecotoxicity of nanoparticles in aquatic and terrestrial regimes showing (**a**) formation of ROS, (**b**) ion release, (**c**) internalization, and (**d**) biological surface coating (Reprinted with permission from Bundschuh et al. 2018)

present in the biosphere may be absorbed or transformed by the nanomaterials, thereby reducing their concentration and reducing the toxic effects associated with them. If the nanoparticle-pollutant is not toxic, then no harmful effects may be seen. However, the reverse is also true in many cases. The toxicity of the herbicide (diuron) and insecticide bifenthrin enhances in the presence of carbon-based nanoparticles. Figure 15.3 depicts the toxicity of nanoparticles in aquatic and terrestrial ecosystems which is primarily due to the formation of ROS that have the capacity to affect biological structures. Silver nanoparticles show toxicity against soil microbial communities, snails, and periphytons. In terrestrial ecosystems, copper oxide nanoparticles (CuO NPs) reduce root length but enhance length of root hair in plants. In aquatic ecosystems, CuO NPs alter protein regulation and gene expression in organisms like marine mussels and zebrafish.

The antibacterial effect of nanomaterials though useful in medical science has a lot to discuss with respect to their effects on environmentally relevant bacterial species. Bacteria are important components of many aquatic and terrestrial food webs where they act as decomposers of organic matter. Therefore, any adverse effects on bacterial populations are likely to have considerable impact on higher tropic levels. Additionally, bacteria are important players in the movement of nanoparticles across the food chain due to their adsorptive or assimilative capacity vis-à-vis nanomaterials. Toxicity studies have shown highly toxic effects of silver and several metal oxide nanoparticles on bacteria (*E. coli*), yeast, and protozoan cells. Metal nanoparticles are also known to retard the growth of several species of phytoplanktons like *Skeletonema marinoi*, *Thalassiosira pseudonana*, *Dunaliella tertiolecta*, and *Isochrysis galbana*.

However, huge knowledge gaps exist with respect to the interaction of nanoparticles with the environmental components. This warrants detailed studies that would assess the stability of nanomaterials in different life-forms so that the threats to different organisms on exposure to nanomaterials can be effectively ascertained.

15.4 Toxicity Toward Plant Forms

Apart from beneficial reports, there have been numerous studies that have implicated nanoparticles in plant toxicity. The phytotoxicity is dependent on different factors like dose, size, and nature of nanoparticles, along with the duration and conditions of exposure. Nanoparticles mainly induce changes in plant physiological traits like chromosomal abnormalities, oxidative damages to biological membranes, decreased photosynthetic rate, growth retardation, reduction in seed germination, biomass, leaf number, root elongation, and even plant death (Fig. 15.4). It is believed that the nanoparticles interact with plant cells that lead to alteration in gene expression and biological pathways which negatively affects plant development. Titanium dioxide nanoparticles have a retarding influence on plant growth since they disrupt cell membrane, damage genomic DNA, and cause photocytotoxicity. Carbon nanotubes have also been reported to reduce shoot and root length, along with cell death and membrane damage in certain plants.

15.5 Ethical Issues

Ethical analysis posed to researchers, policy makers, and users of nanomaterials should be at the core of nanotechnology applications since it may appraise the society about the risks and health hazards associated with the use of nanotechnology. Researchers working on the ethical aspects of nanotechnology have been engrossed with issues as equity, privacy, security, and environmental impact, while issues at the workplace have conveniently taken a backseat. Social and economic research on nanotechnology can enable the manufacturers and policy makers in making judicious decisions when employing the novel technology and maximizing its benefit for human welfare. Apart from this, in-depth research keeping in mind the societal implications will provide the government and the common man with a realistic picture devoid of unwarranted fears. Table 15.3 lists the ethical issues with respect to the personnel engaged in nanotechnology research and industry.

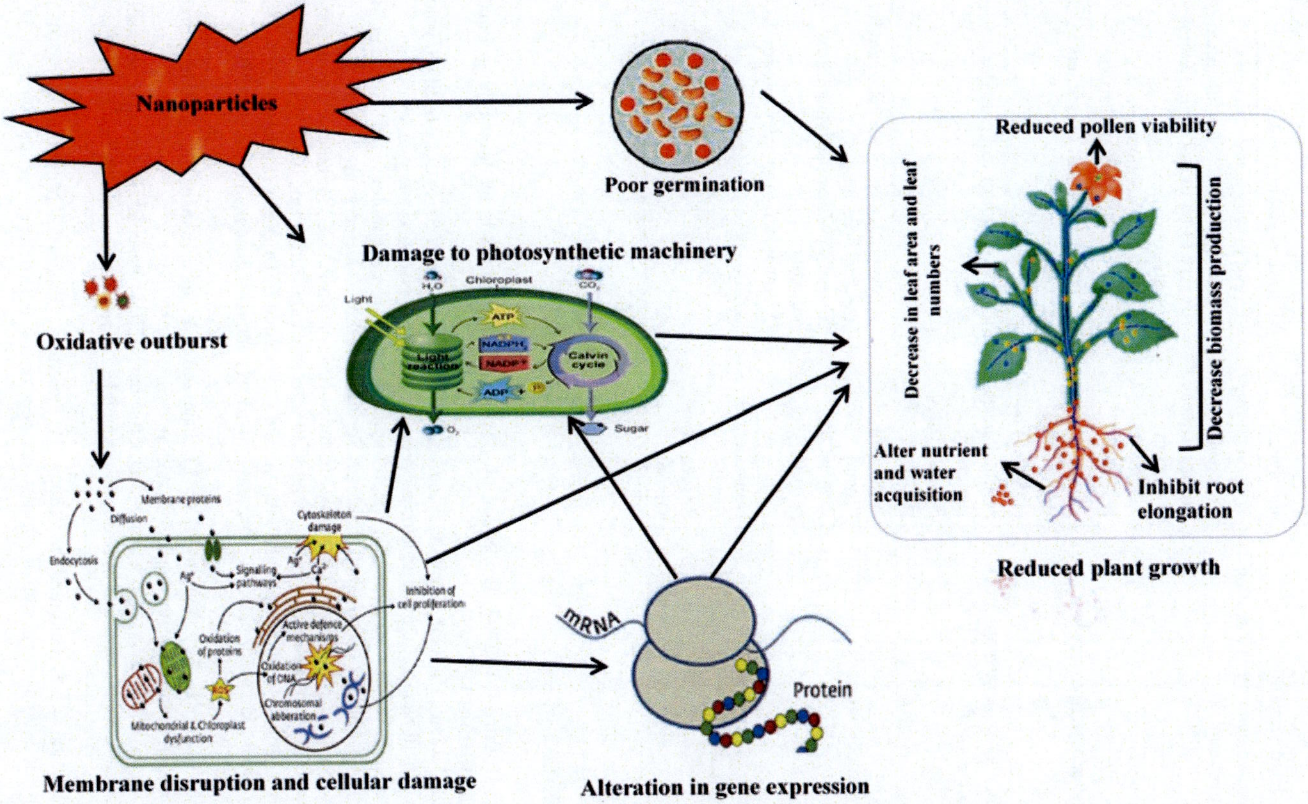

Fig. 15.4 Toxic effects of nanomaterials on plant growth and development (Reprinted with permission from Usman et al. 2020)

Table 15.3 Ethical issues pertaining to workplace situations involving nanomaterials (Reprinted from Schulte and Salamanca-Buentello 2007)

Work-related scenarios	Ethical principles involved	Decision-making issues
Identification and communication of hazards and risks	Responsibilities of scientists Nonmaleficence Autonomy Respect for persons	Extent to which strengths and weaknesses of data are identified Degree of participation in public discussion Accuracy of communications timeliness of communications
Workers' acceptance of risks	Autonomy Respect for persons' justice	Extent of inclusion of workers in decision-making
Selection and implementation of workplace controls	Nonmaleficence Beneficence Respect for persons	Level of control technologies utilized
Medical screening of nanotechnology workers	Autonomy Privacy Respect for persons	Appropriateness of the rationale for medical screening Extent to which participation is voluntary Maintenance of privacy test results
Investment in toxicological and control research	Nonmaleficence Justice Respect for persons	Adequacy of investment

References

Asgharian B, Price OT (2007) Deposition of ultrafine (nano) particles in the human lung. Inhal Toxicol 19(13):1045–1054

Asharani PV, Lianwu Y, Gong Z, Valiyaveettil S (2011) Comparison of the toxicity of silver, gold and platinum nanoparticles in developing zebrafish embryos. Nanotoxicology 5:43–54

Bainbridge WS (2004) Social and ethical implications of nanotechnology. In: Bhushan B (ed) Springer handbook of nanotechnology. Springer, Berlin, Heidelberg, pp 1135–1151

Baky NA, Faddah LM, Al-Rasheed NM, Al-Rasheed NM, Fatani AJ (2013) Induction of inflammation, DNA damage and apoptosis in rat heart after oral exposure to zinc oxide nanoparticles and the cardioprotective role of alpha-lipoic acid and vitamin E. Drug Res (Stuttg) 63:228–236

Baroli B (2010) Skin absorption and potential toxicity of nanoparticulate nanomaterials. J Biomed Nanotechnol 6(5):485–496

Batley GE, Kirby JK, McLaughlin MJ (2013) Fate and risks of nanomaterials in aquatic and terrestrial environments. Acc Chem Res 46:854–862

Bayat N, Lopes VR, Scholermann J, Jensen LD, Cristobal S (2015) Vascular toxicity of ultra-small TiO_2 nanoparticles and single walled carbon nanotubes in vitro and in vivo. Biomaterials 63:1–13

Boraschi D, Italiani P, Palomba R, Decuzzi P, Duschl A, Fadeel B, Moghimi SM (2017) Nanoparticles and innate immunity: new perspectives on host defence. Semin Immunol 34:33–51

Bostan HB, Rezaee R, Valokala MG, Tsarouhas K, Golokhvast K, Tsatsakis AM, Karimi G (2016) Cardiotoxicity of nano-particles. Life Sci 165:91–99

Bundschuh M, Filser J, Lüderwald S, McKee MS, Metreveli G, Schaumann GE, Schulz R, Wagner S (2018) Nanoparticles in the environment: where do we come from, where do we go to? Environ Sci Eur 30(1):6

Cao Y, Gong Y, Liao W, Wu C, Wang M, Yang Q (2018) A review of cardiovascular toxicity of TiO_2, ZnO and ag nanoparticles (NPs). Biometals 31:457–476

Cao Y, Long J, Liu L, He T, Jiang L, Zhao C, Li Z (2017) A review of endoplasmic reticulum (ER) stress and nanoparticle (NP) exposure. Life Sci 186:33–42

Card JW, Zeldin DC, Bonner JC, Nestmann ER (2008) Pulmonary applications and toxicity of engineered nanoparticles. Am J Physiol Lung Cell Mol Physiol 295(3):L400–L411

Cesta MF, Ryman-Rasmussen JP, Wallace DG, Masinde T, Hurlburt G, Taylor AJ, Bonner JC (2010) Bacterial lipopolysaccharide enhances PDGF signaling and pulmonary fibrosis in rats exposed to carbon nanotubes. Am J Respir Cell Mol Biol 43:142–151

Chen R, Huo L, Shi X, Bai R, Zhang Z, Zhao Y, Chang Y, Chen C (2014a) Endoplasmic reticulum stress induced by zinc oxide nanoparticles is an earlier biomarker for nanotoxicological evaluation. ACS Nano 8:2562–2574

Chen T, Hu J, Chen C, Pu J, Cui X, Jia G (2013) Cardiovascular effects of pulmonary exposure to titanium dioxide nanoparticles in ApoE knockout mice. J Nanosci Nanotechnol 13:3214–3222

Chen T, Yan J, Li Y (2014b) Genotoxicity of titanium dioxide nanoparticles. J Food Drug Anal 22(1):95–104

Chen Z, Wang Y, Zhuo L, Chen S, Zhao L, Luan X, Wang H, Jia G (2015) Effect of titanium dioxide nanoparticles on the cardiovascular system after oral administration. Toxicol Lett 239: 123–130

Crosera M, Bovenzi M, Maina G, Adami G, Zanette C, Florio C, Filon LF (2009) Nanoparticle dermal absorption and toxicity: a review of the literature. Int Arch Occup Environ Health 82(9):1043–1055

Danielsen PH, Cao Y, Roursgaard M, Moller P, Loft S (2015) Endothelial cell activation, oxidative stress and inflammation induced by a panel of metal-based nanomaterials. Nanotoxicology 9: 813–824

Demir E, Turna F, Vales G, Kaya B, Creus A, Marcos R (2013) In vivo genotoxicity assessment of titanium, zirconium and aluminium nanoparticles, and their microparticulated forms, in *drosophila*. Chemosphere 93:2304e10

Dhasmana A, Firdaus S, Singh KP, Raza S, Jamal QMS, Kesari KK, Rahman Q, Lohani M (2017) Nanoparticles: applications, toxicology and safety aspects. In: Kesari K (ed) Perspectives in environmental toxicology. Environmental science and engineering. Springer, Cham, pp 47–70

Fennell TR, Mortensen NP, Black SR, Snyder RW, Levine KE, Poitras E, Harrington JM, Wingard CJ, Holland NA, Pathmasiri W, Sumner SC (2017) Disposition of intravenously or orally administered silver nanoparticles in pregnant rats and the effect on the biochemical profile in urine. J Appl Toxicol 37:530–544

Ferreira AJ, Cemlyn-Jones J, Cordeiro CR (2013) Nanoparticles, nanotechnology and pulmonary nanotoxicology. Rev Port Pneumol 19:28–37

Gao J, Mahapatra CT, Mapes CD, Khlebnikova M, Wei A, Sepulveda MS (2016) Vascular toxicity of silver nanoparticles to developing zebrafish (*Danio rerio*). Nanotoxicology 10:1363–1372

Goyal P, Basniwal RK (2017) Toxicity of nanoparticles and their impact on environment. In: Ghorbanpour M, Manika K, Varma A (eds) Nanoscience and plant–soil systems. Soil biology, vol 48. Springer, Cham, pp 531–543

Haberl N, Hirn S, Holzer M, Zuchtriegel G, Rehberg M, Krombach F (2015) Effects of acute systemic administration of TiO_2, ZnO, SiO_2, and ag nanoparticles on hemodynamics, hemostasis and leukocyte recruitment. Nanotoxicology 9:963–971

Halamoda KB, Chapuis BC, Guney-Ayra S, Juillerat-Jeanneret L (2012) Induction of oxidative stress, lysosome activation and autophagy by nanoparticles in human brain-derived endothelial cells. Biochem J 441:813–821

Hashempour S, Ghanbarzadeh S, Maibach HI, Ghorbani M, Hamishehkar H (2019) Skin toxicity of topically applied nanoparticles. Ther Deliv 10(6):383–396

Hassanein KM, El-Amir YO (2017) Protective effects of thymoquinone and avenanthramides on titanium dioxide nanoparticles induced toxicity in Sprague-Dawley rats. Pathol Res Pract 213: 13–22

He D, Garg S, Wang Z, Li L, Rong H, Ma X, Li G, An T, Waite TD (2019) Silver sulfide nanoparticles in aqueous environments: formation, transformation and toxicity. Environ Sci 6: 1674–1687

Holland NA, Becak DP, Shannahan JH, Brown JM, Carratt SA, Winkle L, Pinkerton KE, Wang CM, Munusamy P, Baer DR, Sumner SJ, Fennell TR, Lust RM, Wingard CJ (2015) Cardiac ischemia reperfusion injury following instillation of 20 nm citrate-capped nanosilver. J Nanomed Nanotechnol 6:10–7439

Holland NA, Thompson LC, Vidanapathirana AK, Urankar RN, Lust RM, Fennell TR, Wingard CJ (2016) Impact of pulmonary exposure to gold core silver nanoparticles of different size and capping agents on cardiovascular injury. Part Fibre Toxicol 13:48

Husain M, Wu D, Saber AT, Decan N, Jacobsen NR, Williams A, Yauk CL, Wallin H, Vogel U, Halappanavar S (2015) Intratracheally instilled titanium dioxide nanoparticles translocate to heart and liver and activate complement cascade in the heart of C57BL/6 mice. Nanotoxicology 9:1013–1022

Iannitti T, Capone S, Gatti A, Capitani F, Cetta F, Palmieri B (2010) Intracellular heavy metal nanoparticle storage: progressive accumulation within lymph nodes with transformation from chronic inflammation to malignancy. Int J Nanomedicine 5:955–960

Ihrie MD, Bonner JC (2018) The toxicology of engineered nanomaterials in asthma. Curr Environ Health Rep 5(1):100–109

Keller AA, McFerran S, Lazareva A, Suh S (2013) Global life cycle releases of engineered nanomaterials. J Nanopart Res 15(6):1692

Kumar A, Dhawan A (2013) Genotoxic and carcinogenic potential of engineered nanoparticles: an update. Arch Toxicol 87(11):1883–1900

Landsiedel R, Ma-Hock L, Van Ravenzwaay B, Schulz M, Wiench K, Champ S, Schulte S, Wohlleben W, Oesch F (2010) Gene toxicity studies on titanium dioxide and zinc oxide nanomaterials used for UV-protection in cosmetic formulations. Nanotoxicology 4:364e81

Li JJ, Muralikrishnan S, Ng C-T, Yung L-YL, Bay B-H (2010) Nanoparticle-induced pulmonary toxicity. Exp Biol Med 235(9):1025–1033

Lin CX, Yang SY, Gu JL, Meng J, Xu HY, Cao JM (2017) Theacute toxic effects of silver nanoparticles on myocardial transmembrane potential, INa and IK1 channels and heart rhythm in mice. Nanotoxicology 11:827–837

Lindberg HK, Falck GC, Catalán J, Koivisto AJ, Suhonen S, Järventaus H, Rossi EM, Nykäsenoja H, Peltonen Y, Moreno C, Alenius H, Tuomi T, Savolainen KM, Norppa H (2012) Genotoxicity of inhaled nanosized TiO_2 in mice. Mutat Res 745:58e64

Liu H, Ma L, Zhao J, Liu J, Yan J, Ruan J, Hong F (2009) Biochemical toxicity of nano-anatase TiO_2 particles in mice. Biol Trace Elem Res 129:170–180

Luyts K, Smulders S, Napierska D, Van KS, Poels K, Scheers H, Hemmeryckx B, Nemery B, Hoylaerts MF, Hoet PH (2014) Pulmonary and hemostatic toxicity of multi-walled carbon nanotubes and zinc oxide nanoparticles after pulmonary exposure in Bmal1 knockout mice. Part FibreToxicol 11:61

Magaye R, Zhao J, Bowman L, Ding M (2012) Genotoxicity and carcinogenicity of cobalt-, nickel- and copper-based nanoparticles. Exp Ther Med 4(4):551–561

Mikkelsen L, Sheykhzade M, Jensen KA, Saber AT, Jacobsen NR, Vogel U, Wallin H, Loft S, Moller P (2011) Modest effect on plaque progression and vasodilatory function in atherosclerosis-prone mice exposed to nanosized TiO(2). Part Fibre Toxicol 8:32

Mo Y, Gu A, Tollerud DJ, Zhang Q (2016) Nanoparticle toxicity and environmental impact. Oxidative Stress Biomater 2016:117–143

Monse C, Hagemeyer O, Raulf M, Jettkant B, van Kampen V, Kendzia B, Gering V, Kappert G, Weiss T, Ulrich N, Marek EM, Bunger J, Bruning T, Merget R (2018) Con-centration-

dependent systemic response after inhalation of nano-sized zinc oxide particles in human volunteers. Part Fibre Toxicol 15:8
Munger MA, Radwanski P, Hadlock GC, Stoddard G, ShaabanA FJ, Grainger DW, Deering-Rice CE (2014) In vivo human time-exposure study of orally dosed commercial silver nanoparticles. Nanomedicine 10:1–9
NANODERM (2007) Quality of skin as a barrier to ultra-fineparticles. Final report (project number: QLK4-CT-2002-02678)
Naya M, Kobayashi N, Ema M, Kasamoto S, Fukumuro M, Takami S, Nakajima M, Hayashi M, Nakanishi J (2012) In vivo genotoxicity study of titanium dioxide nanoparticles using comet assay following intratracheal instillation in rats. Regul Toxicol Pharmacol 62:1e6
Ng CT, Li JJ, Bay BH, Yung LY (2010) Current studies into the geno-toxic effects of nanomaterials. J Nucleic Acids 2010:1–12
Nowack B, Bucheli TD (2007) Occurrence, behavior and effects of nanoparticles in the environment. Environ Pollut 150:5–22
Nurkiewicz TR, Porter DW, Hubbs AF, Stone S, Chen BT, Frazer DG, Boegehold MA, Castranova V (2009) Pulmonary nanoparticle exposure disrupts systemic microvascular nitric oxide signaling. Toxicol Sci 110:191–203
Oberdorster G, Stone V, Donaldson K (2007) Toxicology of nanoparticles: a historical perspective. Nanotoxicology 1:2–25
Parry MC, Bhabra G, Sood A, Machado F, Cartwright L, Saunders M, Ingham E, Newson R, Blom AW, Case CP (2010) Thresholds for indirect DNA damage across cellular barriers for orthopaedic biomaterials. Biomaterials 31(16):4477–4483
Raskar S, Laware S (2014) Effect of zinc oxide nanoparticles on cytology and seed germination in onion. Int J Curr Microbiol App Sci 3:467–473
Ray PC, Yu H, Fu PP (2009) Toxicity and environmental risks of nanomaterials: challenges and future needs. J Environ Sci Health C Environ Carcinog Ecotoxicol Rev 27(1):1–35
Roberts JR, McKinney W, Kan H, Krajnak K, Frazer DG, Thomas TA, Waugh S, Kenyon A, MacCuspie RI, Hackley VA, Castranova V (2013) Pulmonary and cardiovascular responses of rats to inhalation of silver nanoparticles. J Toxicol Environ Health A 76:651–668
Roller M (2009) Carcinogenicity of inhaled nanoparticles. Inhal Toxicol 21:144–157
Ryman-Rasmussen JP, Cesta MF, Brody AR, Shipley-Phillips JK, Everitt JI, Tewksbury EW, Moss OR, Wong BA, Dodd DE, Andersen ME, Bonner JC (2009) Inhaled carbon nanotubes reach the subpleural tissue in mice. Nat Nanotechnol 4:747–751
Sadiq R, Bhalli JA, Yan J, Woodruff RS, Pearce MG, Li Y, Mustafa T, Watanabe F, Pack LM, Biris AS, Khan QM, Chen T (2012) Genotoxicity of TiO_2 anatase nanoparticles in B6C3F1 male mice evaluated using pig-a and flow cytometric micronucleus assays. Mutat Res 745:65e72
Schulte PA, Salamanca-Buentello F (2007) Ethical and scientific issues of nanotechnology in the workplace. Environ Health Perspect 115(1):5–12
Seaton A, Tran L, Aitken R, Donaldson K (2010) Nanoparticles, human health hazard and regulation. J R Soc Interface 7:119–129
Sha B, Gao W, Wang S, Li W, Liang X, Xu F, Lu TJ (2013) Nano-titanium dioxide induced cardiac injury in rat under oxidative stress. Food Chem Toxicol 58:280–288
Sharma A, Muresanu DF, Lafuente JV, Patnaik R, Tian ZR, Buzoianu AD, Sharma HS (2015a) Sleep deprivation-induced blood-brain barrier breakdown and brain dysfunction are exacerbated by size-related exposure to ag and cu nanoparticles. Neuroprotective effects of a 5-HT3 receptor antagonist ondansetron. Mol Neurobiol 52:867–881
Sharma A, Muresanu DF, Patnaik R, Sharma HS (2013) Size-and age-dependent neurotoxicity of engineered metal nanoparticles in rats. Mol Neurobiol 48:386–396
Sharma HS, Kiyatkin EA, Patnaik R, Lafuente JV, Muresanu DF, Sjoquist PO, Sharma A (2015b) Exacerbation of methamphetamine neurotoxicity in cold and hot environments: neuroprotective effects of an antioxidant compoundH-290/51. Mol Neurobiol 52:1023–1033
Singh J, Dutta T, Kim KH, Rawat M, Samddar B, Kumar P (2018) Green synthesis of metals and their oxide nanoparticles: applications for environmental remediation. J Nanobiotechnol 16:84

Stapleton PA, Nichols CE, Yi J, McBride CR, Minarchick VC, Shepherd DL, Hollander JM, Nurkiewicz TR (2015) Microvascular and mitochondrial dysfunction in the female F1 generation after gestational TiO_2 nanoparticle exposure. Nanotoxicology 9:941–951

Sycheva LP, Zhurkov VS, Iurchenko VV, Daugel-Dauge NO, Kovalenko MA, Krivtsova EK, Durnev AD (2011) Investigation of genotoxic and cytotoxic effects of micro- and nanosized titanium dioxide in six organs of mice in vivo. Mutat Res 726:8–14

Trouiller B, Reliene R, Westbrook A, Solaimani P, Schiestl RH (2009) Titanium dioxide nanoparticles induce DNA damage and genetic instability in vivo in mice. Cancer Res 69: 8784–8789

Tsaousi A, Jones E, Case CP (2010) The in vitro genotoxicity of orthopaedic ceramic (Al2O3) and metal (CoCr alloy) particles. Mutat Res 697(1–2):1–9

Usman M, Farooq M, Wakeel A, Nawaz A, Cheema SA, Rehman HU, Ashraf I, Sanaullah M (2020) Nanotechnology in agriculture: current status, challenges and future opportunities. Sci Total Environ 721:137778

Xu J, Shi H, Ruth M, Yu H, Lazar L, Zou B, Yang C, Wu A, Zhao J (2013) Acute toxicity of intravenously administered titanium dioxide nanoparticles in mice. PLoS One 8:e70618

Yan Z, Wang W, Wu Y, Wang W, Li B, Liang N, Wu W (2017) Zinc oxide nanoparticle-induced atherosclerotic alterations in vitro and in vivo. Int J Nanomedicine 12:4433–4442

Zhang J, Zou Z, Wang B, Xu G, Wu Q, Zhang Y, Yuan Z, Yang X, Yu C (2018) Lysosomal deposition of copper oxide nanoparticles triggers HUVEC cells death. Biomaterials 161:228–239

Nanobiotechnology: Present Status and Future Prospects

16

Abstract

The nanotechnology market is booming and has led to development of new products, flourishing economy, more jobs, and establishment of new industries in many countries. It has been estimated that the global nanotechnology market is likely to reach $90.5 billion by 2021 at a compound annual growth rate (CAGR) of 18.2% in the 5-year period of 2016–2021. A number of countries have initiated extensive research programs focused on nanotechnology with increasing funding over the years. However, several challenges remain that must be overcome to bring this technology at the doorsteps of the common man.

16.1 Introduction

Nanotechnology is a rapidly growing field that has witnessed significant amount of public interest due to applications of nanomaterials in many areas of human endeavors like communication, public health, textile, food, and agricultural industries. There has been an increasing demand of nanomaterials which has resulted in their increased rate of production. The growth and the present trend in nanotechnology as well as bio-nanotechnology are quite promising and booming. The field has excellent potential to consume more resources attracting further investment, an economy with strong foundations in nanotechnology to development of new products, flourishing economy, more jobs, and establishment of new industries in many countries. This has ignited increased funding in the nanotechnology sector in the last couple of decades. However, research into the possible hazards of nanoparticles with respect to the environment as well as human health is still in its infancy.

S. Srivastava, A. Bhargava, *Green Nanoparticles: The Future of Nanobiotechnology*,
https://doi.org/10.1007/978-981-16-7106-7_16

16.2 Present Status and the Expanding Market

Nanoscience and nanotechnology is one of the fastest growing research and technology areas with the investment in nanotechnology R&D soaring from $432 million in 1997 to about $4.1 billion in 2005. About 2000 tons of nanoparticles were produced in 2004, and the production rate was expected to touch ~58,000 tons by the year 2020. According to a recent report by Business Communications Company, the nanotechnology market is likely to reach $90.5 billion by 2021 at a compound annual growth rate (CAGR) of 18.2% in the 5-year period of 2016–2021. The energy market for nanotechnology is estimated to increase from $5.7 billion in 2018 to $10.0 billion by 2023 at a CAGR of 12.0% for the 5-year period.

The nanotechnology market in the United States was estimated at US$16 billion in the year 2020 accounting for 29.53% of the global market. China, Japan, the European Union, and Canada are other noteworthy players in the nanotechnology arena. After the formation of cross-departmental Unit G4: Nanosciences and Nanotechnologies in 2000, research on nanotechnology in the European Union accelerated between 2002 and 2006 under Framework Programme 6 (FP6) when there was increased funding for nanotechnology-based projects. Also an action plan was formulated for nanotechnology in 2005. Horizon 2020, the European Union research and innovation program, aimed at bridging the gap between nanotechnology research and markets, with major thrust areas being directed toward sustainable development, increased competitiveness, skilled employment, and better quality of life. Horizon 2020 had a program fully dedicated to nanotechnology by the name "Nanotechnologies, Advanced Materials, Biotechnology, and Advanced Manufacturing and Processing" (NMBP).

The National Committee for Direction and Coordination of Nanoscience and Nanotechnology Research was established in 2001 in China with the aim of coordination and coherent planning of nanotechnology research. This was followed by the release of National Medium- and Long-Term Science and Technology Development Plan in 2006 and formulation of a 5-year research plan in 2012 with high focus on nanotechnology. The Department of Science and Technology, Government of India, launched the Programme on Nanomaterials: Science and Devices in 2000 for initiating end-to-end projects keeping in mind the specific societal needs like improvement in water quality, development of alternative energy, and sustainability. Seven years later, the Nano Mission program was launched with greater funding from the government. The main objectives of this program were research infrastructure development, strengthening of public-private partnerships, formation of technology development centers, creation of skilled human resource, and international collaboration (Fig. 16.1).

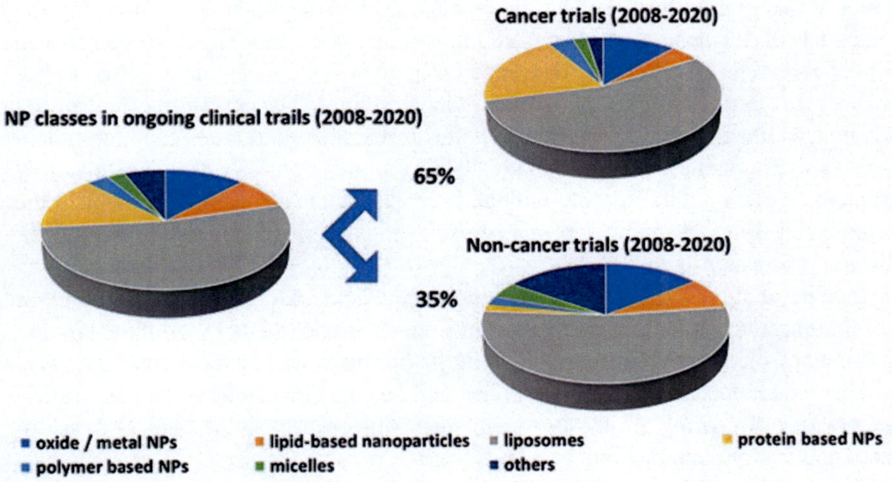

Fig. 16.1 Nanoparticle classes investigated in 409 clinical trials from the year 2008 to 2020 (active, ongoing, or recruiting), identified in the clinicaltrials.gov database in May 2020 (Reprinted with permission from Germain et al. 2020)

16.3 Challenges in Nanobiotechnology

The invisible nature of nanoparticles, toxicity issues, complexity of the properties, environmental concerns, and lack of knowledge are the key challenges faced by nanotechnologists in this field.

The opportunity for nanotechnology to revolutionize diverse technical areas has been well understood. However, nanotechnology is still in its infancy and has an almost unpredictable growth, which seems steady and increasing. The rapid development of nanotechnology and its associated products and services present enormous opportunities as well as challenges. Nanotechnology coupled with green chemistry is an increasingly researched area that ensures better production and utilization of nanomaterials along with safeguarding human health and the environment.

16.3.1 Nanomedicines

One major area of nanotechnology use in biological sciences is in the prevention, diagnosis, and treatment of diseases. It is now considered that nanomaterials cannot be solely considered as a drug delivery system but may become the active therapeutic component. The development of nanoparticulate nanomedicines has been on the rise due to advantages like increase in therapeutic efficacy, decrease in dose, and reduction in the side effects. About 50 nanotechnology-based formulations have been approved for clinical use, while another 400 are in clinical trials. Hybrid nanocrystals or multifunctional nanocrystals, i.e., nanoentities, accomplishing the dual goals of diagnosis and therapy are increasingly becoming popular, especially in cancer research. Nanocrystals comprise entirely of drug molecule, with complete absence of carrier components like polymer or lipid. Nanocrystals offer several advantages like enhanced permeability, greater retention, and facilitation for attachment to many ligands. The poor bioavailability of less soluble drugs is resolved by an increase in the saturation solubility and decrease in the diffusional pathway of the nanocrystal drugs. A number of nanocrystal drugs are available in the market for different disorders (Table 16.1).

In spite of significant breakthrough in nanomedicine, Dr. Kinam Park, Professor of Pharmaceutics at Purdue University, in a recent article in 2019 was of the opinion that nanomedicine was overrated and that its promises will never be realized. This will lead to reallocation of resources, time, and financial investments to other priority sectors. The following are the key challenges with respect to the clinical development and use of nanomedicines (Box 16.1):

Box 16.1. Considerations for the Translational Development of Nanomedicines (Reprinted from Hua et al. 2018)

Nanopharmaceutical design.

Key considerations

- Route of administration
- Reduced complexity in formulation design
- Final dosage form for human use
- Biocompatibility and biodegradability
- Pharmaceutical stability (physical and chemical)

Current obstacles:

- Large-scale production according to GMP standards
- For example, reproducibility, infrastructure, techniques, expertise, and cost
- Quality control assays for characterization
- For example, size and polydispersity, morphology, charge, encapsulation, surface modifications, purity, and stability

Preclinical evaluation

Key considerations

- Need for validated and standardized assays for early detection of toxicity
- Evaluation in appropriate models of disease
- Adequate understanding on in vivo behavior, incl. Cellular and molecular interactions

(continued)

Table 16.1 Nanocrystal products on the market (Reprinted with permission from Shegokar and Müller 2010)

Trade name	Therapeutic use	Applied technology	Pharma company	Administration route
Rapamune® (rapamycin, sirolimus)	Immunosuppressive	élan nanosystems	Wyeth pharmaceuticals	Oral
Emend® (aprepitant)	Antiemetic	élan nanosystems	Merck & co.	Oral
Tricor® (fenofibrate)	Hypercholesterolemia	élan nanosystems	Abbott laboratories	Oral
Triglide® (fenofibrate)	Hypercholesterolemia		Produced by SkyePharma marketed by Sciele pharma, Inc. (Atlanta, CA, USA)	Oral
Megace ES® (megestrol acetate)	Antianorexic	élan nanosystems	Par pharmaceutical companies, Inc. (Spring Valley, NY, USA)	Oral
Avinza® (morphine sulfate)	Psychostimulant drug	élan nanosystems	King Pharmaceuticals	Oral
Focalin® XR (dexmethylphenidate HCl)	Psychostimulant drug	élan nanosystems	Novartis	Oral
Ritalin® LA (methylphenidate HCl)	Psychostimulant drug	élan nanosystems	Novartis	Oral
Zanaflex capsules TM (tizanidine HCl)	Muscle relaxant	élan nanosystems	Acorda	Oral

Box 16.1 (continued)

- Pharmacokinetics (absorption, distribution, metabolism, and excretion)
- Pharmacodynamics (intracellular trafficking, functionality, toxicity, and degradation)

Current obstacles

- Development of more specialized toxicology studies for nanomedicines
- Adequate understanding of the interaction of NNM with tissues and cells
- Adequate structural stability of NNM following in vivo administration
- Limited degree of accumulation of nanomedicines in target organs/tissues/cells

Clinical evaluation for commercialization
Key considerations

- Simplification of developmental pathways from invention to commercialization to minimize time and expense
- Evaluation of safety/toxicity in humans (acute and chronic)
- Evaluation of therapeutic efficacy in humans
- Optimal clinical trial design

Current obstacles

- Lack of clear regulatory guidelines specific for NNMs
- Complexity of NNM patents and IP
- Limited understanding of the biological interaction of NNM with the biological environment (incl. target site) in the body of patients

1. Biological challenges.
2. Large-scale manufacturing at low cost in comparison with currently used therapies.
3. Biocompatibility and safety: this requires filling knowledge gaps with respect to hazards of nanomaterials.
4. Issues in the large-scale production of nanocrystals.
5. Intellectual property and government regulations.
6. Absence of testing methods and international standards for risk assessment of nanomaterials.

16.3.2 Food Industry

Nanotechnology has been utilized for fabrication of innovative products for food industry and in preservation of food items. However, the following challenges remain:

1. Production of efficient delivery systems that can be safely used for edible items.
2. Preventing leaching of nanomaterials from food packaging constituents into the food item.
3. Complete understanding of nanoscale functionalities and toxicities of nanomaterials.
4. Generation of hazardous non-eco-friendly by-products.
5. Potential risks and associated toxicity issues.
6. In-depth studies on nanoparticle interaction with living forms prior to their use in the industry.

16.3.3 Agriculture

Nanobiotechnology has found many applications in agricultural sector like development of nanofertilizers, nanopesticides, and nanobiosensors. However, its use in agriculture is also related with numerous adverse effects. Therefore, development of nanoagriculture in its full swing requires the following:

1. Proper assessment and overcoming the risk issues related to the use of nanoparticles.
2. Validation of the non-harmful dosages of nanoparticles for agriculture within the safety limits.
3. Studies on effective soil management that can mitigate the toxic effect of nanoparticles.
4. Thorough studies on the screening and optimization of nanoparticles for different plant species.
5. Innovative synthesis methods with specific control over composition (shape, size, etc.) of the nanoparticles.
6. Field testing on actual farm conditions.

16.4 Thrust Areas

The use of nanobiotechnology and related items in different industries is likely to benefit most of the developed as well as developing economies. The top five nanobiotechnology applications on the priority list of developing countries include:

1. Energy storage, production, and efficient conversion with minimal losses
2. Increased crop production

3. Water and soil remediation
4. Disease diagnosis
5. Drug delivery systems

References

Cha DS, Chinnan MS (2004) Biopolymer-based antimicrobial packaging: a review. Crit Rev Food Sci Nutr 44(4):223–237

DST (2001) Annual report 2000–2001. Department of Science and Technology, New Delhi

DST (2008) Annual report 2007–2008. Department of Science and Technology, New Delhi

Dupas C, Lahmani M (eds) (2007) Nanoscience: nanotechnologies and nanophysics. Springer, Berlin

Fautz C, Fleischer T, Ma Y, Liao M, Kumar A (2015) Discourses on nanotechnology in Europe, China and India. In: Ladikas M, Chaturvedi S, Zhao Y, Stemerding D (eds) Science and technology governance and ethics. Springer, Cham, pp 125–143

Germain M, Caputo F, Metcalfe S, Tosi G, Spring K, Åslund AKO, Pottier A, Schiffelers R, Ceccaldi A, Schmid R (2020) Delivering the power of nanomedicine to patients today. J Control Release 326:164–171

Hasan A, Morshed M, Memic A, Hassan S, Webster TJ, Marei HE (2018) Nanoparticles in tissue engineering: applications, challenges and prospects. Int J Nanomedicine 13:5637–5655

Hua S, de Matos MBC, Metselaar JM, Storm G (2018) Current trends and challenges in the clinical translation of Nanoparticulate nanomedicines: pathways for translational development and commercialization. Front Pharmacol 9:790

MOST (2001) National programme on the development of nanoscience and nanotechnology (2001–2010). Ministry of Science and Technology of the People's Republic of China

MOST (2006) Outline of national medium- and long-term science and technology development plan (2006–2020). Ministry of Science and Technology of the People's Republic of China

MOST (2012) The 12th five-year national plan for science and technology development. Ministry of Science and Technology of the People's Republic of China

Muller R, Keck C (2012) Twenty years of drug nanocrystals: where are we, and where do we go? Eur J Pharm Biopharm 80:1–3

Muller RH, Gohla S, Keck CM (2011) State of the art of nanocrystals_special features, production, nanotoxicology aspects and intracellular delivery. Eur J Pharm Biopharm 78:1–9

Murthy SK (2007) Nanoparticles in modern medicine: state of the art and future challenges. Int J Nanomedicine 2(2):129–141

Nagarwal RC, Kumar R, Dhanawat M, Das N, Pandit JK (2011) Nanocrystal technology in the delivery of poorly soluble drugs: an overview. Curr Drug Deliv 8:398–406

Pandey G (2018) Challenges and future prospects of agri-nanotechnology for sustainable agriculture in India. Environ Technol Innov 11:299–307

Park K (2019) The beginning of the end of the nanomedicine hype. J Control Release 305:221–222

Roco M (2007) National nanotechnology initiative: past, present, and future. In: Goddard WA, Brenner DW, Lyshevski SE, Iafrate GJ (eds) Handbook of nanoscience, engineering, and technology, 2nd edn. Boca Raton, CRC Press, pp 16–26

Sahoo M, Vishwakarma S, Panigrahi C, Kumar J (2020) Nanotechnology: current applications and future scope in food. Food Front 2:3–22

Shegokar R, Müller RH (2010) Nanocrystals: industrially feasible multifunctional formulation technology for poorly soluble actives. Int J Pharm 399(1–2):129–139